Novelties in Wood Engineering and Forestry

Novelties in Wood Engineering and Forestry

Editors

Aurel Lunguleasa
Florin Dinulica
Camelia Cosereanu

Basel • Beijing • Wuhan • Barcelona • Belgrade • Novi Sad • Cluj • Manchester

Editors
Aurel Lunguleasa
Department of Wood
Engineering, Transilvania
University of Brasov
Brasov, Romania

Florin Dinulica
Department of Silviculture,
Transilvania University of
Brasov
Brasov, Romania

Camelia Cosereanu
Department of Wood
Processing and Design od
Wood Products, Transilvania
University of Brasov
Brasov, Romania

Editorial Office
MDPI
St. Alban-Anlage 66
4052 Basel, Switzerland

This is a reprint of articles from the Special Issue published online in the open access journal *Forests* (ISSN 1999-4907) (available at: https://www.mdpi.com/journal/forests/special_issues/Wood_Engineering_Forestry).

For citation purposes, cite each article independently as indicated on the article page online and as indicated below:

Lastname, A.A.; Lastname, B.B. Article Title. *Journal Name* **Year**, *Volume Number*, Page Range.

ISBN 978-3-0365-9845-1 (Hbk)
ISBN 978-3-0365-9846-8 (PDF)
doi.org/10.3390/books978-3-0365-9846-8

© 2023 by the authors. Articles in this book are Open Access and distributed under the Creative Commons Attribution (CC BY) license. The book as a whole is distributed by MDPI under the terms and conditions of the Creative Commons Attribution-NonCommercial-NoDerivs (CC BY-NC-ND) license.

Contents

Martin Lexa, Roman Fojtík, Viktor Dubovský, Miroslav Sedlecký, Aleš Zeidler and Adam Sikora
Influence of the External Environment on the Moisture Spectrum of Norway Spruce (*Picea abies* (L.) KARST.)
Reprinted from: *Forests* **2023**, *14*, 1342, doi:10.3390/f14071342 1

Răzvan V. Câmpu and Rudolf A. Derczeni
European Beech Log Sawing Using the Small-Capacity Band Saw: A Case Study on Time Consumption, Productivity and Recovery Rate
Reprinted from: *Forests* **2023**, *14*, 1137, doi:10.3390/f14061137 17

Veronica Dragusanu (Japalela), Aurel Lunguleasa, Cosmin Spirchez and Cezar Scriba
Some Properties of Briquettes and Pellets Obtained from the Biomass of Energetic Willow (*Salix viminalis* L.) in Comparison with Those from Oak (*Quercus robur*)
Reprinted from: *Forests* **2023**, *14*, 1134, doi:10.3390/f14061134 29

Alin M. Olarescu, Aurel Lunguleasa, Loredana Radulescu and Cosmin Spirchez
Manufacturing and Testing the Panels with a Transverse Texture Obtained from Branches of Norway Spruce (*Picea abies* L. Karst.)
Reprinted from: *Forests* **2023**, *14*, 665, doi:10.3390/f14040665 49

Florin Dinulica, Adriana Savin and Mariana Domnica Stanciu
Physical and Acoustical Properties of Wavy Grain Sycamore Maple (*Acer pseudoplatanus* L.) Used for Musical Instruments
Reprinted from: *Forests* **2023**, *14*, 197, doi:10.3390/f14020197 69

Luboš Červený, Roman Sloup and Tereza Červená
The Potential of Smart Factories and Innovative Industry 4.0 Technologies—A Case Study of Different-Sized Companies in the Furniture Industry in Central Europe
Reprinted from: *Forests* **2022**, *13*, 2171, doi:10.3390/f13122171 83

Mohammad Hassan Mazaherifar, Hamid Zarea Hosseinabadi, Camelia Coșereanu, Camelia Cerbu, Maria Cristina Timar and Sergiu Valeriu Georgescu
Investigation on *Phoenix dactylifera*/*Calotropis procera* Fibre-Reinforced Epoxy Hybrid Composites
Reprinted from: *Forests* **2022**, *13*, 2098, doi:10.3390/f13122098 111

Anamaria Avram, Constantin Ștefan Ionescu and Aurel Lunguleasa
Some Methods for the Degradation-Fragility Degree Determination and for the Consolidation of Treatments with Paraloid B72 of Wood Panels from Icon-Type Heritage Objects
Reprinted from: *Forests* **2022**, *13*, 801, doi:10.3390/f13050801 131

Gabriela Balea (Paul), Aurel Lunguleasa, Octavia Zeleniuc and Camelia Coșereanu
Three Adhesive Recipes Based on Magnesium Lignosulfonate, Used to Manufacture Particleboards with Low Formaldehyde Emissions and Good Mechanical Properties
Reprinted from: *Forests* **2022**, *13*, 737, doi:10.3390/f13050737 151

Sergiu-Valeriu Georgescu, Daniela Șova, Mihaela Campean and Camelia Coșereanu
A Sustainable Approach to Build Insulated External Timber Frame Walls for Passive Houses Using Natural and Waste Materials
Reprinted from: *Forests* **2022**, *13*, 522, doi:10.3390/f13040522 173

Aurel Lunguleasa, Cosmin Spirchez and Alin M. Olarescu
Calorific Characteristics of Larch (*Larix decidua*) and Oak (*Quercus robur*) Pellets Realized from Native and Torrefied Sawdust
Reprinted from: *Forests* **2022**, *13*, 361, doi:10.3390/f13020361 . . . . **189**

Article

Influence of the External Environment on the Moisture Spectrum of Norway Spruce (*Picea abies* (L.) KARST.)

Martin Lexa [1,*], Roman Fojtík [1], Viktor Dubovský [2], Miroslav Sedlecký [1], Aleš Zeidler [1] and Adam Sikora [1]

[1] Faculty of Forestry and Wood Sciences, Czech University of Life Sciences Prague, Kamýcká 129, 165 00 Prague, Czech Republic; fojtikr@fld.czu.cz (R.F.); sedlecky@fld.czu.cz (M.S.); zeidler@fld.czu.cz (A.Z.); sikoraa@fld.czu.cz (A.S.)

[2] Department of Mathematics, Faculty of Civil Engineering, VSB-TU Ostrava, Ludvíka Podéště 1875/17, 708 00 Ostrava, Czech Republic; viktor.dubovsky@vsb.cz

* Correspondence: lexa@fld.czu.cz

Abstract: The fluctuation of relative humidity and temperature in the surrounding environments of wood products is an important parameter influencing their mechanical properties. The objective of this study was to investigate the complex relationship between the moisture content and mechanical properties of wood as a critical aspect in the design of durable and reliable structures. Over a period of 669 days, a simulated type of experiment was conducted, during which the moisture content and external temperature were continuously measured in a compact profile of Norway spruce (*Picea abies* (L.) KARST.). The data were processed using quadratic and cubic models to establish a predictive model. It was found that the quadratic models slightly outperformed the cubic models when considering time lags greater than six days. The final model demonstrated a significant improvement in explaining the variance of the dependent variable compared to the basic model. Based on these findings, it can be concluded that understanding the relationship between the moisture content and temperature of wood samples plays an important role in wood's efficient use, particularly for timber constructions. This understanding is vital for accurately predicting the mechanical characteristics of wood, which, in turn, contributes to the development of more durable and reliable structures.

Keywords: wood moisture content; temperature–moisture relationship; timber construction; predictive modeling; quadratic models; cubic models

Citation: Lexa, M.; Fojtík, R.; Dubovský, V.; Sedlecký, M.; Zeidler, A.; Sikora, A. Influence of the External Environment on the Moisture Spectrum of Norway Spruce (*Picea abies* (L.) KARST.). *Forests* 2023, 14, 1342. https://doi.org/10.3390/f14071342

Academic Editors: Aurel Lunguleasa, Camelia Cosereanu and Florin Dinulica

Received: 13 June 2023
Revised: 22 June 2023
Accepted: 26 June 2023
Published: 29 June 2023

Copyright: © 2023 by the authors. Licensee MDPI, Basel, Switzerland. This article is an open access article distributed under the terms and conditions of the Creative Commons Attribution (CC BY) license (https://creativecommons.org/licenses/by/4.0/).

1. Introduction

Wood, a material utilized for centuries in the construction industry, offers exceptional mechanical properties and suitability for general building applications and outdoor construction. It is a natural, renewable resource known for its availability, ease of processing, and versatility. Being a carbon-neutral material, it stores atmospheric carbon dioxide during its growth phase and retains it even after being processed into building components [1–6]. These characteristics make wood an environmentally friendly choice that contributes to carbon sequestration and helps to mitigate the impact of carbon emissions from alternative building materials [3].

Wood's excellent strength-to-weight ratio allows for efficient load-bearing capacities, while its cellular structure, consisting of lignocellulosic fibers, provides inherent strength and resilience against bending and compression forces [7–11]. Additionally, the unique fiber arrangement of wood enables it to resist lateral loads [12], making it suitable for constructing durable frameworks capable of withstanding various environmental conditions. However, being a hygroscopic material, moisture in wood directly impacts its structural integrity, dimensional stability, and resistance to decay and degradation [13–19]. An excessive moisture content can lead to undesirable consequences, such as swelling, warping, and loss of strength, compromising the performance and longevity of wood-based products [8,20–22]. Conversely, insufficient moisture content can result in shrinkage, cracking, and brittleness, further compromising the material's mechanical properties [8,10,20,23,24]. Therefore, understanding the intricate relationship between wood's moisture content and its mechanical behavior is vital for designing durable and reliable wood structures. Due to their different structures, individual wood species may differ in their relationship to the joint actions of moisture and temperature. These structural differences also significantly influence the resulting mechanical properties of the wood [25–27].

In the field of timber construction, our study builds upon the findings of two influential research papers. The first paper delved into the primary losses of prestressing force in spruce timber used in transversally prestressed wooden constructions [10]. For 669 days, a simulated experiment was conducted to measure prestressing force, external temperature, and moisture. The findings revealed a significant decrease in prestressing force over time during the primary loss phase, leading to the development of a mathematical model of the losses of prestress force. The paper also highlighted the increasing use of timber in bridge constructions and timber-concrete composites, with specific parameters significantly influencing their long-term behavior. The second paper offered a comprehensive study on the use of transverse prestressing in timber structures, illustrating how environmental conditions can significantly influence the prestressing force and, consequently, the load-bearing capacity and stiffness of these structures [27]. The researchers proposed a mathematical model to predict changes in prestressing force over time based on climatic conditions, underscoring the importance of renewable materials in construction for sustainability.

Building upon these findings, our study utilized the same piece of timber as a sample for measuring temperature and moisture. This decision was based on the strong correlation between temperature and moisture in the aforementioned papers. This investigation was carried out to describe the dependency between the surrounding temperature and the wood moisture of the specimen. Given the complexity of this relationship, special regression techniques were used to handle the multicollinearity inherent in these variables. Specifically, we used principal component analysis and repeated cross-validation methods. By applying these advanced statistical techniques, we aimed to describe the dependency between the surrounding temperature and the wood moisture of the specimen, thereby contributing to the ongoing discourse on the sustainable use of timber in construction [21,27].

2. Materials and Methods

2.1. Materials

The experiment utilized a compact profile of Norway spruce (*Picea abies* (L.) KARST.) wood with cross-sectional dimensions of 138 mm × 138 mm and a length of 273 mm, as described in [10,27]. To determine the essential characteristics and ensure the quality of the timber, we measured its average density (oven-dry density) according to the ČSN 49 0108 standard. The recorded density value was 312.0 ± 14.1 kg·m^{-3}. Additionally, the density profile was evaluated using an X-ray beam on a QTRS-01X Tree Ring Analyzer (Quintek Measurement Systems Inc., Knoxville, TN, USA). The automated sample measurements, utilizing the QTRS-01X software (Quintek Measurement Systems Knoxville, Knoxville, TN, USA) with a step size of 0.01 mm (Figure 1), provided us with the average growth ring width, measuring 6.48 ± 0.80 mm.

In order to mitigate the potential presence of internal defects such as hidden knots, cracks, or other irregular growth patterns, the sample was subjected to a full-volume CT scan using a Siemens Somatom Scope Power CT scanner (Siemens Healthineers, Erlangen, Germany). Figure 2 represents an illustrative image from this scan, specifically focusing on the area where the temperature and humidity sensors were placed. This comprehensive scan revealed no significant issues that could influence the evaluated properties. Furthermore, a microscopic assessment (Figure 3) was conducted to verify the wood's classification as spruce.

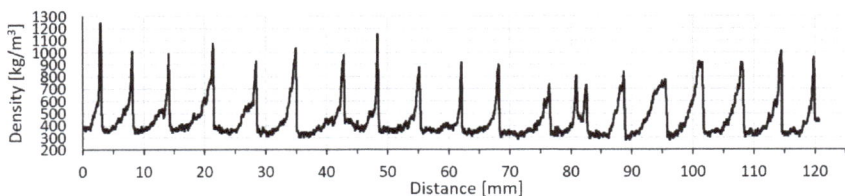

Figure 1. The density profile of the analyzed specimen.

Figure 2. Cross-sectional view of the evaluated timber with no visible defects, as captured using the CT scanner.

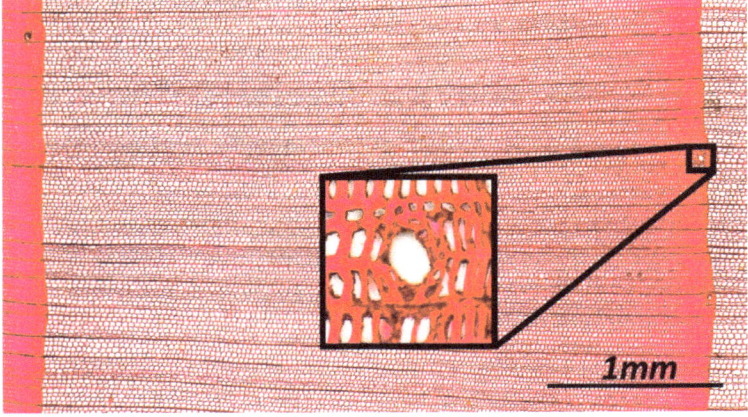

Figure 3. Spruce growth ring traits in an analyzed specimen: a cross-section highlighting resin canals and the gradual transition between earlywood and latewood.

2.2. Data Collection

The sample was exposed to the exterior environment at VSB TU Ostrava, Faculty of Civil Engineering. Wood moisture content measurements were conducted using the ALMEMO FHA 696 MFS1 sensor (Ahlborn, Holzkirchen, Germany), and temperature values were captured with the ALMEMO FPA 686 sensor (Ahlborn, Holzkirchen, Germany). Data collection was facilitated with the ALMEMO 710 data logger (Ahlborn, Holzkirchen, Germany). The experiment was conducted from 15 November 2016 to 14 September 2018, spanning a total duration of 669 days. The moisture content (M %) and external temperature (T °C) were continuously measured and recorded at minute intervals during this period.

2.3. Initial Observation

The most common indicator for measuring the linear relationship between two variables is the correlation coefficient, which is the reason why it is given first here. Similarly, (ordinary) least squares regression is the simplest and easiest way to model such a relationship. Thus, both can be taken as a good starting point.

The graph in Figure 4 shows the daily averages of the temperature and moisture measurements over the entire experiment, i.e., 669 days from 15 November 2016 to 14 September 2018. One can observe a decrease in the moisture content during the days with temperatures above 15 °C and an increase when the temperatures were low. This expected behavior fully agrees with the calculated value of the correlation coefficient, which is $cor(T, M) = -0.7022$.

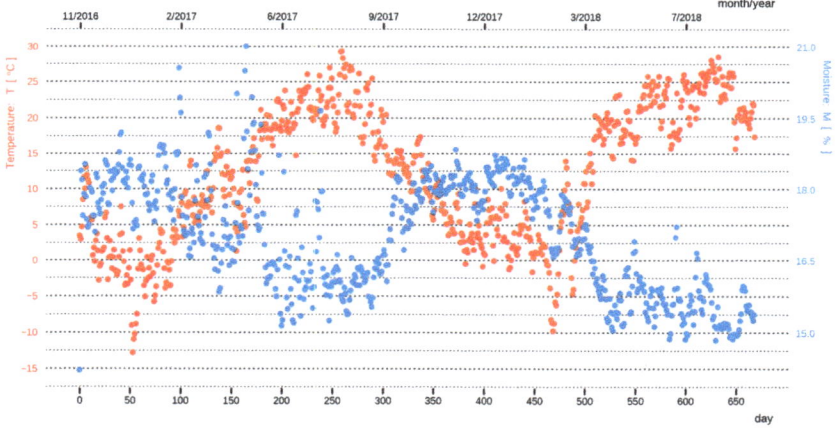

Figure 4. Daily means of measured temperature (red points/values/axis) and moisture (blue points/values/axis) values during entire experiment. (Ordinary) least squares regression leads to the following linear function:

$$M = 18.12 - 0.09013T, \qquad (1)$$

where the negative coefficient of -0.09013 once again indicates decreasing moisture values with increasing temperature. The formula results, compared with the measured values, are shown in the scatter plot in Figure 5. Each point on the plot has coordinates given by the measured temperature and moisture values. The predicted values are represented by the green line. The closer the points on the scatter plot are to the predicted curve, the better the model fits. Here, one group of points above the line (temperatures around 5 °C and moisture around 18%) and another group below the line (temperatures around 20 °C and moisture around 16%) can be seen.

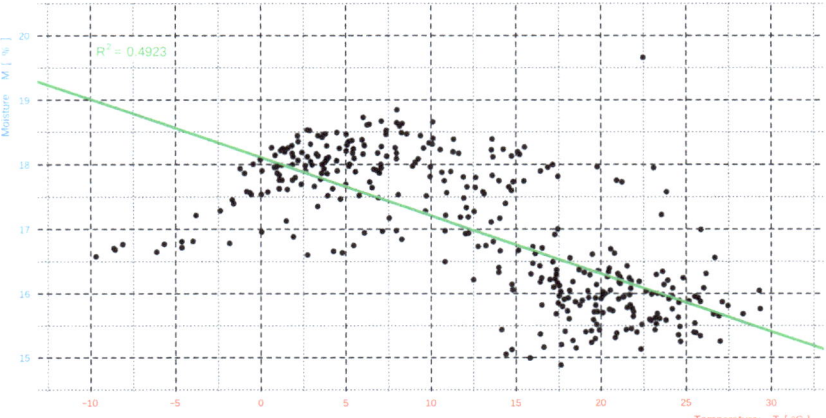

Figure 5. Measured temperature and moisture scatter plot with linear formula.

Several indicators can be used to describe the quality of the model and its performance. Among them, *R-squared*, R^2; the root mean square error, *RMSE*; the mean absolute error, *MAE*; the Nash–Sutcliffe efficiency coefficient, *NSE*; and the performance index, *PI*, are used here. While R^2, the *RMSE*, and the *MAE* are widely and regularly used, efficiency indices and other information criteria are not as common. However, with advances in the use of machine learning methods, these indices are gaining increasing use. In [28], the model and its features were selected based on the *RMSE* and Wilmott's efficiency coefficient *d*, with Wilmott's d being the basis of the performance index, *PI*, used here.

The Nash–Sutcliffe efficiency index is defined in [29], and a recent comparison with other indices can be found in [30]. Its power as a model performance criterion was successfully demonstrated in [31,32].

The *R-squared*, R^2 (or so-called coefficient of determination), can be viewed as the percentage of variance of the dependent variable explained by the model. Its values range is [0, 1], with a value of 1 indicating a perfect fit; obviously, the higher the value is, the better the model is.

Both the root mean square error, *RMSE*, and the mean absolute error, *MAE*, are measures of the error between the observations and the predicted values. Both are nonnegative, with 0 being a perfect fit; of course, the lower the value is, the better the model is. The *RMSE* and *MAE* values are expressed in the same unit terms as the observations, so that they are easy to interpret.

Finally, two quantifiers of model quality are used, namely, the Nash–Sutcliffe coefficient of efficiency, *NSE*, and the performance index, PI. Both characterize the predictive power of the model, the difference being their sensitivity to outliers of the predictors and, hence, the predictions.

The *NSE* values range is $(-\infty, 1]$, with negative values indicating an unsuitable model; NSE = 0 for the predictive power of the average observation and $NSE = 1$ indicates a complete fit. Usually, a model with $NSE > 0.6$ is considered good.

The value of the efficiency index, *PI*, is calculated as the product of the Willmott (efficiency) index d and the Pearson correlation coefficient *r*. The value of *PI* ranges from 0 to 1. This index, *PI*, together with the following evaluation table, was proposed by Camargo in [33]. More information about related topics could be found in [34], where a wide range of model performance indices are discussed. Based on *PI*, the model is considered poor for $PI \leq 0.6$, satisfactory for $0.6 < PI \leq 0.65$, good for $0.65 < PI \leq 0.7$, and, finally, very good for $PI > 0.7$.

The values describing the linear model are:

$$R^2 = 0.4923, \quad RMSE = 0.8499, \quad MAE = 0.6309, \quad NSE = 0.493, \quad PI = 0.5655.$$

These values are not very satisfactory, and therefore, the model needs to be improved.

Note that the values of R^2 and NSE are very close to each other, which is a common property of these measures. Their main difference lies in usage; while R^2 quantifies the goodness of fit of a given statistical model, NSE indicates how well the outcome is predicted by the model simulation. The R^2 is computed on the same dataset on which the model is trained, while the NSE can be used on completely new, unknown data. This is the reason why the NSE is preferred when evaluating model performance. Therefore, the NSE, rather than R^2, will be examined in all the considerations below.

2.4. Data Processing

2.4.1. Reasons for the Insufficiency of the Linear Model

The quality of the linear model may be affected by the length of the experiment. Here, its duration was less than two years. This yielded a disbalance between the observations from different seasons. Specifically, in the Central European climate conditions, the experiment took place in Ostrava, which meant a shortage of autumn measurements.

Thus, in order to deal with this problem, we decided to use values measured across one whole year. The second and more likely reason is the insufficiency of temperature as a predictor. This is because the effect of temperature is not instantaneous, i.e., a time lag must be assumed, and therefore, the moisture models, as a function of previous temperatures, must be verified.

There are (at least) two ways to incorporate previous temperatures into the model. The first is to use the past values themselves; the second is to use the past values in terms of their moving averages.

Polynomial, i.e., quadratic, cubic, and biquadratic models should also be investigated, since the moisture and temperature correspondence need not to be linear.

2.4.2. Year Choice

There are several options for choosing the year in which the model will be trained. Although the first and, in a sense, most natural option was to use 2017, the year starting 1 July 2017 and ending 30 June 2018 was chosen.

The reasons for this decision are illustrated by the box plots in Figure 6 and the associated Tables 1 and 2. Through them, five different time periods are compared, specifically, the entire experimental period and the years starting with the given dates, i.e., 1 January 2017, 1 April 2017, 1 July 2017, and 1 September 2017.

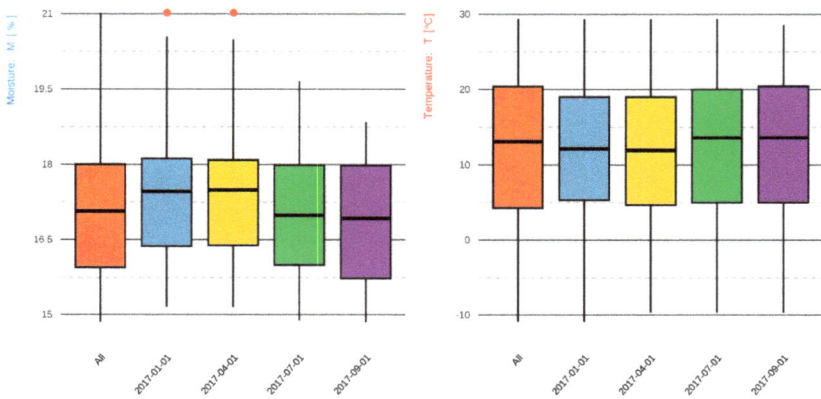

Figure 6. Comparison of different time periods by means of moisture (**left**) and temperature (**right**); outliers represented by red dots.

Table 1. Moisture measurement characteristics.

	All	1 January 2017	1 April 2017	1 July 2017	1 September 2017
Min	14.227	15.151	15.151	14.882	14.851
1st Q	15.947	16.369	16.387	16.002	15.733
Median	17.069	17.462	17.494	16.992	16.931
Mean	17.008	17.326	17.287	17.003	16.864
3rd Q	18.001	18.116	18.087	17.991	17.982
Max	21.017	21.017	21.017	19.66	18.845
IQR	2.054	1.747	1.699	1.989	2.249
sd	1.195	1.073	1.025	1.07	1.183

Table 2. Temperature measurement characteristics.

	All	1 July 2017	1 April 2017	1 July 2017	1 September 2017
Min	−12.746	−12.746	−9.692	−9.692	−9.692
1st Q	4.271	5.306	4.685	5.019	5.019
Median	13.068	12.121	11.91	13.599	13.599
Mean	12.302	11.781	11.745	12.561	12.773
3rd Q	20.392	18.994	18.994	20.003	20.437
Max	29.357	29.357	29.357	29.357	28.577
IQR	16.121	13.688	14.309	14.984	15.419
sd	9.307	8.619	8.42	8.722	8.958

The goal is to find a year that most closely matches the measurement for the entire period. The moisture boxplot, as seen in Figure 6, shows that there are outliers in the cases 1 July 2017 and 1 April 2017 and also that the median moisture values in these years are higher than the median for the whole experiment. The years 1 July 2017 and 1 September 2017 look similar to the whole period. This assessment is supported by the values in Table 1, where the exact differences in the medians and other statistical characteristics can be observed. The final selection of 1 July 2017 was due to its *IQR* (interquartile range) value, which is close to the moisture *IQR* for all the data. This choice was tested with the Kruskal–Wallis (rank sum test) and Dunn (multiple comparison) post hoc tests, which confirmed this decision.

The temperature measurements were examined and tested in a similar manner, and no differences were found between the proposed years. Therefore, the moisture-based selection described above remains valid.

Thus, all the models were trained on the dataset derived from the observations from the year between the dates of 1 July 2017 and 31 August 2018. Surely, the rest of the measurements were not discarded but were used further as model testing data.

2.4.3. Predictor Choice

As mentioned above, one way to improve the power of the model is to add more predictors and increase the degree of the formula, i.e., to also examine quadratic, cubic, etc., polynomial regression.

As an equally important option, one should examine the effect of past temperatures, i.e., describe the current moisture value in terms of lagged temperatures, T_i, where the index i denotes the number of days lagged. For example, T_1 is yesterday's temperature.

Moving averages can also be used. Let $ma(T)_{i,j}$ denote the moving average of the temperature over i days with a lag of j days. Here, for example, $ma(T)_{3,0}$ is the three-day moving average with the current day temperature included. Similarly, $ma(T)_{5,1}$ is a five-day moving average starting with yesterday's temperature.

The plots in Figure 7 indicate the impact of the dependence between humidity and lagged temperature in the correlation coefficients, their powers, or moving averages on the lag length.

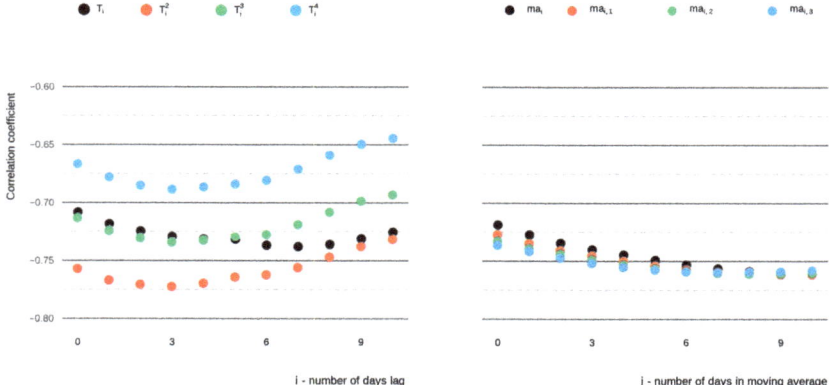

Figure 7. Dependence of the correlation coefficient on the length of the time lag; lagged temperatures and their powers (**left**), moving averages (**right**).

It can easily be observed that adding an increasing number of lag days does not necessarily lead to an increase in the correlation. The highest value for the linear term of temperature is reached with a lag of 7 days, and it is equal to $cor(M, T_7) = -0.738$, while for the quadratic term, a lag of 3 days is the best choice, with the value $cor(M, T_3^2) = -0.7721$.

The same is true for increasing the degrees of the polynomial terms, where the lowest correlation coefficient is obtained for the highest power considered, i.e., for the biquadratic terms of T_i^4. The highest correlation appears in the case of the quadratic terms T_i^2.

The right-hand side of Figure 7 shows the situation for the moving averages, which is different, because the correlation coefficients do not change significantly when time lags are added. In $ma_{i,j}$ (with T for temperature omitted here), the first index i indicates the number of days for the moving average, while the second index j indicates the length of the time shift, e.g., $ma_{5,3}$ is a moving average of 5 days shifted by 3 days.

In all cases, the maximum values are around -0.76 for moving averages of 7 or 8 days. Their differences are of the order of hundredths. This is neither a problem nor an argument for excluding the time-lagged moving averages from further consideration. The reason why time-lagged moving averages are examined is the possibility of their use together with the temperature values of T. Since moving averages are (of course) computed from expressions of T, their joint use causes multicollinearity of the predictors in the regression model.

The above discussion led to the decision to investigate models containing temperature terms with a maximum lag of 7 days and a polynomial of degree 3. The maximum length of the considered moving average was also set to 7 days.

This decision was based not only on the above discussion but also on the natural assumption that temperatures with a time lag of more than 7 days are too distant to affect the actual moisture.

Similar reasoning led to the elimination of the actual temperature terms, since temperature cannot affect moisture instantaneously. Furthermore, a strong relationship between the current (e.g., today's) temperature T and the previous day's (e.g., yesterday's) temperature T_1, expressed by a correlation coefficient equal to $cor(T, T_1) = 0.9688$, must be taken into account.

Although the correlation coefficients of these terms suggest a strong relationship, their rejection seems to be meaningful from a physical point of view. The goal is to construct not a mathematical model with the highest rank but a model corresponding to the real conditions.

2.5. Model Preparation

After the above reduction in the number of possible predictors, 21 predictors still remained, specifically, temperatures with a time lag of a maximum of 7 days and their first, second, and third powers.

Building a model with such a large number of predictors is tedious, as well as time- and power-consuming, and more importantly, because the temperature terms are correlated, there is still the possibility that the resulting model will be overestimated.

The following steps can be taken to address this problem. One starts with a simple least squares regression model with all the predictors and checks not only the useful measures, such as the R-squared or RSS, but also, in particular, the significance of the coefficients obtained. Those predictors that do not meet the criteria (*p*-value and significance level) can be rejected and the model can be rebuilt without them. These steps are repeated until a model with the desired properties is obtained.

The described procedure need not to be performed manually in a loop, since there are special predictor selection methods to speed it up.

Here, the so-called *stepAIC* method was used. This method builds models based on different subsets of the considered predictors and selects the one with the lowest AIC value, where AIC stands for the Akaike Information Criterion. This criterion quantifies the loss of information caused by modifying the set of predictors. This modification can be performed through a forward step, i.e., adding a new predictor; a backward step, i.e., removing a predictor; or the combination of both directions.

The AIC penalizes the model for adding more variables to it. Its absolute value is not important, because the procedure only compares its tendency, i.e., whether it decreases or increases as the predictors change. This is similar to the modified R-squared used in ordinary least squares methods.

This criterion was introduced by Akaike in [35], and its usage was further developed over time. A comprehensive summary of its use in statistical modeling and a selection of regression models can be found in [36].

The procedure of the model's preparation and its final tuning and testing was implemented in the statistical computing language *R* [37] in its IDE (Integrated Development Environment), *RStudio* [38], with the help of the specialized packages *MASS* [39] and *Caret* (Classification and Regression Training) [40] for model preparation, tuning, and final selection and *Tidyverse* [41] for the initial data description, analysis, tidying, and cleansing. Finally, *ggplot2* [42] was used for the visualizations.

Illustration of Modeling Procedure

According to the procedure described above, 30 ordinary least squares models were prepared. These models were divided into three groups: models containing only linear terms (denoted as *Linear*), models containing linear and quadratic terms (denoted as *Quadratic*), and finally, models combining linear, quadratic, and cubic terms (denoted as *Cubic*). There were 10 models in each of these groups, each based on a different number of lagged temperature values.

We denoted these models as L_{ols}^i, Q_{ols}^i, and C_{ols}^i, where L stands for *Linear*, Q for *Quadratic*, and C for *Cubic*; the superscript i denotes the maximum lag length number used here; and the subscript denotes the method used, i.e., *ols* stands for *ordinary least squares* here.

In this notation, Q_{ols}^3 is a quadratic model with a maximum lag of 3 days, i.e., an ordinary least squares model predicting moisture based on T_1, T_1^2, T_2, T_2^2, T_3, and T_3^2. Together with the intercept, 7 coefficients need to be found.

Increasing the number of lagged temperatures and their powers yields longer and, therefore, more complicated formulas. For Q_{ols}^3, considered here, one obtains the following formula:

$$Q_{ols}^3 = 17.82 + 0.03486 T_1 - 0.002875 T_1^2 - 0.02456 T_2 + 0.0004491 T_2^2 + 0.01611 T_3 - 0.00249 T_3^2.$$

The prediction power of this formula is described with these values:

$Training: RMSE = 0.6674, MAE = 0.5177, NSE = 0.61, PI = 0.6776,$
$Testing: RMSE = 0.7451, MAE = 0.5728, NSE = 0.6123, PI = 0.6743,$

where the first set of model characteristics are computed on the *training dataset*, i.e., the year from 1 July 2017 to 30 June 2018, and the second set of characteristics are computed on the entire dataset, i.e., the observations from the entire experimental period, which serve here as the *test dataset*.

The main difference between the training and test evaluations is in the measures of the *RMSE* and *MAE*, which are increased using all the data. The values of the performance indicators *NSE* and *PI* remain almost the same.

After building, exploring, and rating the ordinary least squares model, the *stepAIC* procedure is performed to improve the predictive power of the model and reduce the number of predictors used.

The above model Q^3_{ols} is transformed into a new model Q^3_{aic} in the following form with the following prediction power:

$$Q^3_{aic} = 17.83 + 0.253T_1 - 0.002715T_1^2 - 0.002159T_3^2. \quad (2)$$

$Training: RMSE = 0.6678, MAE = 0.5184, NSE = 0.6096, PI = 0.6772,$
$Testing: RMSE = 0.7446, MAE = 0.5729, NSE = 0.6129, PI = 0.6747.$

As can easily be observed, the *RMSE*, *MAE*, *NSE*, and *PI* values remain the same; thus, the improvement lies in the reduction in the number of predictors. The Q^3_{ols} formula contains 7 terms, namely, the intercept and 6 lagged temperatures, while Q^3_{aic} contains only 4, i.e., the intercept and 3 temperature terms, and only 2 of them are quadratic.

3. Results and Discussion

3.1. Model Comparison and Rating

In order to compare the performances of all the models developed, Table 3 summarizes their *RMSE*, *MAE*, *NSE*, and *PI* values calculated on the test data from the whole experimental period. Here, one can find either the one with the lowest error (*RMSE* or *MAE*) or the one with the highest predictive ability (*NSE* or *PI*).

Table 3. The performance characteristics of proposed ordinary least square models; (L—linear, Q—quadratic, C—cubic; for different time lags).

	RMSE			MAE			NSE			PI		
Lag	L	Q	C	L	Q	C	L	Q	C	L	Q	C
0	0.847	0.786	0.774	0.640	0.601	0.564	0.499	0.568	0.582	0.565	0.632	0.648
1	0.835	0.772	0.765	0.634	0.59	0.558	0.513	0.583	0.591	0.579	0.647	0.657
2	0.823	0.759	0.754	0.625	0.583	0.548	0.527	0.597	0.603	0.592	0.660	0.669
3	0.812	0.745	0.742	0.617	0.573	0.542	0.539	0.613	0.615	0.604	0.675	0.680
4	0.806	0.739	0.737	0.614	0.569	0.541	0.547	0.619	0.621	0.612	0.681	0.685
5	0.803	0.733	0.736	0.611	0.562	0.542	0.550	0.625	0.621	0.615	0.686	0.685
6	0.793	0.728	0.736	0.604	0.559	0.542	0.561	0.630	0.621	0.626	0.691	0.685
7	0.791	0.731	0.736	0.603	0.561	0.542	0.563	0.627	0.621	0.628	0.689	0.685
8	0.790	0.731	0.736	0.603	0.561	0.542	0.565	0.627	0.621	0.629	0.689	0.685
9	0.791	0.731	0.736	0.603	0.561	0.542	0.563	0.627	0.621	0.628	0.689	0.685
10	0.791	0.731	0.739	0.603	0.561	0.545	0.563	0.627	0.619	0.628	0.689	0.683

The use of Table 1 leads to the rejection of all the linear models because none of them satisfy either the *NSE* or the *PI* criterion for a good model, namely, *NSE* < 0.6 and *PI* < 0.65 for all of them. For the same reason, the quadratic models with a lag of less than 3 days and cubic models with a lag of less than 2 days are also rejected.

In addition, Table 3 also shows the changes in the values of the performance characteristics depending on the change in the type of model, i.e., from quadratic to cubic, or the change in the number of predictors, i.e., the maximum time lag used. These changes are very small, being of the order of thousandths, for time lags longer than 2 days. Therefore, there are again natural questions as to whether the use of multiple predictors will yield the intended goal, i.e., a formula that is not only highly ranked but also easy to use and understand.

The highest values of the *NSE* and *PI* are obtained with formula Q^6_{aic}, which is given and described as follows:

$$Q^6_{aic} = 17.83 + 0.03612T_1 - 0.002586T_1^2 - 0.001022T_2^3 - 0.002586T_5^2 - 0.001022T_6^2. \quad (3)$$

$Training: RMSE = 0.6442, MAE = 0.4961, NSE = 0.6367, PI = 0.7019,$
$Testing: RMSE = 0.7281, MAE = 0.5593, NSE = 0.6298, PI = 0.6909.$

The differences in the *RMSE*, *MAE*, *NSE*, and *PI* between the best model Q^6_{aic} and the model Q^3_{aic} proposed above are:

$$\Delta RMSE = -0.01644, \Delta MAE = -0.01362, \Delta NSE = 0.0169, \Delta PI = 0.01613,$$

Thus, one could conclude that although there is no doubt that Q^6_{aic} outperformed Q^3_{aic} in all the parameters, the level of improvement is not high enough to justify a higher number of predictors.

The dependence of the *RMSE*, *MAE*, *NSE*, and *PI* on the number of predictors, i.e., the time lag used, is also shown in Figures 8–11. These figures visually provide the same information as that shown in Table 3. It was possible to observe an improvement in all the characteristics with incrementing time lags until the best value was reached, after which they started to stagnate or even decrease.

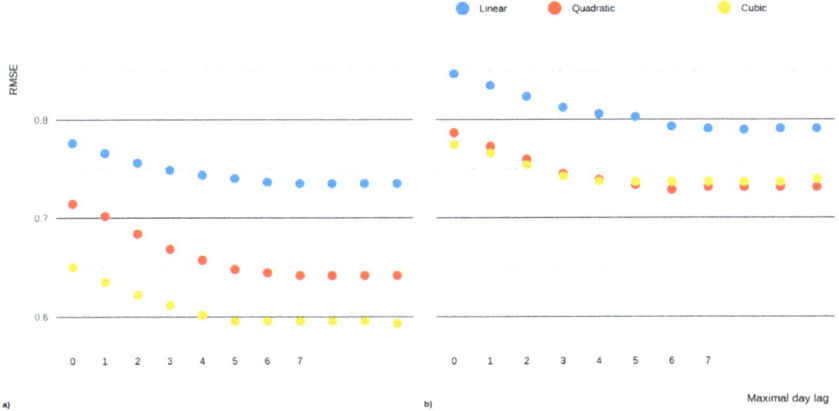

Figure 8. RMSE values for models with different time lags used: (**a**) training data; (**b**) all data.

The figures show the results for both datasets, i.e., the training and the testing parts. In the training part, the cubic models performed significantly better than the quadratic models. However, the performance examined for testing, i.e., for new and unknown data, is more important in deciding which model to choose. Additionally, at this point, the figures show a noticeable increase in the closeness between the performances of the quadratic and cubic models on the test dataset. Furthermore, note that for the *RMSE*, *NSE*, and *PI*, the quadratic models slightly outperform the cubic models for time lags greater than 6 days.

Finally, note that Table 3 and Figures 8–11 confirm the correctness of the assumptions made in the previous sections, particularly the rejection of the models with time lags longer than 7 days. For emphasis, results are also presented for the models with time lags longer than 7 days and up to 10 days.

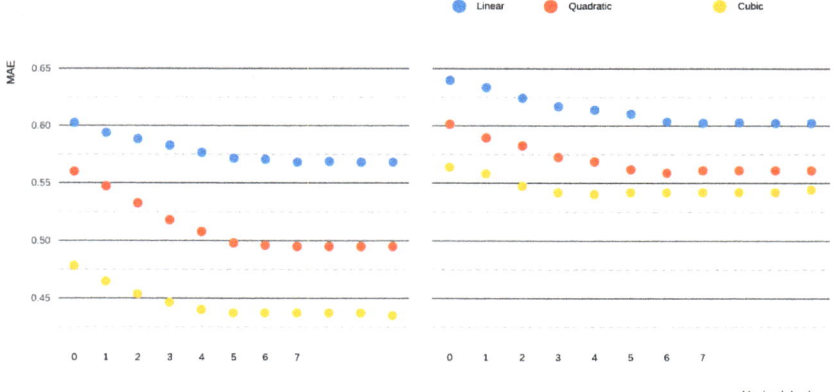

Figure 9. MAE values for models with different time lags used: (**a**) training data; (**b**) all data.

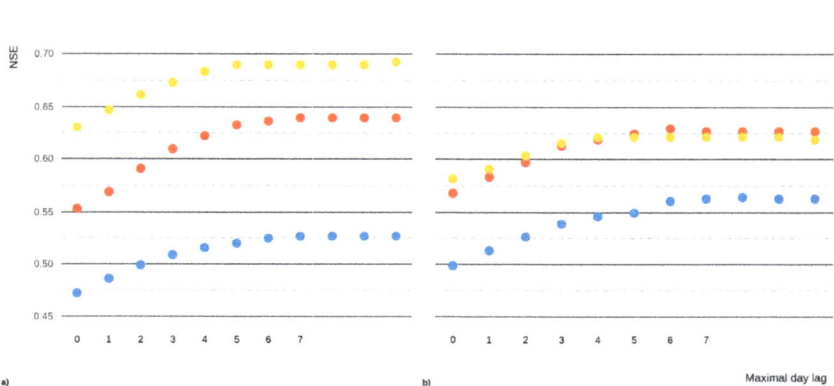

Figure 10. NSE values for models with different time lags used: (**a**) training data; (**b**) all data.

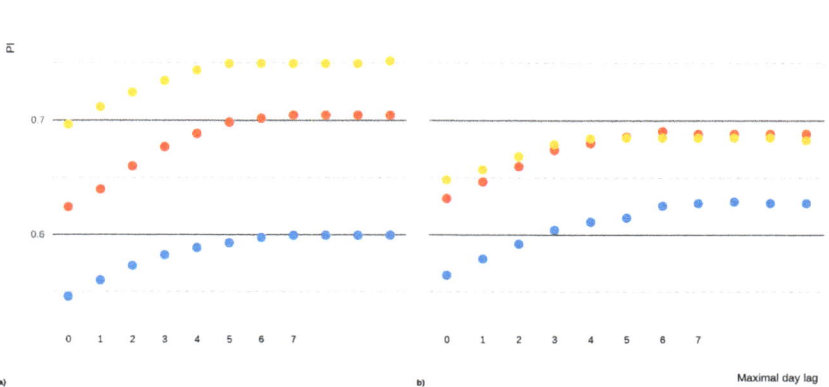

Figure 11. PI values for models with different time lags used: (**a**) training data; (**b**) all data.

3.2. Final Tuning

The choice of the quadratic model as the final formula introduces a new problem. The formula Q_{aic}^3 is a quadratic function of two variables and as such reaches its global maximum. Specifically, for $T_1 = 4.67\ °C$ and $T_3 = 0\ °C$, the maximum moisture predicted by this model is $M_{max} = 17.89\%$.

The existence of a global maximum is true for all the quadratic and cubic models obtained by means of the method presented above.

In order to solve this problem, the entire model underwent a new tuning process. In this second run, moving averages of different lengths and different time delays were added. Under this new setup, ordinary least squares models were again built and then adjusted using the stepAIC methods.

Then, their *RMSE*, *MAE*, *NSE*, and *PI* characteristics were calculated and examined, and the models were ranked based on them. Finally, the final model was selected. This was a model derived from the Q_{aic}^3 model (2) described in the previous section.

The final model had the form:

$$M = 17.91 + 0.08295 T_1 - 0.004101 T_1^2 - 0.07851 ma(T)_{7,1}, \qquad (4)$$

where the $ma(T)_{7,1}$ is a moving average of 7 days, shifted by 1 day.

The form of the final model (4) is very simple because the quadratic term T_3^2 was suppressed through the stepAIC procedure. The current moisture is therefore given as a linear function of the temperatures of the previous seven days, except for the term representing the previous day, which is quadratic, and such a function cannot be bounded. Thus, the problem of the global maxima and, hence, the bounded predictions is solved.

The predictive power of the final model is given by the following values:

Training : $RMSE = 0.6676$, $MAE = 0.5145$, $NSE = 0.6099$, $PI = 0.6775$,
Testing : $RMSE = 0.7355$, $MAE = 0.5617$, $NSE = 0.6222$, $PI = 0.6834$,

which are slightly better than those of the base model Q_{aic}^3, as seen in Equation (2), and slightly worse than those of the model Q_{aic}^6, in Equation (3), with the best rank among all the models.

In addition to the above statistics, we computed the coefficient of determination of the final model. This index was mentioned in the Introduction in connection to the basic linear formula in Equation (1). Recall that its value can be understood as a measure of dependent variable variance, which is explained by the model.

The final model (4) reached a value of 0.7016, whereas for the basic model (1), it was only 0.4923. This is a significant improvement, which is also true in the case of all the measures of the final model compared to the basic model.

However, as mentioned above, the model characteristics are not the only criteria that justify the choice of the final model. Its simplicity also needs to be taken into account.

In addition to the numerical characteristics and indices, the performance of the final model is illustrated in two graphs.

The graph in Figure 12 shows the measured and predicted moisture throughout the entire experiment. The significant descriptive values of the measurements and predictions are indicated by the markers on the y-axis; in particular, their minimum, maximum, first and third quartiles, and median are shown.

The graph in Figure 13 is a scatter plot that provides a representation of the agreement between the moisture measured and the moisture predicted by the final model. Each point on this graph has coordinates given as a pair (*prediction, measurement*); the closer the predicted value is to the measured value, the closer it is to the $x = y$ line. The line $x = y$ capturing this ideal state is plotted as a red dashed line in Figure 13. Note that the closer the points are to the $x = y$ line, the higher the *NSE* value is.

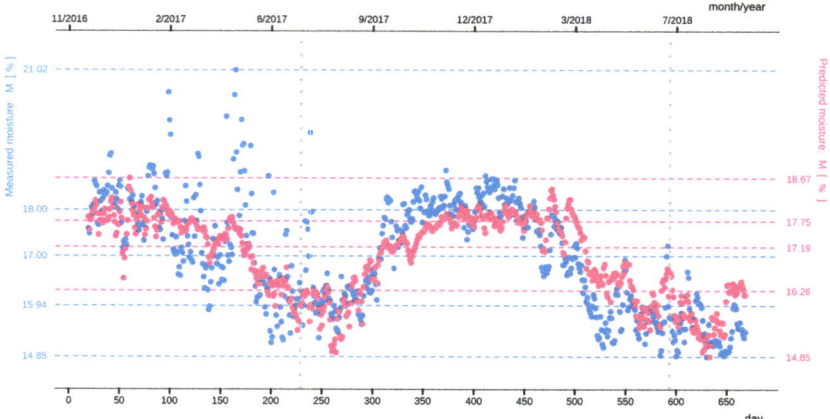

Figure 12. Measured and predicted moisture during the entire experiment.

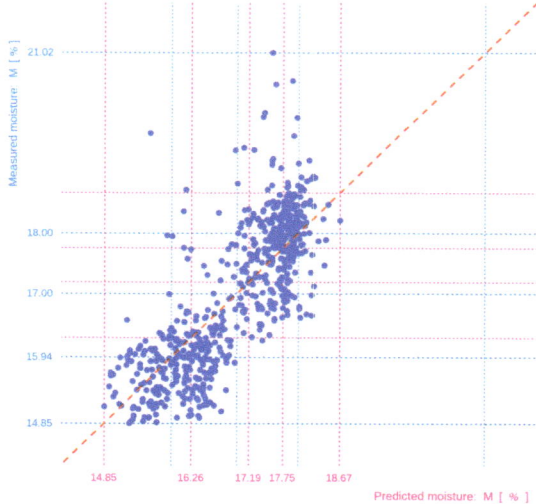

Figure 13. Measured and predicted moisture values scatter plot.

The graph in Figure 12 depicts higher differences between the measurements and simulations for the first half of 2017, and the scatter plot in Figure 13 contains points that are further from the desired $x = y$ line. Thus, it can be seen that the model could not accurately perform predictions in all circumstances. This is due to the fact that it is only a temperature model, which does not take into account other meteorological parameters such as precipitation in the form of either rain or snow in winter.

However, as a simple temperature-based model, the proposed final model is very effective.

4. Conclusions

A simulation experiment was conducted from 2016 to 2018, and the prestressing force, external temperature, and moisture content of wood samples were measured. The main objective of this simulation was to propose a formula describing the dependence of prestress on time, temperature, and the moisture content.

Since temperature and moisture are strongly related, their measurements were re-examined together to obtain a description of their relationship. This was the aim of the work presented in this paper.

Our examination of the data led to the proposal of a model for predicting the actual moisture content as a function of temperature values with different time lags. Decisions regarding the exact length of the time lag and the maximal power of the polynomial temperature term were made based on the results of stepAIC regression methods, which are designed to reduce the number of predictors while preserving (or improving) the predictive power of the regression model. As the proposed model is only a temperature model, it cannot capture all the moisture changes and fluctuations perfectly; to increase its power, other external influences also need to be taken into account. However, based on the evaluation of the final model using multiple statistical indicators, its high predictive ability was proven.

Author Contributions: Conceptualization, R.F., M.L. and V.D.; methodology, M.L., V.D., R.F., A.Z., M.S. and A.S.; software, V.D.; validation, V.D., M.L. and R.F.; formal analysis, V.D., M.L., R.F., A.Z. and M.S.; investigation, R.F.; resources, R.F., M.S., A.Z. and M.L.; data curation, V.D.; writing—original draft preparation, M.L., V.D., R.F., A.Z., M.S. and A.S.; writing—review and editing, M.L.; visualization, V.D.; supervision, R.F.; project administration, R.F.; funding acquisition, R.F., M.S. and A.S. All authors have read and agreed to the published version of the manuscript.

Funding: This research was funded by EVA 4.0.—Advanced research supporting the forestry and wood-processing sector's adaptation to global change and the 4th industrial revolution, OP RDE (grant number CZ.02.1.01/0.0/0.0/16_019/0000803).

Data Availability Statement: Data are available on request.

Conflicts of Interest: The authors declare no conflict of interest.

References

1. Buchanan, A.H.; Levine, S.B. Wood-Based Building Materials and Atmospheric Carbon Emissions. *Environ. Sci. Policy* **1999**, *2*, 427–437. [CrossRef]
2. Karjalainen, T.; Kellomäki, S.; Pussinen, A. Role of Wood-Based Products in Absorbing Atmospheric Carbon. *Silva Fennica.* **1994**. *28*, 67–80. [CrossRef]
3. Mantau, U. Wood Flow Analysis: Quantification of Resource Potentials, Cascades and Carbon Effects. *Biomass Bioenergy* **2015**, *79*, 28–38. [CrossRef]
4. Belgacem, M.N.; Pizzi, A. (Eds.) *Lignocellulosic Fibers and Wood Handbook: Renewable Materials for Today's Environment*; Scrivener Publishing: Beverly, MA, USA; Wiley: Beverly, MA, USA; Hoboken, NJ, USA, 2016; ISBN 978-1-118-77352-9.
5. Dinwoodie, J.M. *Timber*; CRC Press: Boca Raton, FL, USA, 2000; ISBN 978-1-135-80810-5.
6. Mitterpach, J.; Fojtík, R.; Machovčáková, E.; Kubíncová, L. Life Cycle Assessment of a Road Transverse Prestressed Wooden–Concrete Bridge. *Forests* **2022**, *14*, 16. [CrossRef]
7. Record, S.J. *The Mechanical Properties of Wood: Including a Discussion of the Factors Affecting the Mechanical Properties and Methods of Timber Testing*; Publishers Printing Company: New York, NY, USA, 1914.
8. Bodig, J.; Jayne, B.A. *Mechanics of Wood and Wood Composites*; Krieger Pub Co.: Malabar, FL, USA, 1993; ISBN 978-0-89464-777-2.
9. Vasiliev, V.V.; Morozov, E.V. *Advanced Mechanics of Composite Materials and Structures*, 4th ed.; Elsevier: Amsterdam, The Netherlands, 2018; ISBN 978-0-08-102209-2.
10. Fojtík, R.; Dubovský, V.; Kozlová, K.; Kubíncová, L. Prestress Losses in Spruce Timber. *Wood Res.* **2020**, *65*, 645–652. [CrossRef]
11. Wenzel, H. *Health Monitoring of Bridges*; Wiley: Chichester, UK, 2009; ISBN 978-0-470-03173-5.
12. Wadi, H.; Amziane, S.; Taazount, M. The Lateral Load Resistance of Unclassified Cross-Laminated Timber Walls: Experimental Tests and Theoretical Approach. *Eng. Struct.* **2018**, *166*, 402–412. [CrossRef]
13. Wang, J.; Cao, X.; Liu, H. A Review of the Long-Term Effects of Humidity on the Mechanical Properties of Wood and Wood-Based Products. *Eur. J. Wood Prod.* **2021**, *79*, 245–259. [CrossRef]
14. Startsev, O.V.; Makhonkov, A.; Erofeev, V.; Gudojnikov, S. Impact of Moisture Content on Dynamic Mechanical Properties and Transition Temperatures of Wood. *Wood Mater. Sci. Eng.* **2017**, *12*, 55–62. [CrossRef]
15. Vololonirina, O.; Coutand, M.; Perrin, B. Characterization of Hygrothermal Properties of Wood-Based Products—Impact of Moisture Content and Temperature. *Constr. Build. Mater.* **2014**, *63*, 223–233. [CrossRef]
16. Buell, T.W.; Saadatmanesh, H. Strengthening Timber Bridge Beams Using Carbon Fiber. *J. Struct. Eng.* **2005**, *131*, 173–187. [CrossRef]

17. Lukowsky, D.; Wade, D. *Failure Analysis of Wood and Wood-Based Products*; Mc Graw-Hill education: New York, NY, USA, 2015; ISBN 978-0-07-183937-2.
18. Obućina, M.; Kitek Kuzman, M.; Sandberg, D. *Use of Sustainable Wood Building Materials in Bosnia and Herzegovina, Slovenia and Sweden*; University of Sarajevo, Mechanical Engineering Faculty, Department of Wood Technology: Sarajevo, Bosnia and Herzegovina, 2017.
19. Fojtík, R.; Dědková, K. Analysis of Diagnostic Methods for Detecting the Presence of *Gloeophyllum* spp. *Wood Res.* **2018**, *63*, 479–486.
20. Thybring, E.E.; Fredriksson, M. Wood Modification as a Tool to Understand Moisture in Wood. *Forests* **2021**, *12*, 372. [CrossRef]
21. Fojtík, R.; Kubíncová, L.; Dubovský, V.; Kozlová, K. Moisture at Contacts of Timber-Concrete Element. *Wood Res.* **2020**, *65*, 917–924. [CrossRef]
22. Sandberg, K.; Pousette, A.; Nilsson, L. Moisture Conditions in Coated Glulam Beams and Columns during Weathering. In Proceedings of the International Conference on Durability of Building Materials and Components, Porto, Portugal, 12–15 April 2011; pp. 1–8.
23. Almeida, G.; Hernández, R.E. Changes in Physical Properties of Tropical and Temperate Hardwoods below and above the Fiber Saturation Point. *Wood Sci. Technol.* **2006**, *40*, 599–613. [CrossRef]
24. Brischke, C.; Rapp, A.O.; Bayerbach, R. Measurement System for Long-Term Recording of Wood Moisture Content with Internal Conductively Glued Electrodes. *Build. Environ.* **2008**, *43*, 1566–1574. [CrossRef]
25. Reiterer, A.; Sinn, G.; Stanzl-Tschegg, S.E. Fracture Characteristics of Different Wood Species under Mode I Loading Perpendicular to the Grain. *Mater. Sci. Eng. A* **2002**, *332*, 29–36. [CrossRef]
26. Zhang, S.Y. Effect of Growth Rate on Wood Specific Gravity and Selected Mechanical Properties in Individual Species from Distinct Wood Categories. *Wood Sci. Technol.* **1995**, *29*, 451–465. [CrossRef]
27. Fojtík, R.; Dubovský, V.; Kubíncová, L.; Stejskalová, K.; Machovčáková, E.; Lesňák, M. Probabilistic Expression of the Function of the Change in Prestressing Force of Timber Elements Depending on Climatic Conditions in Situ. *Constr. Build. Mater.* **2023**, *377*, 130955. [CrossRef]
28. Fransson, E.; Eriksson, F.; Erhart, P. Efficient Construction of Linear Models in Materials Modeling and Applications to Force Constant Expansions. *NPJ Comput. Mater.* **2020**, *6*, 135. [CrossRef]
29. Nash, J.E.; Sutcliffe, J.V. River Flow Forecasting through Conceptual Models Part I—A Discussion of Principles. *J. Hydrol.* **1970**, *10*, 282–290. [CrossRef]
30. Duc, L.; Sawada, Y. A Signal-Processing-Based Interpretation of the Nash–Sutcliffe Efficiency. *Hydrol. Earth Syst. Sci.* **2023**, *27*, 1827–1839. [CrossRef]
31. Dubovský, V.; Dlouhá, D.; Pospíšil, L. The Calibration of Evaporation Models against the Penman–Monteith Equation on Lake Most. *Sustainability* **2020**, *13*, 313. [CrossRef]
32. Dlouhá, D.; Dubovsky, V.; Pospíšil, L. Optimal Calibration of Evaporation Models against Penman–Monteith Equation. *Water* **2021**, *13*, 1484. [CrossRef]
33. Camargo, A.P.; Sentelgas, P.C. Avaliação Do Desempenho de Diferentes Métodos de Estimativa Da Evapotranspiração Potencial No Estado de São Paulo, Brasil. *Rev. Bras. Meteorol.* **1997**, *5*, 89–97.
34. Sobenko, L.R.; Pimenta, B.D.; Camargo, A.P.D.; Robaina, A.D.; Peiter, M.X.; Frizzone, J.A. Indicators for Evaluation of Model Performance: Irrigation Hydraulics Applications. *Acta Sci. Agron.* **2022**, *45*, e56300. [CrossRef]
35. Akaike, H. A New Look at the Statistical Model Identification. *IEEE Trans. Automat. Contr.* **1974**, *19*, 716–723. [CrossRef]
36. Sakamoto, Y.; Ishiguro, M.; Kitagawa, G. *Information Criteria and Statistical Modeling*; Springer: New York, NY, USA, 2006; ISBN 978-0-387-71886-6.
37. R Core Team. *R: A Language and Environment for Statistical Computing*; R Foundation for Statistical Computing: Vienna, Austria, 2023.
38. RStudio, PBC. *RStudio Team RStudio: Integrated Development Environment for R*; RStudio, PBC: Boston, MA, USA, 2020.
39. Venables, W.N.; Ripley, B.D. *Modern Applied Statistics with S*, 4th ed.; Springer: New York, NY, USA, 2002.
40. Kuhn, M. Building Predictive Models in R Using the Caret Package. *J. Stat. Softw.* **2008**, *28*, 1–26. [CrossRef]
41. Wickham, H.; Averick, M.; Bryan, J.; Chang, W.; McGowan, L.D.; François, R.; Grolemund, G.; Hayes, A.; Henry, L.; Hester, J.; et al. Welcome to the Tidyverse. *J. Open Source Softw.* **2019**, *4*, 1686. [CrossRef]
42. Wickham, H. *Ggplot2: Elegant Graphics for Data Analysis*; Springer: New York, NY, USA, 2016; ISBN 978-3-319-24277-4.

Disclaimer/Publisher's Note: The statements, opinions and data contained in all publications are solely those of the individual author(s) and contributor(s) and not of MDPI and/or the editor(s). MDPI and/or the editor(s) disclaim responsibility for any injury to people or property resulting from any ideas, methods, instructions or products referred to in the content.

Article

European Beech Log Sawing Using the Small-Capacity Band Saw: A Case Study on Time Consumption, Productivity and Recovery Rate

Răzvan V. Câmpu * and Rudolf A. Derczeni

Department of Forest Engineering, Forest Management Planning and Terrestrial Measurements, Transilvania University of Braşov, Şirul Beethoven No. 1, 500123 Braşov, Romania; derczeni@unitbv.ro
* Correspondence: vasile.campu@unitbv.ro; Tel.: +40-729-123-450

Abstract: In rural, isolated areas, sawmills are often equipped with one or more small-capacity hand-fed band saws. Even in this situation, the productivity of the band saw must be viewed through the factors that influence it, namely the characteristics of logs and the optimization of the stages and activities carried out. Therefore, time consumption, the structure of working time and the recovery rate in sawing logs into lumber provide important information for users. The structure of the sawing operation for a work team made up of an operator and an assistant was divided into six work stages. The sawing pattern used involves sawing the log up to approximately half of the diameter, then rolling the log with 180° and continuing the sawing, aiming to obtain lumber with a thickness of 40 and 50 mm from the central part of the log. The productivity was 2.45 $m^3 \cdot h^{-1}$, the recovery rate was 70.84% and the working time real-use coefficient was 0.37. Research has highlighted the positive correlation between working time and the middle diameter of the logs (R^2 = 0.84). The feeding speed was also determined along with the quality of cuts, which was expressed by the thickness uniformity of the lumber and the presence of cutting teeth traces on the newly created surfaces.

Keywords: lumber; band saw; productivity; working time structure; recovery rate

1. Introduction

In Romania, European beech (*Fagus sylvatica* L.) is the most widespread species, occupying about 33.4% of the national forest surface which amounts to 6.604 million hectares, representing 27.7% of the country's surface [1]. Beech forests are found in the Carpathian Mountains, starting from altitudes of 300 m to altitudes of 1400 m [2]. In 2020, 19.7 million cubic meters of wood were harvested from Romanian forests, of which 6.110 million cubic meters (31%) was beach wood [1]. The wood is harvested by 5185 accredited logging companies. In 2020, about 64.1% was harvested from forests owned by the state, 32.3% from private forests and 3.6% from forest vegetation located on lands outside the forest [1]. Individual owners may harvest at most 20 $m^3 \cdot year^{-1}$ of the total forest that they own. Private properties represent 51.9% of the national forest surface [1]. In 2011, there were approximately 830,000 private properties, of which 828,000 had a surface smaller than 10 hectares and 2200 had a surface above 10 hectares. The number of private properties is on the rise because the restitution process is not over yet and there are still numerous lawsuits taking place. It may be estimated though that when the restitution process is over, state-owned forests will represent about 40% of the national forest surface [3,4].

In [5], it is mentioned that there are about 7000 wood processing factories that produce about 4.47 million m^3 of lumber annually, many of them being of small capacity, with a production of 8–10 m^3 per day. Taken together, the forestry and wood processing industries have a contribution of 3.5% to the country's GDP, representing an important branch of the economy in which approximately 128,000 employees are working, while another 186,000 are working in related sectors [6]. Sawmills that have a small sawing capacity of up to

Citation: Câmpu, R.V.; Derczeni, R.A. European Beech Log Sawing Using the Small-Capacity Band Saw: A Case Study on Time Consumption, Productivity and Recovery Rate. *Forests* **2023**, *14*, 1137. https://doi.org/10.3390/f14061137

Academic Editor: Gianni Picchi

Received: 30 April 2023
Revised: 22 May 2023
Accepted: 29 May 2023
Published: 31 May 2023

Copyright: © 2023 by the authors. Licensee MDPI, Basel, Switzerland. This article is an open access article distributed under the terms and conditions of the Creative Commons Attribution (CC BY) license (https://creativecommons.org/licenses/by/4.0/).

5000 m^3·year^{-1} of beech logs often use one or more band saws with a sawing capacity of 4–8 m^3·shift^{-1}. In Romania, small sawmills are important components of the industry because they substitute large sawmills in the market conditions in which the latter could not operate feasibly [7]. They use logs with unproductive dimensions and characteristics for large factories, they supply the local market, substitute imports, provide jobs and bring revenues to local budgets through the sale of manufactured products [8,9]. Moreover, the purchase cost is low, which is an important aspect especially in rural areas. One such piece of equipment is the band saw FBO–03–CUT, used in the present article and made in Romania by S.C. Cutean Company Srl.

In general, in the literature of the field, the articles published address two main aspects: (i) the study of the productivity of machinery and the factors on which the latter depends [5,7,10,11]; and (ii) worker health and safety, risk factors, accidents and occupational disease analysis [7,12–15]. The purpose of the present article falls within the first direction of research and is represented by the determination of time consumption and productivity of the FBO–03–CUT band saw when sawing beech logs into lumber, in order to optimize activities and increase work productivity.

2. Materials and Methods

2.1. Field Data Collection and Equipment

The research was carried out in a beech lumber factory from Buzău county, which has as its object of activity for the processing of beech wood, the final product being beech lumber. The measurements took place at the end of February 2021, the beech logs being partially frozen.

To achieve the goal of this paper, the research methodology had to respond to the following challenges: (1) description of the band saw used for sawing logs into lumber; (2) description of the work team and the work tasks of each worker; (3) establishment of a work time structure by stages and activities that allow for the grouping of activities carried out when sawing logs into lumber; and (4) the determination of productivity, recovery rates, coefficient of use of the working time and feeding speed of the band saw.

The band saw FBO–03–CUT is a horizontal one, with manual advance, intended for sawing logs into lumber and into beams up to 160 mm thick, with a maximum length of 7.5 m (Figure 1).

FBO–03–CUT is operated by two workers, an operator and an assistant. The operator carries out the following activities: together with the assistant, he or she feeds the band saw platform with logs and fixes the log with clamps; adjusts the lumber thickness; ensures the advance of the saw; gives instructions to the assistant so that the rolling logs correspond to the sawing scheme used; changes the cutting blade at regular time intervals or when the situation requires it; and informs the technical team if the band saw is not working properly. The assistant participates in feeding the platform, rolls the logs according to the instructions received from the operator, fixes the logs on the platform and manually removes lumber and sawdust from the sawing area.

In the research undertaken, the sawing pattern used involves sawing the log up to approximately half of the diameter, then rolling the log with 180° and continuing the sawing, aiming to obtain lumber with a thickness of 40 and 50 mm from the central part of the log. Toward the periphery of the section, the log was sawed into lumber with a thickness of 25 mm (Figure 2).

Figure 1. The band saw FBO–03–CUT.

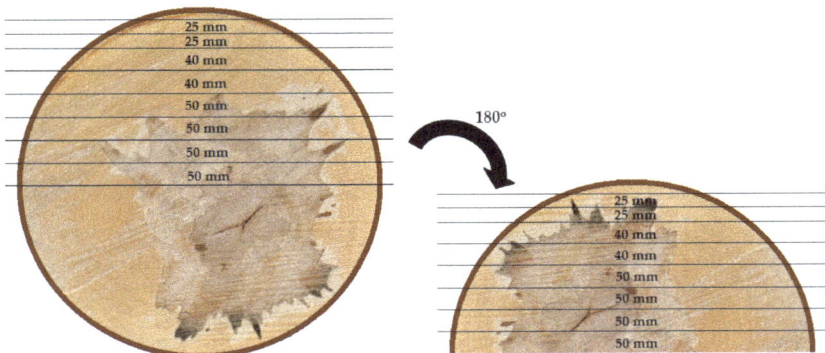

Figure 2. The sawing pattern used for sawing beech logs.

The consumption of working time and its structure was analyzed at the level of stages and activities. Activities which were strictly necessary from a technological point of view for the normal development of the production process were considered work stages. To these, a series of activities which were not absolutely necessary from a technological point of view was added. Their acceptance was justified in order to ensure the conditions imposed by work safety norms, ergonomic, physical and physiological requirements [16].

The structure of the sawing operation for a piece of lumber, for the work team made up of an operator and an assistant, was divided into six work stages according to Table 1. The activities specific to each stage are presented in Table 2.

Table 1. Stages in the sawing operation.

Stage	Symbol	Begins	Ends
1. Loading and fixing the log on the band saw platform	LFL	When the operator and assistant start feeding a new log to the band saw platform	When the operator and assistant have finished clamping the log to the band saw platform
2. Slabs sawing	SS	When the operator manually moves the band saw	When the slabs are detached from the log
3. Setting the thickness of the lumber/slabs	STL	When the operator sets the lumber thickness adjustment mechanism	When the operator has finished setting the thickness of the lumber
4. Lumber sawing	LS	When the operator manually moves the band saw	When the piece of lumber is completely detached from the log
5. Lumber evacuation	LE	When the assistant removes the sawdust from the piece of lumber	When the piece of lumber reaches the conveyor that feeds the circular edging saw
6. Returning the band saw to the start position	RBS	When the operator begins to move the band saw to the starting position	When the band saw reaches the starting position and the sawing cycle can be resumed

Table 2. Working time structure for sawing beech logs (adapted from [17]).

Working Time Structure					Operation	Stage	Activity
WP		NT				-	Delays caused by meals, breaks, rest, personal needs and organization
	WT	PW	MW		Log sawing	SS	Slab sawing
						LS	Lumber sawing
						STL	Setting the thickness of the lumber/slabs
			CW			LE	Removal of sawdust from the piece of lumber; Lumber evacuation
						RBS	Returning the band saw to the start position
		SW	PT	SU		LFL	Loading and fixing the log on the band saw platform; Rolling the log corresponding to the sawing pattern and fixing it with clamps
			ST	MT		-	Changing the cutting blade

Note: WP—workplace time; NT—non-working time; WT—working time; PW—productive working time; SW—supportive working time; MW—main working time; CW—complementary working time; PT—preparatory time; SU—set-up time; ST—service time; MT—maintenance time.

Time was measured in seconds, by using the continuous time study method. A stopwatch was used to measure time by recording the beginning and the ending of each stage or activity. The time for delays and their causes (change of cutting blade, adjustments of the band saw, removal of sawdust from the area of the band saw, meal break, personal needs, rest, etc.) were also measured.

After sawing each piece of lumber, the minimum width of each piece on the narrowest face and the thickness at the ends of the piece and in the middle were measured. The minimum width was measured with a tape measure, and it corresponded to the width of the edged lumber. The thickness was measured with a caliper, both measurements being expressed in millimeters. Thus, based on the width and the average thickness of the lumber pieces, their volume was calculated. The quality of the cut was also evaluated, observing the presence of cutting teeth marks on the faces of the timber pieces and the uniformity of the thickness of timber pieces.

2.2. Data Analysis

Ensuring the statistical representativeness of the research conducted is the first step in data analysis. For this, the economic agent was asked about the size of the lot of logs that were to be sawed using the above-mentioned sawing pattern. The answer to the question was about 50 m^3, representing 94 beech logs with mid-diameters between 35 and 70 cm, and lengths between 2.70 and 3.60 m according to SR 2024-1993 (the national standard regarding the dimensions and quality of beech logs for lumber). In order to determine the minimum number of logs subjected to measurements, it was necessary to determine the coefficient of variation of the mid-diameter. The mid-diameters were measured for 30 logs, the calculated coefficient of variation being 22.95% (Table 3).

Table 3. Characteristics of logs and resulting lumber. Statistical indicators of the mid-diameter of average logs and the recovery rate.

No. Log	Log Characteristics			Log Volume	Number of Cuts	Lumber Volume	Recovery Rate
	Length	Mid-Diameter					
	m	cm	Statistical Indicators	m^3	no.	m^3	%
1	3.6	47		0.625	9	0.396	63.39
2	3.0	53		0.662	11	0.464	70.07
3	2.9	68		1.053	14	0.846	80.29
4	3.3	45		0.525	10	0.414	78.85
5	3.0	40		0.377	9	0.278	73.82
6	3.0	45		0.477	11	0.370	77.61
7	3.4	35		0.327	9	0.184	56.35
8	2.7	36		0.275	8	0.160	58.24
9	3.1	66		1.061	14	0.894	84.33
10	2.8	42		0.388	9	0.260	66.94
11	4.0	49	Mean mid-diameter (cm)—47.03;	0.754	10	0.641	85.01
12	3.0	62	Standard deviation—10.794;	0.906	11	0.750	82.81
13	3.0	40	Variation coefficient (%)—22.95;	0.377	7	0.259	68.71
14	3.0	52	Minimum (cm)—35;	0.637	11	0.484	76.01
15	3.2	52	Maximum (cm)—68;	0.680	11	0.460	67.66
16	3.1	39	Mean recovery rate (%)—70.84;	0.370	9	0.278	75.03
17	3.2	36	Standard deviation—9.891;	0.326	9	0.174	53.31
18	3.0	61	Variation coefficient (%)—13.96;	0.877	14	0.750	85.52
19	3.4	45	Minimum (%)—53.31;	0.541	9	0.368	68.12
20	2.9	35	Maximum (%)—85.52;	0.279	8	0.173	62.11
21	3.2	35		0.308	8	0.185	60.24
22	3.0	40		0.377	9	0.278	73.75
23	3.2	38		0.363	9	0.292	80.53
24	3.0	66		1.026	14	0.832	81.05
25	2.7	66		0.924	14	0.778	84.27
26	3.2	35		0.308	9	0.169	54.83
27	3.2	52		0.680	11	0.461	67.89
28	2.8	37		0.301	8	0.165	54.74
29	3.1	41		0.409	9	0.273	66.70
30	3.0	53		0.662	11	0.424	64.05
			Total	16.873	-	12.461	73.86

Further, the number of necessary measurements was established by using the relation suggested by [18]:

$$n = \frac{u^2 \cdot S_\%^2 \cdot N}{N \cdot \Delta_\%^2 + u^2 \cdot S_\%^2} \quad (1)$$

n—the minimum number of logs that will be sawed;
$u = 1.96$—the standard deviation of normal distribution, corresponding to the transgression probability $\alpha = 5\%$;

$S\% = 22.95\%$—the variation coefficient of the middle diameter for the logs analyzed;
$\Delta = \pm 10\%$—the limit error;
$N = 94$—the total number of logs in the lot intended for sawing.

Knowing the parameters that come into play when establishing the number of sample pieces, by applying the above formula, a log number—$n = 17$—was obtained. Because $n < 30$, the result obtained was considered a temporary value n', n being recalculated with the same formula where u is replaced by t (t Student distribution) [18]. The value of t is determined according to the number of freedom degrees $f = n' - 1$ and according to α. At 16 freedom degrees and $\alpha = 5\%$ it yields that $t = 2.120$.

By applying the formula again, a number of 19 logs was obtained. A great number of measurements were made (30 logs, those for which the coefficient of variation of the mid-diameter was determined) in order to normalize the distribution of the values measured and minimize the Hawthorne effect.

Productivity defined as the ratio of input to output [19–21], is a synthetic indicator which defines the production capacity level of use in a system under certain work conditions and it is expressed, in log sawing, usually in the following form $m^3 \cdot h^{-1}$, respectively:

$$W = \frac{V}{TU} \quad (2)$$

W—the productivity;
V—the log volume sawed in a time unit (m^3);
TU—the time unit taken into consideration (hour, work shift, etc.).

Furthermore, by dividing the working time corresponding to the LS and SS stages by the workplace time, the working time real-use coefficient (K) when sawing the logs could be established [10]:

$$K = \frac{TLS + TSS}{WP} \quad (3)$$

K—the working time real-use coefficient;
TLS—the working time consumed in the LS stage;
TSS—the working time consumed in the SS stage;
WP—workplace time.

The recovery rate, when logs are sawed into lumber, was expressed as a percentage representing the ratio of the volume of lumber obtained to the volume of sawed logs [11,22,23]:

$$RR = \frac{V_{lumber}}{V_{logs}} \cdot 100 \quad (4)$$

RR—the recovery rate (%);
V_{lumber}—the volume of lumber obtained (m^3);
V_{logs}—the volume of sawed logs (m^3).

The feeding speed was calculated in two ways: (1) as a ratio of the cumulative length of the lumber pieces resulting from a log to the working time TLS, corresponding to the LS stage being expressed in $m \cdot min^{-1}$; (2) as a ratio of the cumulative area of the sawing cuts to the working time TLS, being expressed in $m^2 \cdot min^{-1}$.

Further, the correlation between the working time and mid-diameter of the logs was tested by using the simple linear regression. The statistical interpretation of the regression was done using an ANOVA.

3. Results and Discussion

Based on the working time consumed for the realization of each work stage, the structure of the working time for sawing logs was established (Table 4). Thirty logs totaling to a volume of $16.87\ m^3$ were sawed. The total working time WP required for sawing the logs was 24,796 s.

Table 4. Working time structure for sawing beech logs.

Number of Logs	Volume	WP									
		WT								NT	
		PW				SW					
		MW		CW		PT (SU)		ST (MT)			
	m³	s·m⁻³	%	s·m⁻³	%	s·m⁻³	%	s·m⁻³	%	s·m⁻³	%
30	16.87	542	36.85	463	31.53	260	17.72	143	9.72	61	4.18
Total		1005 s·m⁻³				403 s·m⁻³					
Overall Total		1469 s·m⁻³									

The working time structure, by time elements, is presented in Figure 3.

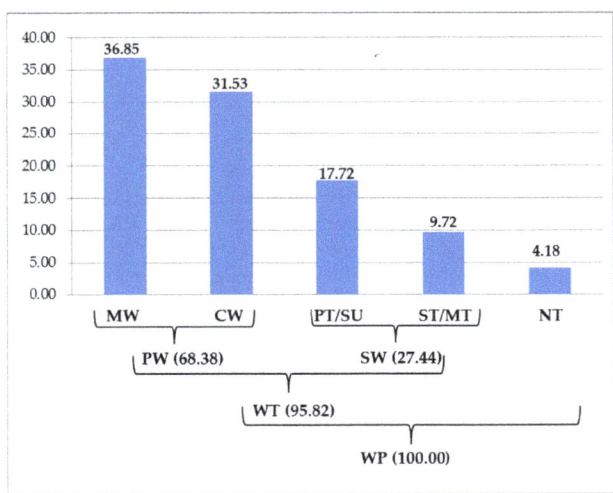

Figure 3. Working time structure according to time elements (%): *WP*—workplace time; *NT*—non-working time; *WT*—working time; *PW*—productive working time; *SW*—supportive working time; *MW*—main working time; *CW*—complementary working time; *PT*—preparatory time; *SU*—set-up time; *ST*—service time; *MT*—maintenance time.

It is observed that the *WT–NT* ratio was 96%:4%, indicating the main share of working time. The distribution of working time, by stages, is presented in Figure 4. It can be seen that the *LS* and *LFL* stages have the largest share; together they represent 47.9% of the working time. The *LS* stage represents 30.18% and is explained by the fact that the advance of the band saw is ensured manually through pushing by the operator. Additionally, *LFL* represents 17.72%, this operation being performed manually. In [24], it was stated that the operating time required to feed the saw with logs, turn the logs and fix them in the clamps, can be reduced by the proper training of the workers. A large share is also represented by delays (13.9%) being caused by technical and personal needs.

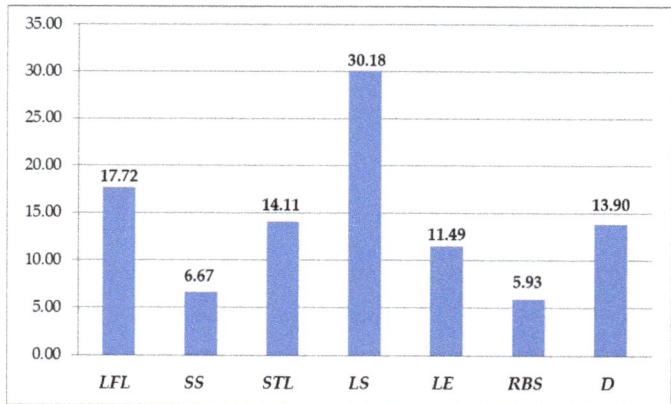

Figure 4. Distribution of working time by stages (%): *LFL*—loading and fixing the log on the band saw platform; *SS*—slab sawing; *STL*—setting the thickness of the lumber; *LS*—lumber sawing; *LE*—lumber evacuation; *RBS*—returning the band saw to the start position; *D*—delays.

Among the dendrometric characteristics of logs, working time best correlates with the mid-diameter ($R^2 = 0.84$). In [24], similar results were also obtained, where $R^2 = 0.702$. The link between the two variables was established by the means of simple linear regression, the results being presented in Table 5. Thus, a strong positive correlation was highlighted between the studied variables ($0.75 < r < 0.95$). In this situation, the coefficient of determination R^2 shows that the dependence of working time on the middle diameter is 84%, 16% of the variation of the working time being caused by other factors.

Table 5. Linear regressions analysis of *WT* in relation to the *dm*.

	ANOVA			The Significance of the Variable Coefficient				
R^2	Standard Error	Degrees of Freedom	F	Variable	Coefficient	Standard Error	t Statistic	p-Value
0.84	144.43	Regression 1 Residual 28	141.50 ***	Constant	−563.54	119.80	−4.70	<0.001 ***
				dm	29.56	2.49	11.90	<0.001 ***

Note: Asterisks denote *F* significance and significant correlations: *p*-value *** < 0.001; *WT*—working time; *dm*—middle cross-section diameter.

The productivity of the FBO–03–CUT band saw was 2.45 m$^3 \cdot$h^{-1} of the sawed logs, which means that 38.73 m$^2 \cdot$h^{-1} of lumber was obtained. In a similar study, [10] obtained a productivity of 2.90 m$^3 \cdot$h^{-1}. In [7], it was established that productivity is often influenced by the human factor through its preparation and the decisions made, and by the variable characteristics of the logs with respect to their dimensions and quality, which limit the operation of the band saw at its maximum capacity. The productivity obtained 2.45 m$^3 \cdot$h^{-1}, which is generally higher than in manual mobile band saws 0.64–0.86 m$^3 \cdot$h^{-1} [25]. Productivity is also influenced by the thickness of the logs. The larger the thickness of the log, the smaller the number of cuts for sawing a log and the higher the productivity [24].

The working time real-use coefficient *K* was 0.37 and falls within the range of 0.3–0.5, specific to the band saws with manual advance [26,27]. According to [10], an increase in the *K* coefficient can be obtained by automating the phases of loading, rolling and fixing the logs and by increasing the speed of the *RBS* stage. By increasing the feeding speed, respectively, by increasing the coefficient *K*, an increase in the productivity of the band saw definitely occurs. However, this increase is not linear; [10] show that for a 100% increase in feeding speed, productivity increased by about 20%. The variable characteristics of the logs in terms of their dimensions and quality influence the feeding speed and thus the working time and the productivity of the machine [28].

In the case of the band saw analyzed, the feeding speed is not constant; it is ensured by pushing the band saw manually and it varies, as previously shown, with the following sawing conditions: the presence of wood defects, the frozen state of the wood, the wear of the cutting blade, etc. Analysis of the results obtained (Table 6) showed that the variation of the feeding speed is higher when it is expressed in $m \cdot min^{-1}$, the coefficient of variation being 39.46%. When the feeding speed is expressed in $m^2 \cdot min^{-1}$, the coefficient of variation is 22.41%. Thus, it can be said that the width of lumber pieces, conditioned by the diameter of the logs, influences the feeding speed. The larger the width of the lumber, the greater the effort made by the operator to manually ensure that the feeding speed becomes lower. Therefore, even though logs of a larger diameter are sawed with a reduced feeding speed, the log band saw achieves a greater capacity and vice versa. The effect of log volume on productivity is greater than the influence of the adjustments in the log feeding speed [24].

Table 6. Descriptive statistics of the variation in the feeding speed.

Descriptive Statistics	Feeding Speed	
	$m \cdot min^{-1}$	$m^2 \cdot min^{-1}$
Mean	9.826	2.339
Standard deviation	3.877	0.524
Variation coefficient (%)	39.46	22.41
Minimum	3.904	1.506
Maximum	16.150	3.696

The average feeding speed obtained in the present case (9.826 $m \cdot min^{-1}$) is intermediate between the feeding speed obtained by [24] (15.40 $m \cdot min^{-1}$) when sawing with a band saw equipped with a hydraulic carriage and [5] (4.74 $m \cdot min^{-1}$) when cutting with a manual mobile band saw.

The recovery rate obtained (70.84%) was high, being determined by the thickness of the lumber (40 and 50 mm thick lumber predominates) and the width of the kerf of 3 mm. Ref. [5] obtains a similar recovery rate between 62 and 74% in beech logs sawed into lumber with a thickness of 30 and 50 mm. If other factors are kept constant, a reduced kerf should increase the recovery rate, since fewer fibers are lost in the form of sawdust [29]. The larger the thickness of the lumber is, the fewer the cuts needed for sawing a log and thus, the higher the recovery rate. Moreover, the log size and quality have a major impact to recovery rate [30]. Losses can also be caused by log quality expressed by the severity of wood defects.

Regarding the quality of the cuts, the presence of the cutting teeth traces was found on the faces of the lumber pieces starting from 3% and reaching up to 10% of the surface of their faces. This is caused by the deviation of the cutting teeth from the set of cutting blade or the vibrations of the blade [31] as a result of the presence of wood defects, the frozen state of the logs, the presence of foreign bodies, low blade tension force etc., associated with manual, uneven feeding speed of the saw when cutting. In general, any asymmetry in the cutting teeth caused by mounting, grinding or damage may result in the generation of a lateral force [32,33] which can cause cutting deviations. This shows that the percentage of the ST time element in the working time structure cannot be reduced. By replacing and maintaining the cutting blade. An appropriate quality of the cut must be ensured, thus reducing the manufacturing defects for the elimination of which lumber pieces must be subjected to in additional processing operations [27]. Regarding the accuracy of the cuts expressed by obtaining pieces of lumber of the same thickness, it was found that a percentage of 5% of the pieces of lumber presented a thickness variation of more than 3 mm between the thicknesses measured at the ends of the piece and those measured in the middle. In general, the cutting precision depends on the size of the cutting forces caused by feeding speed, the thickness of logs [34], the rotation speed, the sharpening of the blade, the setting system of the lumber thickness being sawed [10], the vibrations and the wear of the cutting blade [31]. A decrease in blade stability causes a vibration that leads to side deflection resulting in cutting deviations, surface roughness, wear and the deformation of the blade [35]. Further, the sawing of frozen logs is difficult, because, at temperatures below $-10\ °C$, the sawdust freezes on the cut surfaces and causes severe dimensional inaccuracies [36]. The cutting blade deviation from the sawing plane during the cut also leads to obtaining lumber with an uneven thickness that requires additional processing operations, reducing the recovery rate and increasing production costs [37]. The debarking of logs, the constant feeding speed of the saw and the avoidance of sawing frozen logs would probably have considerably reduced the non-conformities related to the quality of cuts and the variation in the thickness of lumber pieces.

4. Conclusions

In this paper, time consumption, productivity and recovery rate in the process of sawing beech logs into lumber were studied. The results showed that the time consumed when sawing logs into lumber was 1469 s·m^{-3}, depending mainly on the mid-diameter of the log ($R^2 = 0.84$).

The productivity obtained (2.45 m^3·h^{-1}) was influenced by the thickness of the logs, the feeding speed and the operator's decisions and skills. Productivity could be increased by reducing the time spent on manual activities through adequate operator training. Although increasing the feeding speed leads to increased productivity, this possibility is limited to manual band saws.

The recovery rate obtained (70.84%) is high, being determined by the thickness of the lumber, the width of the kerf and by the quality of the logs.

Regarding the quality of the cuts, the presence cutting teeth traces was found on the faces of lumber pieces starting from 3% and reaching up to 10% of the surface of their faces. Additionally, it was found that a percentage of 5% of the pieces of lumber presented a thickness variation of more than 3 mm. Literature in the field, mentions numerous factors that can determine the cutting blade deviation from the sawing plane during the cut or the blade vibration. Among them, specific to this research, is the partially frozen state of the logs.

For beech sawmills that use low-capacity band saws, the results obtained regarding time consumption, productivity, recovery rate and quality of cuts can be used for a better understanding of the factors that influence the above-mentioned characteristics and thus, make the best decisions regarding the design and organization of technological flows in sawmills.

Author Contributions: Conceptualization, R.V.C. and R.A.D.; methodology, R.V.C. and R.A.D.; validation, R.V.C. and R.A.D.; formal analysis, R.V.C.; investigation, R.V.C.; resources, R.V.C.; data curation R.V.C. and R.A.D.; writing—original draft preparation, R.V.C.; writing—review and editing, R.V.C.; visualization, R.V.C. and R.A.D.; supervision, R.V.C. and R.A.D. All authors have read and agreed to the published version of the manuscript.

Funding: This research received no external funding.

Data Availability Statement: Not applicable.

Acknowledgments: The authors would like to thank Eng. Radu Florin Robert for the contributions brought to the collection of data and for facilitating the relations with the factory where the research was carried out.

Conflicts of Interest: The authors declare no conflict of interest.

References

1. Institutul Național de Statistică. *Statistica Activităților din Silvicultură în Anul 2020*; Institutul Național de Statistică: Bucharest, Romania, 2021; p. 30.
2. Stănescu, V.; Șofletea, N.; Popescu, O. *Flora Forestieră Lemnoasă a României*; Editura Ceres: Bucharest, Romania, 1997; p. 516.

3. World Bank. Analiza Funcțională a Administrației Publice Centrale Din România. Analiza Funcțională a Sectorului Mediu și Păduri în România. Final Report. Project Cofinanced by the European Social Fund through the Operational Program—Development of Administartive Skills, World Bank, Washington, DC, USA. 2011; p. 55. Available online: https://sgg.gov.ro/docs/File/UPP/doc/rapoarte-finale-bm/etapa-II/MMP_FR_Environment_Water_Forestry_Vol_1_Main_Report_ROM_FINAL.pdf (accessed on 12 February 2023).
4. Ciubotaru, A.; Câmpu, V.R. Delimbing and Cross-Cutting of Coniferous Trees–Time Consumption, Work Productivity and Performance. *Forests* **2018**, *9*, 206. [CrossRef]
5. Gligoraș, D.; Borz, S.A. Factors affecting the effective time consumption, wood recovery and feeding speed when manufacturing lumber using a FBO-02 cut mobile bandsaw. *Wood Res.* **2015**, *60*, 329–338.
6. PWC. Industria Silvică și de Prelucrare a Lemnului din România. Comunicat de Presă. 2016. Available online: https://www.pwc.ro/en/press_room/assets/2016/wood-industry-ro.pdf (accessed on 14 February 2023).
7. Borz, S.A.; Oghnoum, M.; Marcu, M.V.; Lorincz, A.; Proto, A.R. Performance of Small-Scale Sawmilling Operations: A Case Study on Time Consumption, Productivity and Main Ergonomics for a Manually Driven Bandsaw. *Forests* **2021**, *12*, 810. [CrossRef]
8. FAO. *Small and Medium Sawmills in Developing Countries*; Forestry Paper N. 28; FAO: Rome, Italy, 1981.
9. Cheța, M.; Marcu, M.V.; Iordache, E.; Borz, S.A. Testing the capability of low-cost tools and artificial intelligence techniques to automatically detect operations done by a small-sized manually driven bandsaw. *Forests* **2020**, *11*, 739. [CrossRef]
10. Sachelarescu, V.; Demetrescu-Gîrbovi, S.T.; Ursulescu, S.T. Cercetări privitoare la debitarea lemnului de fag în cherestea cu ajutorul fierăstraielor panglică. *Analele ICAS* **1970**, *24*, 257–268.
11. Guallpa, M.; Suatunce, J.; Canchignia, H. Tiempos y rendimiento en el proceso de aserrado de Eucalyptus globulus Labill, con sierra circular y de cinta. *Enfoque UTE* **2019**, *10*, 126–143. [CrossRef]
12. Thepaksorn, P.; Thongjerm, S.; Incharoen, S.; Siriwong, W.; Harada, K.; Koizumi, A. Job safety analysis and hazard identification for work accident prevention in para rubber wood sawmills in southern Thailand. *J. Occup. Health* **2017**, *59*, 542–551. [CrossRef]
13. Thepaksorn, P.; Koizumi, A.; Harada, K.; Siriwong, W.; Neitzel, R.L. Occupational noise exposure and hearing defects among sawmill workers in the south of Thailand. *Int. J. Occup. Saf. Ergon.* **2019**, *25*, 458–466. [CrossRef]
14. Fidan, M.S.; Yasar, S.S.; Komut, O.; Yasar, M. A study on noise levels of machinery used in lumber industry enterprises. *Wood Res.* **2020**, *65*, 785–796. [CrossRef]
15. Irsath, M.M.; Jayan, T.J. Occupational Health and Hygiene Study of Sawmill Workers in Sathyamangalam. *Int. Res. J. Eng. Technol.* **2022**, *9*, 1445–1449.
16. Câmpu, V.R.; Ciubotaru, A. Time consumption and productivity in manual tree felling with a chainsaw—A case study of resinous stands from mountainous areas. *Silva Fenn.* **2017**, *51*, 1657. [CrossRef]
17. Björheden, R.; Thompson, M.A. An international nomenclature for forest work study. In Proceedings of the IUFRO 1995 S3:04 Subject Area: 20th World Congress, Tampere, Finland, 6–12 August 1995; Field, D.B., Ed.; Miscellaneous Report 422; University of Maine: Orono, ME, USA, 1995; pp. 190–215.
18. Giurgiu, V. *Metode ale Statisticii Matematice Aplicate în Silvicultură*; Editura Ceres: Bucharest, Romania, 1972; p. 566.
19. Björheden, R. Basic time concepts for international comparisons of time study reports. *J. For. Eng.* **1991**, *2*, 33–39. [CrossRef]
20. Kanawaty, G. *Introduction to Work Study*, 4th ed.; International Labour Office: Geneva, Switzerland, 1991; p. 524.
21. Lindroos, O. Scrutinizing the theory of comparative time studies with operator as a block effect. *J. For. Eng.* **2010**, *21*, 20–30. [CrossRef]
22. Ene, N.; Bularca, M. *Fabricarea Cherestelei*; Editura Tehnică: Bucharest, Romania, 1994; p. 448.
23. Brandstetter, M.; Ispas, M.; Campean, M. Conversion efficiency of fir sawlogs into lumber. *Pro Ligno* **2020**, *16*, 68–74.
24. Ištvanić, J.; Lučić, R.B.; Jug, M.; Karan, R. Analysis of factors affecting log band saw capacity. *Croat. J. For. Eng.* **2009**, *30*, 27–35.
25. Lolila, N.J.; Mchelu, H.A.; Mauya, E.W.; Madundo, S.D. Lumber Recovery and Production Rates of Small-Scale Mobile Sawmilling Industries in Northern Tanzania. *Tanzan. J. For. Nat. Conserv.* **2021**, *90*, 74–83.
26. Ciubotaru, A. *Sortarea și Prelucrarea Lemnului*; Editura Lux Libris: Brașov, Romania, 1997; p. 194.
27. Ciubotaru, A.; Câmpu, V.R.; David, E.C. *Exploatarea și Prelucrarea Lemnului*; Editura Universității Transilvania din Brașov: Brașov, Romania, 2012; p. 119.
28. De Lasaux, M.J.; Spinelli, R.; Hartsough, B.R.; Magagnotti, N. Using a small-log mobile sawmill system to contain fuel reduction treatment cost on small parcels. *Small-Scale For.* **2009**, *8*, 367–379. [CrossRef]
29. Wade, M.W.; Bullard, S.H.; Steele, P.H.; Araman, P.A. Estimating hardwood sawmill conversion efficiency based on sawing machine and log characteristics. *For. Prod.* **1992**, *42*, 21–26.
30. Wang, S.J. A new dimension sawmill performance measure. *For. Prod. J.* **1988**, *38*, 64–68.
31. Ulsoy, A.G.; Mote, C.D., Jr.; Szymani, R. Principal developments in band saw vibration and stability research. *Holz als Roh-und Werkstoff* **1978**, *36*, 273–280. [CrossRef]
32. Loehnertz, S.P.; Cooz, I.V. *Saw Tooth Forces in Cutting Tropical Hardwoods Native to South America*; Research Paper FPL-RP-567; United States Department of Agriculture, Forest Service, Forest Products Laboratory: Madison, WI, USA, 1998; p. 16.
33. Moradpour, P.; Doosthoseini, K.; Scholz, F.; Tarmian, A. Cutting forces in bandsaw processing of oak and beech wood as affected by wood moisture content and cutting directions. *Eur. J. Wood Wood Prod.* **2013**, *71*, 747–754. [CrossRef]
34. Orlowski, K.A.; Chuchala, D.; Stenka, D.; Przybylinski, T. Assessment of wear of the bandsaw teeth in industrial conditions. *Acta Fac. Xylologiae Zvolen* **2022**, *64*, 69–76. [CrossRef]

35. Krilek, J.; Kuvik, T.; Kováč, J.; Meliherčík, J. Research on the side deflection of the saw band of a joinery band saw influenced by the selected technical and technological factors. *Manag. Syst. Prod. Eng.* **2020**, *28*, 148–153. [CrossRef]
36. Brandstetter, M.; Campean, M. Evaluation of lumber y22ield as function of the sawing equipment. *Pro Ligno* **2022**, *18*, 50–56.
37. Rasmussen, H.K.; Kozak, R.A.; Maness, T.C. An analysis of machine-caused lumber shape defects in British Columbia sawmills. *For. Prod. J.* **2004**, *54*, 47–56.

Disclaimer/Publisher's Note: The statements, opinions and data contained in all publications are solely those of the individual author(s) and contributor(s) and not of MDPI and/or the editor(s). MDPI and/or the editor(s) disclaim responsibility for any injury to people or property resulting from any ideas, methods, instructions or products referred to in the content.

Article

Some Properties of Briquettes and Pellets Obtained from the Biomass of Energetic Willow (*Salix viminalis* L.) in Comparison with Those from Oak (*Quercus robur*)

Veronica Dragusanu (Japalela) [1], Aurel Lunguleasa [1,*], Cosmin Spirchez [1] and Cezar Scriba [2]

[1] Wood Processing and Design Wooden Product Department, Transilvania University of Brasov, 29 Street Eroilor, 500036 Brasov, Romania; tamara.dragusanu@unitbv.ro (V.D.); cosmin.spirchez@unitbv.ro (C.S.)
[2] Forestry Operations, Forest Management and Land Surveying Department, Transilvania University of Brasov, 29 Street Eroilor, 500036 Brasov, Romania; caesarus@unitbv.ro
* Correspondence: lunga@unitbv.ro

Abstract: Fast-growing species have been increasingly developed in recent years, and among them, those cultivated to obtain combustible woody biomass have shown rapid development. The purpose of this research study is to highlight the properties of the briquettes and pellets obtained from energetic willow compared to the briquettes and pellets obtained from oak biomass. Methodologies have been based on international standards and were used to find the physical, mechanical, and calorific properties of the two types of briquettes and pellets. The results did not highlight a significant difference between the two categories of briquettes and pellets obtained from the two hardwood species (energetic willow and oak). Characteristics such as the calorific value were 20.7 MJ/kg for native pellets and 21.43 MJ/kg for torrefied pellets of energetic willow, as well as the compressive strength of 1.02 N/mm^2, surpassed the same characteristics of briquettes and pellets obtained from oak biomass. Other characteristics of energetic willows, such as energetic density of 18.0×10^3 MJ/m^3, splitting strength of 0.08 N/mm^2, shear strength of 0.86 N/mm^2, and abrasion of 1.92%, were favorably related to the oak biomass. The ecological analysis highlighted the high potential of the ecological willow in a period when the quantities of carbon dioxide released into the atmosphere by human activities are very high, and its sequestration by existing forests is insufficient. As a general conclusion of this research study, it can be stated that the two categories of briquettes and pellets obtained from the woody biomass of the energetic willow and oak species have similar characteristics, which can be used separately or together in ecological and sustainable combustion.

Keywords: briquette; pellet; CIELab space; density; calorific value; energetic density; ash content

Citation: Dragusanu, V.; Lunguleasa, A.; Spirchez, C.; Scriba, C. Some Properties of Briquettes and Pellets Obtained from the Biomass of Energetic Willow (*Salix viminalis* L.) in Comparison with Those from Oak (*Quercus robur*). *Forests* **2023**, *14*, 1134. https://doi.org/10.3390/f14061134

Academic Editor: Stefano Grigolato

Received: 9 May 2023
Revised: 27 May 2023
Accepted: 29 May 2023
Published: 31 May 2023

Copyright: © 2023 by the authors. Licensee MDPI, Basel, Switzerland. This article is an open access article distributed under the terms and conditions of the Creative Commons Attribution (CC BY) license (https://creativecommons.org/licenses/by/4.0/).

1. Introduction

The main use of lignocellulosic biomass, regardless of its nature and origin, is in the field of energy production. The sources for obtaining it are varied, such as the woody remains from the exploitation of forests and wood processing factories, but also from forestry and agricultural crops dedicated exclusively to obtaining energetic biomass. Among these specially dedicated crops for biomass are listed woody plants with a short rotation cycle (willow, poplar, acacia, etc.), with a high potential to obtain a secure income in the short term [1] but also to take over and sequester of carbon from the atmosphere within a short period of time. A plantation with ordinary woody species of trees needs, on average, 90–160 years to reach maturity, but a plantation with fast-growing species needs only 25–35 years. Therefore, in addition to the fact that plantations with fast-growing species recover their investment faster, they could obtain a (72–78%) better economic efficiency [1]. In addition to the fact that the short-rotation crops are a renewable source of ecological energy [2,3], some fast-growing species also have a role in the remediation of soil areas degraded by heavy metals [4–6] and/or with high moisture content in the soil, improving environmental pollution problems, recycling of wastewater [7], and sustaining the development of rural localities. This category of energetic crops could include energetic willow (*Salix viminalis*), elephant grass (*Mischantus giganteum*), Chinese reed (*Miscanthus sinensis*), Pampa's grass (*Cortaderia Rosea*), sorghum (*Sorghum halepense*), Sudan grass (*Sorghum sudanense*), etc. [6].

Worldwide, there are over 300 species of willow. The energetic willow (*Salix viminalis*) is an agricultural crop plant that is woody and shrubby, located in agricultural areas and less often in forest habitats. In addition to the role of restoring the productive circuit of some highly degraded lands such as tailings dumps, former sites of chemical plants, and heavily eroded, saline or sandy soils, these woody species produce considerable biomass (about 35 t/year/ha wet biomass, and when some specific irrigation and fertilization treatments are applied, the biomass increase up to 60 t/ha/year) which is available in each growing year [6]. In addition, this woody species has a fast growth cycle with height gains of 3–3.5 cm/day. Its calorific value is 18,224 kJ/kg for 3-year-old sticks (with an average height of 7 m and a diameter at the base of 8 cm) and 18,265 MJ/kg for the bark of the same sticks [6]. The energetic willow contains a large amount of salicylic acid, which helps in a rapid loss of moisture up to 14%–16%, with advantages in its processing without artificial drying and open storage for a long period of time without biological degradation. The harvesting of thin stems is conducted with current agricultural machinery only during the period of vegetative rest after November, such as combine harvesters, tractors, and trailers, when the park of agricultural machinery is under conservation, and the willow leaves have fallen to the ground and can create a fertilizing layer for the soil [8,9]. This woody biomass, in the form of wood chips, chops, briquettes, and pellets, represents a cheaper alternative to ordinary firewood. The energy obtained from energetic willow is a renewable one, the vegetation period of a plantation being 25–30 years, starting with the third year after planting [10–12]. A calorific value of 19–21 MJ/kg, equal to 5.5–6.1 kWh/kg, is ideal for obtaining briquettes and pellets [6,7]. Many authors have analyzed impacts on climate change [13], financial analysis [14], energetic impact [15], biofuel potential [16], yield [17–20], establishing of surface cover [21,22], biomass production [23], trait and genome [24,25], potential [26], wood quality [27], briquetting [28], genetic structure and diversity [29,30], structure [31], and prognosis [32], all of these being obtained on energetic willow plantation. The energetic willow is planted on the ground in such a way that two adjacent rows have a distance of 750 mm between them, and a greater distance of 900 mm between four rows is necessary for moving the harvesting installations [6,21].

Energetic crops represent an alternative to fossil fuels; it is estimated that in the EU-27 will be a potential of 47 Mtoe/year before 2025, and after 2025 this value will exceed 138 Mtoe/year, i.e., namely 14% of the total energy consumption at European level. It is also stated that an area of 13.2 Mha is available for energetic crops before 2025, and after 2025 this will exceed 26.2 Mha [33].

The biomass obtained when harvesting the energetic willow could be stored and used in the form of briquettes and pellets [34]. Going further, other authors [35,36] have improved the energetic and hydrophobic performances of biomass through a torrefaction thermal process. Based on the results obtained at temperatures above 200 °C, the torrefaction process of the woody biomass contributed to an increase in the calorific value by up to 60% and a decrease in the hydrophilicity by up to 20%. It was also specified that at the European level, renewable energy represents 24% of total energy consumption, and the consumption of wood and wood waste represents over 68% of biomass. A small part, namely less than 1% of this woody biomass, is derived from energetic willow crops [33].

Lignocellulosic briquettes and pellets are solid fuel products whose main advantage is the densification of small lignocellulosic biomass up to a density of 800–1250 kg/m^3 [37–39]. They maintain the advantages of lignocellulosic biomass, including those related to renewability [40], environmental friendliness [41], and neutrality of carbon dioxide emission [42]. In addition, briquettes obtained from energetic crops (miscanthus, energy willow, sorghum, etc.) [43] bring an important contribution to the natural environment by eliminating oxygen in the atmosphere [44] and sequestering carbon dioxide in each vegetative year [6,45]. Other economic and ecological effects of biomass briquetting are presented by other researchers [46,47].

Objectives: If previous studies in the field refer to particular aspects of the energetic willow, the main objective of the research is to correlate its ecological, physical, and calorific properties. Particularly, the purpose of this research is to analyze the properties of the briquettes and pellets obtained from the biomass of the energetic willow in order to use it effectively. The physical-mechanical properties, such as the density, breaking resistance of the briquettes and shearing resistance of the pellets, and calorific properties, such as calorific value, calorific density, and ash content, will be analyzed. Some ecological aspects of energetic willow will also be emphasized. In addition, to observe the position of the energetic willow biomass within the whole lignocellulosic biomass, a comparison was made between the briquettes and pellets obtained from the energetic willow biomass and that of the oak waste, both in the native and torrefied state.

2. Materials and Methods

2.1. Ecological Aspects

Similar to any tree or forest, the cultivation of the energetic willow brings multiple additional ecological benefits, two of which are more important, namely the sequestration of carbon dioxide from the atmosphere and the elimination of oxygen through the photosynthesis process [13,48,49]. The method for determining the amount of carbon dioxide sequestered in energetic willow crops is similar to that for trees in forests (as oak species could be) and contains several steps [13], as follows:

— Determining the weight of the green biomass cutting from the part above the ground, using the following relationship for the group of twigs:

$$M1 = \frac{\pi \cdot d^2}{4} \times H \times De \times n \ [\text{kg}] \quad (1)$$

where: D is the average diameter of a stem of the shoot, in m; H is the average height of the shoot stem, in m; De is the density of green wood, in kg/m^3; n is the number of sticks resulting from a shoot.

— The addition of the woody part corresponding to the root, which corresponds to about 10% of the area above the ground, respectively, the total value of the woody part will be:

$$M2 = 1.1 \times M1 \ [\text{kg}] \quad (2)$$

— Determination of the absolute dry mass of the woody mass resulting from a stump group, taking into account that 72.5% is dry mass and 27.5% is water in different forms (liquid, vapor, and chemically dissociated), that is, this will be:

$$M3 = 1.2 \times M1 \times 0.725 \ [\text{kg}] \quad (3)$$

— Determination of the carbon mass in the wood of the energetic willow, considering that the carbon content for the energetic willow is 48.4% [50,51], which means a mass of:

$$M4 = 1.2 \times M1 \times 0.725 \times 0.484 \ [\text{kg}] \quad (4)$$

— Determination of the sequestered carbon dioxide content in the woody part of the biomass. It is taken into account that CO_2 has one carbon molecule and two other oxygen molecules, the atomic mass of carbon is 12, and that of oxygen is 16. Consequently, the mass of CO_2 will be a ratio between the atomic mass of all CO_2 and the atomic mass of carbon, respectively, 44/12 = 3.67. Therefore, to determine the mass of carbon dioxide sequestered in trees, the previous mass of carbon M4 will be multiplied by 3.67, which will be:

$$M5 = 3.67 \times M4 \ [\text{kg}] \quad (5)$$

— For an energetic willow crop, the amount of sequestered carbon dioxide per surface unit is calculated with the following relationship:

$$M6 = n1 \times M5 \ [kg\ CO_2/ha] \quad (6)$$

where: n1 represents the number of cuttings existing on one hectare of energetic willow culture.

So, taking into account all variables, the general relationship will be:

$$Mg = \pi \times D^2 \times H \times De \times n \times 0.725 \times 0.484 \times n1 \times 3.67 \times 1.2 \times 0.25 \ [kg\ CO_2/ha] \quad (7)$$

2.2. Granulometry of the Energetic Willow and Oak Crushed Material

The willow biomass was taken from an energetic willow plantation, and the oak biomass was taken from a circular saw in the form of sawdust. The granulometry of the small material characterizes its dimensions, depending on which the obtained pellets and briquettes could have better or worse physical–mechanical characteristics. Both types of small materials were sorted with a 5 × 5 mm sieve in order to have appropriate and homogeneous sizes. The main purpose of this determination was to determine the different fractions of the small material because an increased percentage of fractions with large sizes will determine a low density and a high breaking strength of the briquettes and pellets, and an increased percentage of the fraction with small sizes will lead to obtaining products with high densities and reduced resistance. For this test, an electrical vibrating device was used with sieve sizes of 4 × 4, 3.13 × 3.13, 2 × 2, 1.25 × 1.25, 0.8 × 0.8 and 0.4 × 0.4. The granulometry of the material was determined by the sieving method. What passed through the last lower sieve was called "Rest." A total of 6 randomly extracted samples of 30 g were used from each type of biomass (oak and energetic willow), and 12 samples were made. In order to obtain the values of the masses (determined with EWJ 600-2M Kern, Merck KGaA, Darmstadt, Germany balance) and the participation percentages for each sieve, the arithmetic mean was performed for the values obtained for each of the 6 samples. For example, the participation percentage of the resulting fraction above the 2 × 2 mm sieve was calculated with the following calculation relationship:

$$P_{2\times 2} = \frac{\frac{1}{6}\sum_{i=1}^{6} m_{2\times 2}}{m_p} \times 100 \ [\%] \quad (8)$$

where: $m_{2\times 2}$ is the mass of the fraction that remained above the mesh sieve 2 × 2 mm; m_p—the mass of the sample taken into consideration for the test, in g.

2.3. Obtaining Briquettes and Pellets

The shredded material had an absolute moisture content of 12 ± 1%, determined by the gravimetric weighing-drying-weighing principle [52,53]. Briquetting was carried out on a Gold Star type hydraulic piston machine (Brasov, Romania) with a capacity of 500 kg/h, and pelletizing was carried out on a Sarras mechanical press (Sarras group, Brasov, Romania) with a capacity of 40 kg/h. For a better choice of briquettes and pellets, the briquettes and pellets obtained in the first 5 min of operation were removed from this study. In addition, for better identification of each piece during testing, each sample had an identification number and was placed on flat white support in an order that was maintained throughout the all-testing period.

2.4. Unit Density of Briquettes and Pellets

Before this determination, all briquettes and pellets were conditioned at a temperature of 20 °C and 55% air humidity for 24 h in order to stabilize the moisture content at 10 ± 1%. The individual density of briquettes and pellets was determined as the ratio between their mass and volume (DIN 51731: 1996) [54]. Taking into account the cylindrical shape of the briquettes and pellets, the density determination relationship was the following (Equation (9)):

$$\rho = \frac{4 \cdot m}{\pi \cdot d^2 \cdot l} \times 10^6 \ [kg/m^3] \quad (1)$$

where: m is the mass of the sample, in g; d is the diameter of the sample, in mm; l is the length of the sample, in mm.

The bulk density of the pellets was determined according to EN 15103:2009 [55] by using a cylindrical vessel with a known internal volume and weighing the vessel filled with pellets. Eight replicates of this test were used to obtain a consistent mean and standard deviation of statistical results.

2.5. The Pellet Torrefaction Treatment

The torrefaction treatment was applied to energetic willow and oak pellets, using temperatures of 180, 200, and 220 °C and 3 treatment times of 1, 2, and 3 h [35]. The main purpose of this determination was to improve the calorific characteristics of the pellets. During torrefaction, part of the wood hemicelluloses is damaged, thereby increasing the calorific value but especially the energetic density through the loss of mass. Additionally, the torrefied pellets become more stable (moisture absorption is reduced) and are sterilized (degradation is more difficult). The higher temperature and duration, the more advantages of torrefaction [41,42]. A Memmert-type laboratory oven (Carbolite Gero Ltd., Hope Valley, UK) was used with the air inlet valve closed in order to eliminate the possibility of oxidation and self-ignition of the samples. Prior to treatment, all the samples were dried for 4 h in the same laboratory oven at a temperature of 105 °C. After drying, the samples were weighed with a precision electronic balance type EWJ 600-2M Kern (Merck KGaA, Darmstadt, Germany), recording their initial and final mass at the end of the treatment period. Based on the two weights, it was possible to determine the mass loss during torrefaction with the help of the following relationship (Equation (10)):

$$ML = \frac{M_i - M_f}{M_i} \times 100 \ [\%] \quad (10)$$

where: ML is the mass loss in %; M_i is the initial mass of samples in g; M_f is the final mass of samples in g.

Eight valid samples for each type of pellet were used to determine the statistical parameters of this test. The biomass briquettes were not torrefied due to their low compressibility and compatibility, which led to their disintegration during the treatment.

2.6. Color Determination of Native and Torrefied Pellets with CIELab Colorimetric Space

This colorimetric space is quantified by 3 distinct parameters, L*, a*, and b*. The L* axis represents lightness and has the value zero for black and one hundred for white; between these values, there were a series of shades of gray. The a* axis refers to the green-red opposition, with negative values towards green and positive values towards red. The b* axis quantifies the blue–yellow opposition, with negative values towards the blue zone and positive values towards the yellow zone [56].

2.7. Calorific Value and Energetic Density of Briquettes and Pellets

The calorific value of briquettes and pellets of *Salix viminalis* and oak was determined on dry and compact samples with a weight of 0.8 ± 0.1 g, taken from briquettes or pellets [57]. An XRY-1C oxygen bomb calorimeter (Shanghai Geological Ltd., Shanghai, China), provided with its own test software, was used. Before testing, the calorimeter was calibrated by using a pill of 1 g of benzoic acid with a known calorific value of 26,454 kJ/kg, thus finding the calorimetric coefficient k from the relationship for determining the high calorific value (*HCV*) (Equation (11)):

$$HCV = \frac{k \cdot (T_f - T_i) - Q_{ct} - Q_{nw} - Q_a}{m} \ [\frac{kJ}{kg}] \quad (11)$$

where: T_f is the final temperature in °C; T_i is the initial temperature in °C; Q_{ct} is the amount of heat released during combustion by the cotton thread in kJ; Q_{nw} is the amount of heat of the nickel wire in kJ; Qa is the amount of heat given by the nitric acid produced during combustion, in kJ.

Eight valid tests were carried out for each type of material [6,28,33] in order to obtain acceptable and significant values. Each test provided values with a high calorific value, low calorific value, burning time, and evolution of the temperature in the 3 stages of the test (fore, main, and after).

Energetic density expresses the calorific energy obtained from each cubic meter of the briquettes or pellets. It was obtained by multiplying the calorific value by briquette/pellet density with the next relationships (Equation (12)):

$$ED = HCV \times \rho_{b/p} \ [MJ/m^3] \qquad (12)$$

where: ED is energetic density in MJ/m^3; HCV is the high calorific value in MJ/kg; $\rho_{b/p}$ is the unit density of briquette or pellet in kg/m^3.

2.8. Ash Content

In order to determine the ash content, 3–5 g of crushed material was taken, which was obtained during the sorting procedure, with the help of the 1 × 1 mm sieve. This material was dried in an oven at 105 °C for 30 min up to a constant mass in order to obtain an absolutely dry mass. Next, this dry material was deposited on a crucible and weighed with an analytical balance type EWJ 600-2M Kern (Merck KGaA, Darmstadt, Germany) with a precision of 2 decimals. The crucibles with the dried material were placed in a calcination furnace at a temperature of 750 °C (ASTM E1755-01: 2020; ISO 2171: 2007) [58,59] for about 40 min. Calcination was considered complete when the ash became light gray, without traces of sparks or non-calcified material. At that moment, the crucible with calcined ash was extracted from the furnace, cooled in a desiccator, and weighed with the same high-precision balance. The ash content was determined with the following calculation relationship:

$$A_c = \frac{mc_f - mc}{mc_i - mc} \times 100 \ [\%] \qquad (13)$$

where: m_{ci} is the initial mass of the crucible with the sample of dry crushed material in g; m_{cf} is the final mass of the crucible with calcined ash in g; m_c with the mass of the empty crucible in g.

At least 6 valid samples were taken into consideration for each type of sample, obtained from energetic willow and oak biomass.

2.9. The Compressive Strength of Briquettes

Before the determination, all briquettes were conditioned at a temperature of 20 °C and a relative air humidity of 65% until obtaining a moisture content of 10 ± 1%. Then, their average length and average diameter were determined. For this determination, the briquettes made of energetic willow and oak were inserted between the two plateaus of the universal WA testing machine (TE Force speed Corporation, Jinan, China) and subjected to the action of compression until they broke. The maximum breaking force was recorded for each briquette, after which the compression breaking resistance was determined as a ratio between the force and the breaking area (Equation (14)):

$$\sigma_c = \frac{F_{max}}{d \cdot l} \ [\frac{N}{mm^2}] \qquad (14)$$

where: F_{max} is the maximum force of briquette breakage in N; d is the diameter of briquettes in mm; l is the length of briquette samples in mm.

Fifteen valid tests were taken into consideration in order to obtain statistical parameters with a confidence interval of 95%.

2.10. Splitting Resistance of Briquettes (Perpendicular and Parallel to the Length)

Even if the briquettes are reconstituted engineering products, similar to the splitting resistance of wood along the longitudinal plane of minimum strength, the splitting resistance of the briquettes was taken into consideration. This resistance was made perpendicular and parallel to the length of the briquettes. To perform the splitting test, the WA-type universal testing machine (TE Force Speed Corporation, Jinan, China) was used, with a specially designed splitting device. The device consisted of a knife with a tip having an angle of 76 degrees and a radius of roundness of 1 mm (in order not to cut the briquette but only to split it). The relations for determining the splitting resistance were the following (Equation (15)):

$$\sigma_{spar} = \frac{F_{max1}}{d \cdot l} \left[\frac{N}{mm^2}\right] \sigma_{sper} = \frac{4 \cdot F_{max2}}{\pi \cdot d^2} \left[\frac{N}{mm^2}\right] \quad (15)$$

where: σ_{spar} is the resistance to splitting parallel to the length of the briquettes in N/mm^2; σ_{sper} is the resistance to splitting perpendicular to the length of the specimen in N/mm^2; F_{max1} is the maximum force of the parallel resistance in N; d is the diameter of briquettes in mm; l is the length of briquettes in mm; F_{max2} is the maximum force of splitting strength perpendicular to the length of briquettes in N.

A total of 10 briquette specimens were tested for this determination, both for the longitudinal and transverse splitting of briquettes.

2.11. The Pellet Shearing Strength

Shearing of pellets is a current problem encountered during storage, transport, and use. In this way, during the combustion process, they become shorter, with serious repercussions on the management of the combustion process. For this test, a universal WA-type testing machine and two shearing devices (metal plates) fixed on the two arms of the compression machine were used. The upper device had the role of shearing, having an angle at the top of about 80 degrees so as not to cut the pellets, and the lower one was provided with five holes with a diameter of $8^{+0.2}$ mm to support the pellets during shearing. The test was considered finished when the shear force dropped suddenly. The shear strength relationship was determined by the next relationship (Equation (16)).

$$\tau_s = \frac{4 \cdot F_{max}}{5 \cdot \pi \cdot d^2} \left[\frac{N}{mm^2}\right] \quad (16)$$

where: F_{max} is the maximum force of breaking in N/mm^2; d is the diameter of the pellet in mm.

2.12. The Briquettes Abrasion

The briquettes were abraded by placing them on the electrical vibrating device (Xinxiang Gaofu Machinery Co., Ltd., Xinxiang, China) with a 2 × 2 mm sieve. The briquettes that were used had a constant moisture content of 10 ± 1% and a maximum height of 30 mm in order to obtain sufficient space for the movement of the briquettes on the wear and sorting sieve. The briquette abrasion was determined by the next relationship (Equation (17)):

$$Ab = \frac{m_{us}}{m_s} \times 100 \, [\%] \quad (17)$$

where: Ab is the abrasion of briquettes in %; m_{us} is the mass of sawdust sifted and collected under the 2 × 2 mm sieve in g; m_s is the mass of initial samples in g.

A total of 8 laboratory tests were performed.

2.13. Statistical Analysis

An arithmetic mean and a standard deviation were calculated for each group of tested values. In addition, the standard deviations, the regression equations, and the coefficient of determination R^2 were identified on the Microsoft Excel 2019 (Microsoft Corp., Redmond, WA, USA) graphs. With the help of the statistical analysis program Minitab 18 (Penn State

University, State College, PA, USA) and its specific graphs, the upper and lower limits of the calculated values were identified for a confidence interval of 95% or an alpha error of 0.05.

3. Results
3.1. Ecological Aspects of the Willow/Oak Plantation

Following the calculations made according to the methodology presented in Section 2.1 above, a mature cutting aged 10 years could absorb about 20 kg of carbon dioxide per year from the atmosphere. Taking into account that saplings under the age of 3 years have reduced foliage and will absorb less CO_2, it results that in an average life of 100 years, four saplings could absorb about one ton of CO_2. Reported at the surface of a plantation, the sequestering of CO_2 was about 4000 tons of CO_2. This value is compared with the amount of carbon dioxide eliminated by human activity of, about 40 billion tons of CO_2 every year [6]. Extrapolating to the level of the planet's three thousand billion trees, it is found that each human life would need about four hundred twenty trees. Relating it to the world's forests and the number of world's trees, a negative ratio is obtained, i.e., for an emission of 40 billion tons of carbon dioxide, the forests will absorb only 30 billion tons of carbon dioxide, with 10 billion tons of carbon dioxide remaining in the atmosphere and leading to global warming through the greenhouse effect. Therefore, another thousand billion trees would be needed, or taking into account the existence of 5000 trees per hectare, a forested area of about 500 million hectares would be needed. The essential problem of the planet nowadays is that every year about 5% of the world's forests are lost [40–43] by returning deforested lands to agriculture, and the amount of carbon dioxide increases yearly. The energetic willow is considered an agricultural crop, which is why it can compensate for the shortage of trees. In addition to the CO_2 emissions that was described above, it could be added a considerable amount of methane gas, eliminated in particular by agricultural farms all over the world [46,47], but also by the degradation of fallen trees due to storms and other natural weather/fire problems [47]). Methane gas released into the atmosphere has a much greater influence than CO_2 on global warming.

Regarding the release of oxygen during the photosynthesis process, the analysis is similar to that of any fast-growing woody species. The crown of a mature cutting produces, on average, 117.9 kg of oxygen each year. It is known that a person needs 9.5 tons of air per year, or considering the composition of air with 23% oxygen and human breathing uses only 25% of this oxygen, it means that a person needs 3.85 t oxygen per year. From these considerations, it follows that a person needs about 30 trees to release the oxygen necessary for breathing. Hence, the slogan that every man must plant a tree in his life must be multiplied 30 times nowadays. It is observed that from this point of view, in total—without taking into account the diversity of the fauna and the different climates from one area of the globe to another—the world's forests still provide the oxygen needed for human life.

3.2. Bulk Density of the Sawdust of the Two Types of Biomasses

Bulk density was determined as a ratio between the mass and the volume of the crushed material disposed into the vessel. The volume was determined using a cylindrical vessel with an inner diameter of 20.17 mm and a height of 42.67 mm, respectively, with a volume of 1342.11 cm^3. The mass of the material contained in the cylindrical vessel was determined as the difference between the mass of the cylinder with material and the mass of the empty vessel, with the help of an analytical balance with a precision of two decimals. For a good placement of the material in the analysis vessel, it was vibrated for 3 min. Based on the 10 tests performed both on the small material of energetic willow and oak, bulk densities of 480.2 kg/m^3 for oak and 209.9 kg/m^3 for energetic willow were obtained. Taking into account the effective density of the two wood species of 675 kg/m^3 for oak and 400 kg/m^3 for willow [47], compacting coefficients of 1.41 for oak and 1.97 for energetic willow were obtained. Therefore, the shredded material of willow was considered much looser than that of oak due to the low density of the species, which had repercussions on the increase in compaction in briquettes and pellets.

3.3. Granulometry of Wood Particles

Due to the use of the same 5 × 5 mm sorting sieve of sawdust before briquetting and pelletizing, the two granulometry curves were almost similar, as can be seen in Figure 1. A slight shift to the right in the granulometric curve of the energetic willow, or the upward shift in the oak curve, was due to the greater weight of the oak chips.

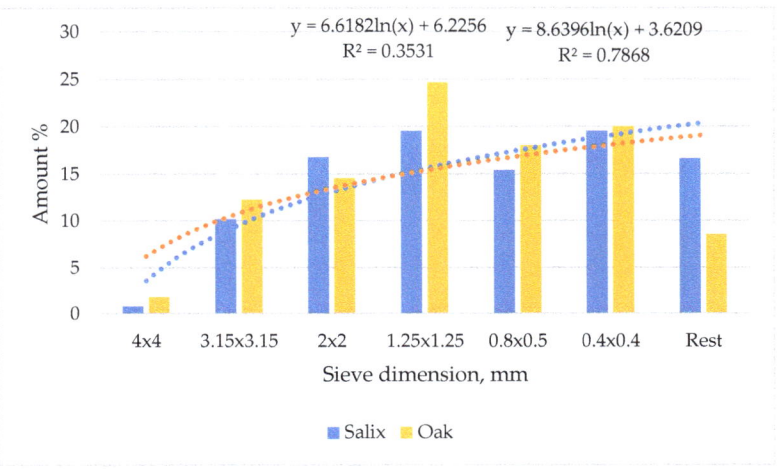

Figure 1. Granulometry of the crushed material for the two types of biomasses.

3.4. Dimensions of Briquettes and Pellets

Based on the values of the 38 samples that were analyzed, average diameter values of 40.86 mm were found in the case of briquettes and 6.35 mm in the case of pellets. The percentage coefficient of expansion after evacuation from the press was 2.1% in the case of briquettes and 5.8% in the case of pellets. The different expansions or enlargements of the pellets were explained by their increased density compared to briquettes. The length of briquettes was 22.78 mm in the case of energetic willow and 52.12 mm in the case of oak biomass, meaning that the oak biomass is more compactable.

In the case of oak, the average diameter for the 38 samples was 41.34 mm for briquettes and 6.26 mm for the pellets. The coefficient of expansion in the case of oak sawdust was 3.3% for briquettes and 4.3% for pellets. The small differences between the expansions for oak and energetic willow were due to the fact that both briquettes and pellets of energetic willow and oak were made on the same briquetting and pelletizing installations. The length of pellets was 12.0 mm in the case of energetic willow and 22.0 mm in the case of oak biomass. Related to the higher length of oak pellets, the same explanation as in the case of briquettes was identified.

3.5. Unit Density of Briquettes and Pellets

The unit density of the briquettes differed between those obtained from energetic willow (766.7 kg/m^3) and oak (877.8 kg/m^3); the oak briquettes were 14.5% denser than the willow ones. These differences were due to the density differences between the two wood species, the oak having a density of 675 kg/m^3 and the energetic willow having a density of 400 kg/m^3 [42]. A unit density increase of 30% was obtained in the case of oak and 91.6% in the case of energetic willow. Therefore, the densification–compaction coefficient in briquettes was 1.3 in the case of oak and 1.91 in the case of energetic willow compared to the density of the woody species and about double the density of the crushed material.

At the same pressing pressure, the crushed material with lower density will compress somewhat more, but it will not compensate for the very large difference in density between the two species. However, the density of the briquettes was very low compared to other briquettes [50], all due to the briquetting machine with a hydraulic drive with a pressure of 20 atm, with the help of which some densities much lower were always obtained related to mechanical ones with helical screw or pressure hammer. It could also be seen that the range of limiting variation of the values in the case of the energetic willow is almost double that of the oak, thus demonstrating a greater inhomogeneity.

The densities of the pellets obtained from the biomass of the energetic willow (1101 kg/m^3) and the oak (1296.3 kg/m^3) were different, with 17.7% higher in the case of the oak compared to that of the energetic willow. The explanations regarding this difference, as well as the range of variation of the different values in the case of the two types of biomasses, were similar to those in the case of briquettes, respectively; it is due to the different densities of the two wood species.

3.6. Pellet Torrefaction

The main parameter of pellet torrefaction was the mass loss, which increased a little with the treatment period of 1, 2, and 3 h but increased a lot with the increase in temperature from 180 to 220 °C. Therefore, maximum mass loss values were obtained for temperatures of 220 °C, between (16.78%–23.47) % for energetic willow and between (20.77–27.47) % for oak (Figure 2).

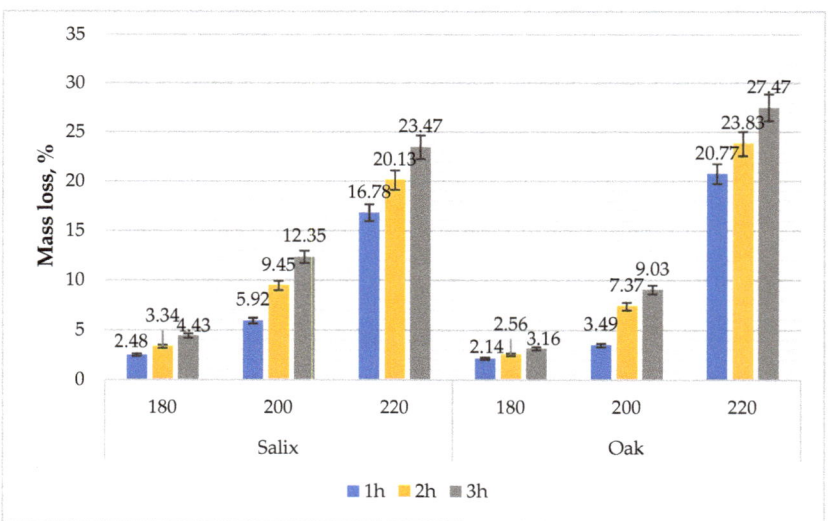

Figure 2. Loss of masses from torrefaction of energetic willow and oak; temperature and time are variable.

It was observed that oak was torrefied better than energetic willow, the losses being somewhat higher due to the chemical composition of the two analyzed species.

3.7. Calorific Value and Energetic Density

Figure 3 shows the low and high calorific values for the two analyzed species, depending on the torrefaction treatment used, respectively, depending on the three torrefaction temperatures, 180 °C, 200 °C, and 220 °C.

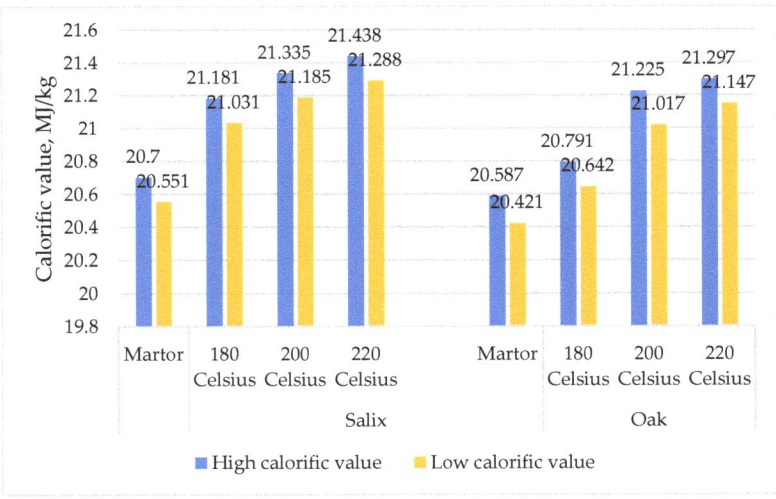

Figure 3. The calorific value of energetic willow and oak pellets depends on the applied thermal treatments; temperature and species are variable.

It is observed that the willow biomass had a higher calorific value than the oak biomass, regardless of the treatment applied; the increase in the calorific value of the energetic willow compared to the calorific value of the oak for the maximum treatment applied was 141 kJ/kg or a percentage of only 0.6%. In total, following the torrefaction process, the calorific value of energetic willow biomass increased by 3.5%, and that of oak increased by 3.4%.

Energetic density, definite with Equation (13), has values of 15.8×10^3 and 18.0×10^3 MJ/m^3 in the case of native Salix and Quercus briquettes and values of 22.7×10^3 MJ/m^3 and 26.67×10^3 MJ/m^3 in the case of native Salix and Quercus pellets. When pellets are torrefied, the energetic density slightly decreases because of higher mass losses related to the value of the calorific increase.

3.8. The Pellets Color after the Torrefaction Process

The torrefaction time had very little influence on the color of the pellets; the most obvious color change was observed when the torrefaction temperature was changed (Table 1).

Table 1. The color of the pellets in the CIELab coloristic space.

Pellet Specie	Treatment	CIELab		
		L*	a*	b*
Salix viminalis	Control	36.5	−15.3	5.6
	180/3	28.7	−20.1	6.5
	200/3	22.8	−21.5	7.1
	220/3	19.4	−22.7	8.4
Quercus robur	Control	38.1	−21.8	9.1
	180/3	36.2	−24.5	9.7
	200/3	32.5	−36.9	10.1
	220/3	28.5	−47.5	10.9

It can be seen that the values of the parameter L* have increased, which means that the pellets change their color from a light gray to black; the values of the parameter a* are negative, which means that the color remains in the dominant green range, and the values of the parameter b* increase slightly during torrefaction, remaining in the domain of the dominant yellow color.

3.9. Compressive Strength of Briquettes

As can be seen in Figure 4, the compressive strength of the two types of briquettes was very different from one to another, with willow briquettes having a compressive strength of 1.02 N/mm^2 and oak briquettes a compressive strength of only 0.33 N/mm^2. This difference highlights the fact that oak briquettes will break much faster during transport and storage in multi-stored bags.

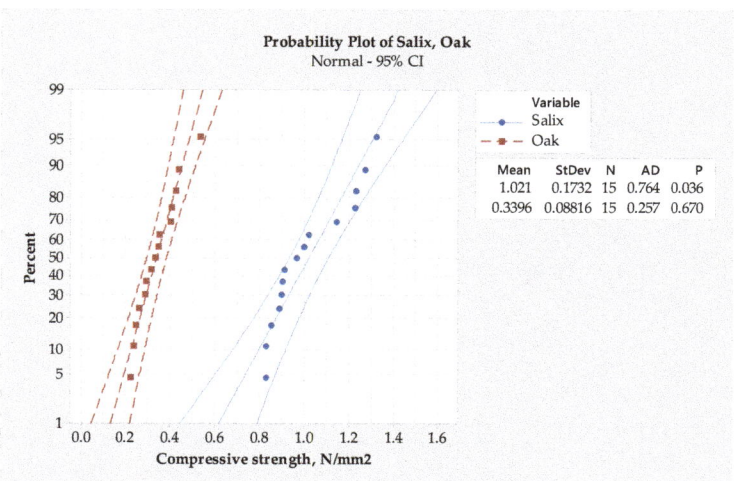

Figure 4. The compressive strength of briquettes: StDev-standard deviation; N-number of samples; AD-Anderson-Darling coefficient; *p*-statistic coefficient.

Statistical significance of the 15 values from Figure 4, represented by *p*-value, falls within the limiting value of 0.05 only for the energetic willow with a value of 0.036, the value of 0.67 for the oak exceeding the imposed limit. This proves that the energetic willow is more homogeneous from the point of view of compressibility. The other Anderson–Darling coefficient shows the same statistical trend.

3.10. The Splitting Strength of Briquettes

The splitting resistance of briquettes (similar to the splitting resistance of wood) was divided into two parts, namely, one perpendicular to the length of the briquette and the other parallel to the length of the briquette. The splitting strength perpendicular to the length of the briquette had small values, about 0.08 N/mm^2 in the case of oak and 0.05 N/mm^2 in the case of energetic willow. The standard deviation of these values was 0.009 N/mm^2 for oak briquettes and 0.006 N/mm^2 for energetic willow (Figure 5). A large variety of values is also observed.

Regarding the splitting of the briquettes, parallel to the length of the specimen (Figure 6), small average values were obtained, namely 1.0 N/mm^2 in the case of oak briquettes and about 0.85 N/mm^2 in the case of energetic willow.

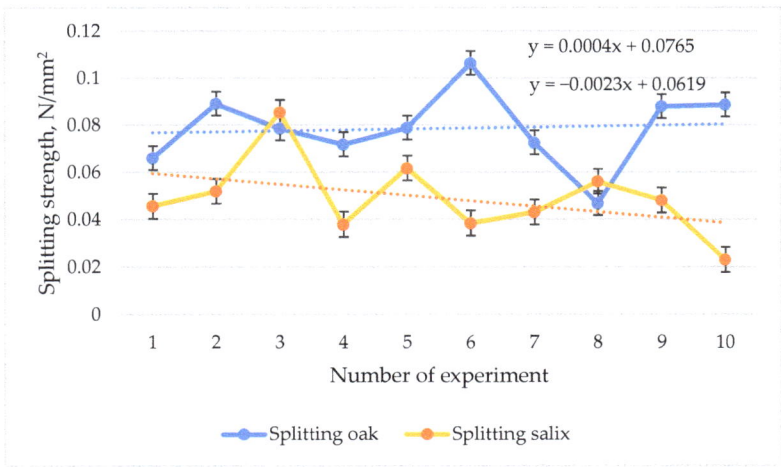

Figure 5. Resistance to splitting perpendicular to the length of the briquette.

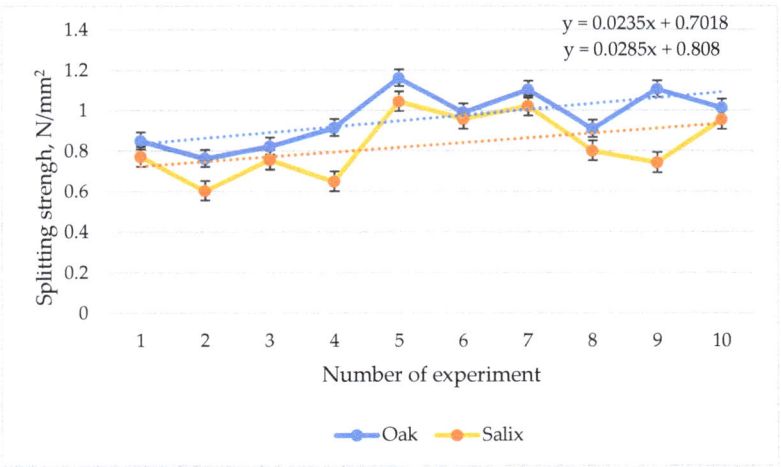

Figure 6. Resistance to splitting parallel to the length of the specimen.

Overall (parallel and perpendicular to the length), the splitting resistance of the energetic willow was slightly lower than that of the oak briquettes.

3.11. Shear Strength of Pellets

The average values of pellet shear were 0.74 N/mm² for willow pellets and 0.86 N/mm² for oak pellets, with standard deviations of 0.045 N/mm² and 0.18 N/mm², respectively. The increase in the shear resistance of the pellets in the case of oak by about 16.2% was determined by the higher density of the oak wood by 68%, from which the pellets were produced, but mainly due to the increase in the density of the pellets by about 17.7%.

Since the order of the experimental values could influence the regression equation, only the relative position of the regression equations for the two types of pellets was analyzed in Figure 7. Due to the parallelism of the two linear equations, it can be concluded that there are no statistically significant differences between the two groups of values.

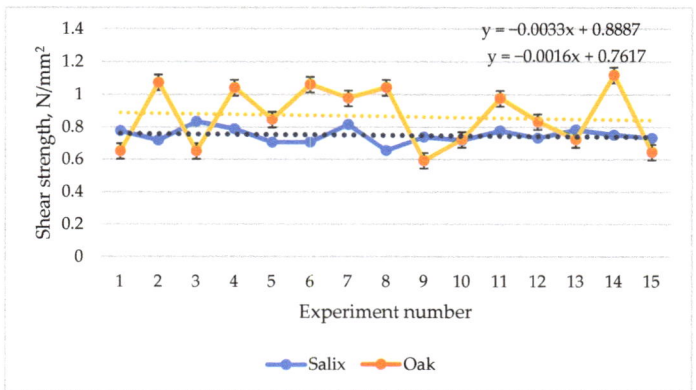

Figure 7. Shear resistance of energetic willow and oak pellets. Salix and oak pellets are variables.

Taking into account the average values and standard deviations of shear strength for a confidence interval of 95% or an alpha-type error of 0.05%, a value range of 0.65–0.83 N/mm^2 was obtained in the case of energetic willow pellets and of 0.5–1.22 N/mm^2 in the case of oak pellets. It is clearly observed that the range of variation of oak pellets was much wider than that of energetic willow pellets; that is, oak pellets were much more inhomogeneous. This can also be seen from the study of the standard deviation, its value in the case of willow pellets being 75% lower than in the case of oak pellets.

3.12. The Briquette Abrasion

The willow briquettes had an abrasion of 1.92%, and the oak briquettes had an abrasion of 4.22% (Figure 8). The 55% lower value of the abrasion of the energetic willow was due to its lower density, which caused less intensity and force when rubbed by the sieve on which the abrasion was made. Even if of lower density, the briquettes of energetic willow had a favorable result on the abrasion.

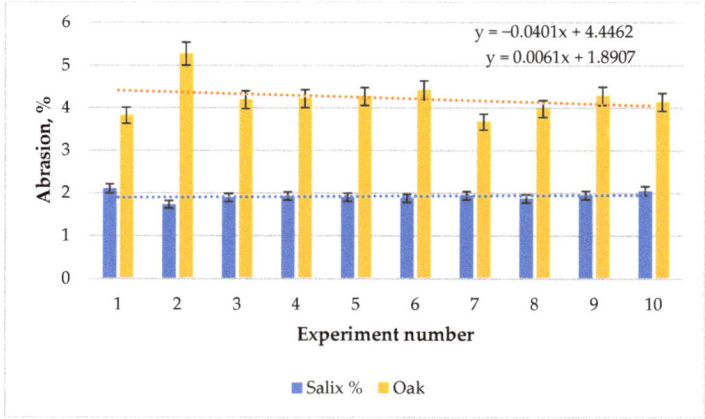

Figure 8. Abrasion of energetic willow and oak briquettes. Salix and oak biomass are the variables.

3.13. Ash Content

The black ash (Figure 9) obtained after the end of the flame had some values of over 10% for both energetic willow and oak sawdust, slightly higher in the case of oak due to substances in the form of oxides in a larger amount inside of the structure of this species.

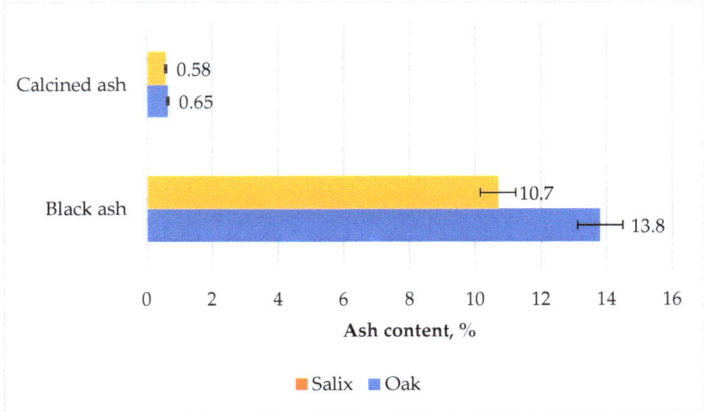

Figure 9. The content of calcined ash and black ash in the case of oak and energetic willow. The variable is the ash content.

The calcined ash of the energetic willow (as well as of the oak) has kept the current values of the woody species, i.e., fell under 1%. Effectively, the calcined ash content of the energetic willow was 12% lower than that of the oak, which brings an added value to this species.

3.14. Modeling the Calorific Power Depending on the Content of Chemical Components

This modeling starts from the premise that all lignocellulosic materials have a different calorific value of the chemical constituents, respectively, of 24.4 MJ/kg for lignin, 18.6 MJ/kg for cellulose, 16.1 MJ/kg for hemicelluloses, 34.5 MJ/kg for extractives, and 0.2 MJ/kg for ash [60–69]. The influence of the torrefaction process is also considered [35,69–71], along with other original research-specific correlations. If the different participation percentage of the chemical components for each wood species is taken into account, the following calculation relationship (Equation (18)) is obtained:

$$CV = 34.5 \times \frac{Ext}{100} + 24.4 \times \frac{Lig}{100} + 18.6 \times \frac{Cel}{100} + 16.1 \times \frac{Hem}{100} + 0.2 \times \frac{Ash}{100} \left[\frac{MJ}{kg}\right] \quad (18)$$

where: *Ext* is the percentage of extractives from the species in %; *Lig* is the lignin percentage of the species in wt%; *Cel* is the percentage of cellulose of the species in % wt; *Hem* is the percentage of hemicelluloses of the species in % wt; *Ash* is the percentage of ash content of the species in % wt.

It is also taken into account that the degradation of hemicelluloses starts at over 180 °C, of cellulose at over 240 °C, and lignin at over 280 °C [46,50]. Therefore, during the heat treatment process by torrefaction, the hemicelluloses are degraded first, and finally, the lignin ash content remains unchanged.

Following this methodology, a calorific value was obtained for native pellets of 20.46 MJ/kg for willow and 20.33 MJ/kg for oak, with a small difference compared to the research values (Table 2). Next, taking into account the loss of mass during torrefaction over the entire thermal treatment area of 23.4% for willow and 27.4% for oak. Taking into account these mass losses, the percentages of cellulose and hemicellulose were changed. Specifically for the energetic willow, 23.4 % was subdivided into 17.09% hemicellulose, with the difference of 6.31% coming from the cellulose. Therefore, there will be a new division of the components from 100%, namely "%ref" in Table 2. The 23.4% will be divided between the other components depending on the initial percentages of the chemical compounds.

Table 2. Modeling the calorific power during torrefaction.

Constituents		Extractive 34.5 MJ/kg	Lignin 24.4	Cellulose 18.6	Hemicelluloses 16.1	Ash 0.2	Total -	MJ/kg, Equation (18)
Salix-native	%	6.07	27.31	49.11	17.09	0.58	100	-
	KJ/kg	209.41	666.364	913.446	257.149	0.118	2046.48	20.46
Salix-torrefied	%	6.07	27.31	42.80	0	0.58	77.6	-
	%ref.	7.47	33.74	54.09	3.99	0.71	100	-
	kJ/kg	257.025	823.256	1006.074	64.239	0.142	21150.736	21.15
Oak-native	%	8.20	20.05	46.08	25.12	0.65	100	-
	kJ/kg	282.9	489.22	857.088	404.43	0.13	2033.76	20.33
Oak-torrefied	%	8.20	20.05	43.8	0	0.65	(−27.4)	-
	%ref.	10.44	25.54	56.42	6.87	0.82	100	-
	kJ/kg	360.18	623.176	1047.412	110.607	0.164	2141.539	21.41

These new percentages of the chemical components will provide a new calorific value of the torrefied pellets of 21.15 MJ/kg for energetic willow and 21.41 for oak. These values were appropriated to the tested value, meaning the modeling was correctly predicted.

4. Discussion

In order to be able to have a discussion on the properties of the analyzed briquettes and pellets, the average values from the paper were centralized together with those of other authors and standards [34,50,56] in Table 3. In addition to these data, other data were taken and analyzed from the research studies of other authors [7,27,35,41–43,57,70–72].

Table 3. Synthesis of properties.

No.	Property	Own Values of Research		Other Value	References
		Willow	Oak		
1.	Unit density of briquettes, kg/m^3	766	877	620–720	CRI-R0415
2.	Unit density of pellets, kg/m^3	1101	1296	Min 1000	ONORM M7135
3.	Mass loss of torrefaction, %	23	27	5–22	Tumuluru et al. [34]
4.	Lightness (L*) of native pellets	36.5	38.1	62 (beech)	Mitani and Barboutis [56]
5.	Lightness (L*) of torrefied pellets	19.4	28.5	45 (beech)	Mitani and Barboutis [56]
6.	Calorific value for native biomass, MJ/kg	20.7	20.5	17.5–19.5	DIN 51731
7.	Calorific value after torrefaction, MJ/kg	21.4	21.2	22.9	Bi et al. [67]
8.	Compressive strength, N/mm^2	1.02	0.33	0.53	Brozek et al. [68,69]
9.	Abrasion of briquettes, %	4.22	1.92	(1.5–3)%	SS 18 17 20
10.	Ash content, %	0.58	0.62	Max 6%	ONORM M7135

The unit density of energetic willow briquettes (766 kg/m^3) was 12.5% lower than that of oak; this difference is due to the 68.7% increased density of oak wood compared to that of energetic willow biomass. Referring to the limiting values of 620–720 kg/m^3 of the Italian Standard CRI-R0415, it can be seen that the experimental values fall within these limits. In addition, Brozek et al. [68] found a density of poplar briquettes of 776 kg/m^3 and 692 kg/m^3 for briquettes from birch biomass. The unit density of pellets, as 1001 kg/m^3 for Salix and 1269 kg/m^3 for oak, falls within the limiting provisions of the Austrian Standard ONORM M7135 (minimal 1000 kg/m^3). Similar values were found by other authors [27,28].

The experimental values of 23% and 27% mass loss by torrefaction are also confirmed by Tumuluru et al. [34], who found values of 5%–22% (for temperatures of 230–270 °C) for willow and 4%–14% for rice straw and wheat. The torrefaction treatment destroys OH groups in the wood, forming a non-polar unsaturated chemical structure, which will protect the pellets against biological degradation, similar to charcoal [56].

The values of the luminance (L*) expressed in the CIELab space of the native pellets obtained from the biomass of the energetic willow and the oak were different, with a difference of 9.1 color units. From this point of view, the energetic willow had a color closer to white than the oak. After torrefaction, the luminance of the two categories of pellets is very close, the difference being only 1.6 color units, in a shade of gray much closer to black than native pellets. Mitani and Barboutis [56] found that the beech species has a decrease in luminance during the torrefaction treatment from 60 to 45 color units, the difference depending on the treatment temperature but also on the structural direction of the wood (longitudinal, radial, and tangential) which is taken into consideration.

The calorific value of native energetic willow (20.7 MJ/kg) and native oak biomass (20.5 MJ/kg) fell within the provisions of all European reference standards, respectively, higher than 16.9 MJ/kg (SS 18 17 20), higher than 18 MJ/kg (ONORM M 7135), between 17.5–19.5 (DIN 51731:2000) and higher than 16.2 MJ/kg (CTI-R0615). Moreover, the calorific value of torrefied pellets, higher than that of native pellets, falls within the above limiting values of European standards. The values obtained in this study are slightly higher than those obtained by other researchers [69] for the clone of the energetic willow *Salix shwerinii* of 20.02 MJ/kg. The same authors found other calorific values of 20.2 N/mm^2 for *Alnus glutinosa* and 19.4 MJ/kg for *Betula pendula*. Balaban and Uçar [60] found an average high calorific value of 20 MJ/kg. Krajnic [42] has stated typical values of calorific value lower than 19.2 for coniferous and 19 MJ/kg for deciduous species. Referring to the calorific value for torrefied pellets, Bi et al. [67] found a maximum value of 22.9 MJ/kg, higher than the one from our own research due to the no-oxygen content during torrefaction. In addition, Hu et al. [69] found a difference between the HCV value before and after torrefaction of 2.12 MJ/kg, a value higher than the one found in the research of 0.8 MJ/kg, explained by the duration and temperature of the heat treatment performed.

Regarding the compressive strength, Brozek et al. [69] have established the compressive strengths of briquettes appropriate to those found in the research (1.02 N/mm^2 for energetic willow and 0.33 N/mm^2 for oak). For example, taking into account the diameter of the briquettes of 60 mm, for briquettes obtained from poplar biomass, a value of 1.35 N/mm^2 was found, and for biomass from the bark of the same species, a value of 0.53 N/mm^2 was also obtained.

The abrasion determines the amount of dust resulting from the transport and handling of briquettes, established by the Swedish Standard SS 18 17 20 in the form of "fines" smaller than 3 mm at a range of (1.5%–3%). The values of 4.22% for the energetic willow exceed the maximum value of 3% due to the low density of the briquettes. The abrasion value of 1.92% for the oak biomass briquettes falls within the limits of the standard, especially due to the density of 877 kg/m^3, 14.4% higher than that of the energetic willow.

The limits of the calcined ash content are different from one standard to another, being (0.7%–1.5%) for woody biomass (SS 18 17 20, DIN 51731 and CTI-R0415) and a maximum of 6% for agricultural biomass (ONORM M7135). Agricultural biomass has a higher mineral content than wood, which is why the ash content is also higher [67–69]). The experimental values of 0.58% for *Salix viminalis* and 0.62% for *Quercus robur* fall within the limits of European standards. Other researchers found values higher than the standardized limits for *Populus alba* (3.61%) and its bark (1.72%) [67], and others [69] found a value of 0.44% for the *Salix shwerinii* clone, 2.17% for *Populus tremula*, 0.78% for *Alnus glutinosa*, and 0.75% for *Betula pendula*.

5. Conclusions

- Energetic willow has a calorific value of 20.7 MJ/kg and an energy density of 22.7 × 10^3 MJ/m^3, higher than those of oak of 20.58 MJ/kg and 26.6 × 10^3 MJ/m^3, respectively. So, the energetic willow has very good calorific behavior, its properties being better than those of oak and other wood species used in the energetic field. Moreover, the research demonstrated why willow is considered one of the deciduous species with the highest calorific value.

- The torrefaction treatment at maximum regime led to a better calorific value of 21.43 MJ/kg in the case of energetic willow, compared to only 21.29 MJ/kg in the case of oak.
- The calcined ash content was lower in the case of energetic willow, with a value of 0.59%, compared to 0.65% in the case of oak.
- Ecologically, energetic willow has the same positive effects of sequestering carbon dioxide from the air and releasing oxygen as any other fast-growing woody species used in combustion.
- Future research will be focused on increasing the carbon content of the energetic willow biomass during the torrefaction process.

Author Contributions: Conceptualization, A.L. and V.D.; methodology, C.S. (Cosmin Spirchez); software, C.S. (Cezar Scriba); validation, A.L., C.S. (Cosmin Spirchez) and C.S. (Cezar Scriba); formal analysis, C.S. (Cezar Scriba); investigation, C.S. (Cosmin Spirchez); resources, C.S. (Cezar Scriba); data curation, C.S. (Cezar Scriba); writing—original draft preparation, A.L.; writing—review and editing, A.L.; visualization, C.S. (Cezar Scriba); supervision, A.L.; project administration, A.L.; funding acquisition, C.S. (Cosmin Spirchez) All authors have read and agreed to the published version of the manuscript.

Funding: This research received no external funding.

Data Availability Statement: Not applicable.

Conflicts of Interest: The authors declare no conflict of interest.

References

1. Manzone, M.; Balsari, P. Planters' performance during a Very Short Rotation Coppice planting. *Biomass Bioenergy* **2014**, *67*, 188–192. [CrossRef]
2. Wilkinson, J.M.; Evans, E.J.; Bilsborrow, P.E.; Wright, C.; Wewison, W.O.; Pilbeam, D.J. Yield of willow cultivars at different planting densities in a commercial short rotation coppice in the north of England. *Biomass Bioenergy* **2007**, *31*, 469–474. [CrossRef]
3. Amichev, B.Z.; Hangs, R.D.; van Ress, K.C.J. A novel approach to simulate the growth of multi-stem willow in bioenergy production systems with a simple process-based model (3PG). *Biomass Energy* **2011**, *35*, 473–488. [CrossRef]
4. Mleczek, M.; Rutowski, P.; Rissman, I.; Kaczmarek, Z.; Golinski, P.; Szentner, K.; Strazynska, K.; Stachowiak, A. Biomass productivity and phytoremediation potential of *Salix alba* and *Salix viminalis*. *Biomass Bioenergy* **2010**, *34*, 1410–1418. [CrossRef]
5. Mleczek, M.; Gąseckaa, B.; Waliszewska, Z.; Magdziak, M.; Szostek, P.; Rutkowskid, J.; Kaniuczakc, M.; Zborowskab, S.; Budzyńskaa, P.; Mleczek, P.; et al. *Salix viminalis* L.—A highly effective plant in phytoextraction of elements. *Chemosphere* **2018**, *212*, 67–78. [CrossRef]
6. Scriba, C.; Lunguleasa, A.; Spirchez, C.; Ciobanu, V. Influence of INGER and TORDIS Energetic Willow Clones Planted on Contaminated Soil on the Survival Rates, Yields and Calorific Value. *Forests* **2021**, *12*, 826. [CrossRef]
7. Stelte, W.; Holm, J.K.; Sanadi, A.R.; Barsberg, S.; Ahrenfeldt, J.; Henriksen, U.B. A study of bonding and failure mechanisms in fuel pellets from different biomass resources. *Biomass Bioenergy* **2011**, *35*, 910–918. [CrossRef]
8. Rebina 2022, Rebina Agrar. Available online: https://balkangreenenergynews.com/energy-willow-salix-viminalis-biomass-where-you-want-it/ (accessed on 2 October 2022).
9. Albertsson, J.; Verwijst, T.; Hansson, D.; Bertholdsson, N.-O.; Åhman, I. Effects of competition between short-rotation willow and weeds on performance of different clones and associated weed flora during the first harvesting cycle. *Biomass Bioenergy* **2014**, *70*, 364–372. [CrossRef]
10. Bergante, S.; Manzone, M.; Facciotto, F. Alternative planting method for short rotation coppice with poplar and willow. *Biomass Bioenergy* **2016**, *87*, 39–45. [CrossRef]
11. Dimitrious, I.; Rosenqvist, H.; Berndes, G. Slow expansion and low yields of willow short rotation coppice in Sweden; implications for future strategies. *Biomass Bioenergy* **2011**, *35*, 4613–4618. [CrossRef]
12. Buchholz, T.; Volk, T.A. Improving the profitability of willow crops-identifying opportunities with a crop budget model. *Bioenergy Res.* **2011**, *4*, 85–95. [CrossRef]
13. Balloffet, N.; Deal, R.; Hines, S.; Larry, B.; Smith, N. Ecosystem Services and Climate Change 2012; U.S. Department of Agriculture, Forest Service, Climate Change Resource Center. Available online: www.fs.usda.gov/ccrc/topics/ecosystem-services (accessed on 15 March 2023).
14. El Kasmioui, O.; Ceulemans, R. Financial analysis of the cultivation of short rotation woody crops for bioenergy in Belgium: Barriers and opportunities. *BioEnergy Res.* **2013**, *6*, 336–350. [CrossRef]
15. Fiala, M.; Bacenetti, J. Economic, energetic and environmental impact in short rotation coppice harvesting operations. *Biomass Bioenergy* **2012**, *42*, 107–113. [CrossRef]

16. Ray, M.; Brereton, N.B.; Shield, I.; Karp, A.; Murphy, R. Variation in cell wall composition and accessibility in relation to biofuel potential of short rotation coppice willows. *Bioenergy Res.* **2012**, *5*, 685–698. [CrossRef]
17. Serapiglia, M.; Cameron, K.; Stipanovic, A.; Abrahamson, L.; Volk, T.; Smart, L. Yield and woody biomass traits of novel shrub willow hybrids at two contrasting sites. *Bioenergy Res.* **2012**, *6*, 533–546. [CrossRef]
18. Volk, T.A.; Abrahamson, L.P.; Cameron, K.D.; Castellano, P.; Corbin, T.; Fabio, E. Yields of willow biomass crops across a range of sites in North America. *Asp. Appl. Biol.* **2011**, *112*, 67–74.
19. Larsen, S.U.; Jørgensen, U.; Lærke, P.E. Willow yield is highly dependent on clone and site. *BioEnergy Res.* **2014**, *7*, 1280–1292. [CrossRef]
20. Mitsui, Y.; Seto, S.; Nishio, M.; Minato, K.; Ishizawa, K.; Satoh, S. Willow clones with high biomass yield in short rotation coppice in the southern region of Tohoku district (Japan). *Biomass Bioenergy* **2010**, *34*, 467–473. [CrossRef]
21. Nissim, W.G.; Laberque, M. Planting micro cuttings: An innovative method for establishing a willow vegetation cover. *Ecol. Eng.* **2016**, *91*, 472–476. [CrossRef]
22. Wang, Z.; MacFarlane, D.W. Evaluating the biomass production of coppiced willow and polar clones in Michigan, USA, over multiple rotations and different growing conditions. *Biomass Bioenergy* **2012**, *46*, 380–388. [CrossRef]
23. Tahvanainena, L.; Rytkönena, V.-M. Biomass production of *Salix viminalis* in southern Finland and the effect of soil properties and climate conditions on its production and survival. *Biomass Bioenergy* **1999**, *16*, 103–117. [CrossRef]
24. Hallingbäck, H.; Berlin, S.; Nordh, N.-E.; Weih, M.; Rönnberg-Wästljung, A.-C. Genome Wide Associations of Growth, Phenology, and Plasticity Traits in Willow (*Salix viminalis* (L.)). *Front. Plant Sci.* **2019**, *10*, 753. [CrossRef]
25. Almeida, P.; Proux-Wera, E.; Churcher, A. Genome assembly of the basket willow, *Salix viminalis*, reveals earliest stages of sex chromosome expansion. *BMC Biol.* **2020**, *18*, 78. [CrossRef]
26. Fromm, J.; Spanswick, R. Characteristics of Action Potentials in Willow (*Salix viminalis* L.). *J. Exp. Bot.* **1993**, *44*, 1119–1125. [CrossRef]
27. Gao, Y.; Jebrane, M.; Terziev, N.; Daniel, G. Evaluation of Wood Quality Traits in *Salix viminalis* Useful for Biofuels: Characterization and Method Development. *Forests* **2021**, *12*, 1048. [CrossRef]
28. Marreiro, H.; Peruchi, R.; Lopes, R.; Andersen, S.; Eliziário, S.; Rotella Junior, P. Empirical Studies on Biomass Briquette Production: A Literature Review. *Energies* **2021**, *14*, 8320. [CrossRef]
29. Rönnberg-Wästljung, A.C. Genetic structure of growth and phenological traits in *Salix viminalis*. *Can. J. For. Res.* **2001**, *31*, 276–282.
30. Munshi, L.A.; Dar, A.R. Genetic Diversity in *Salix viminalis* in the Kashmir Valley, India. *Am. J. Biochem. Mol. Biol.* **2011**, *1*, 178–184. [CrossRef]
31. Lascoux, M.; Thorsén, J.; Gullberg, U. Population structure of a riparian willow species, *Salix viminalis* L. *Genet. Res.* **1996**, *68*, 45–54. [CrossRef]
32. European Bioenergy Outlook. 2013. Available online: https://www.pfcyl.es/sites/default/files/biblioteca/documentos/european_bioenergy_outlook_2013.pdf (accessed on 3 June 2020).
33. Griu, T.; Lunguleasa, A. *Salix viminalis* vs. Fagus sylvatica—Fight for renewable energy from woody biomass in Romania. *Environ. Eng. Manag. J.* **2016**, *15*, 413–420.
34. Tumuluru, J.S.; Sokhansanj, S.; Wright, C.; Boardman, R.; Hess, J.R. Review on Biomass Torrefaction Process and Product Properties and Design of Moving Bed Torrefaction System Model Development. In Proceedings of the 2011 ASABE Annual International Meeting, Louisville, KY, USA, 7–10 August 2011.
35. Chen, W.H.; Cheng, W.Y.; Lu, K.M.; Wuang, Y.P. An Evaluation on Improvement of Pulverized Biomass Property for Solid through Torrefaction. *Appl. Energy* **2011**, *11*, 3636–3644. [CrossRef]
36. Ecomatcher 2022. How to Calculate CO2 Sequestration. Available online: https://www.ecomatcher.com/how-to-calculate-co2-sequestration/ (accessed on 7 September 2022).
37. FAO. Environmental Aspects of Natural Resources: Forestry. 2022. Available online: https://www.fao.org/3/ad905e/AD905E02.htm (accessed on 4 April 2022).
38. Ens, J.; Farrell, R.; Bélanger, N. Early effects of afforestation with willow (*Salix purpurea*, "Hotel") on soil carbon and nutrient availability. *Forests* **2013**, *4*, 137–154. [CrossRef]
39. Nurek, T.; Gendek, A.; Dąbrowska, M. Influence of the Die Height on the Density of the Briquette Produced from Shredded Logging Residues. *Materials* **2021**, *14*, 3698. [CrossRef] [PubMed]
40. Kumar, S.; Kumar, R.; Pandey, A. *Current Developments in Biotechnology and Bioengineering: Waste treatment Processes for Energy Generation*; Elsevier: Amsterdam, The Netherlands, 2019.
41. Francescato, V.; Antonin, E.; Bergomi, L.Z. *Wood Fuels Handbook*; Print House AIEL—Italian Agri forestry Energy Association: Legnaro, Italy, 2008.
42. Krajnic, N. *Wood Fuel Handbook*; Food and Agriculture Organization (FAO) of the United Nations: Pristina, Kosovo, 2015.
43. Demirbas, A.; Sahin-Demirbas, A. Briquetting Properties of Biomass Waste Materials. *Energy Source Part A* **2004**, *26*, 83–91. [CrossRef]
44. Dhillon, R.S.; von Wuelhlisch, G. Mitigation of Global Warming through Renewable Biomass. *Biomass Bioenergy* **2013**, *48*, 75–87. [CrossRef]
45. Kongsager, R.; Napier, J.; Mertz, O. The carbon sequestration potential of tree crop plantations. *Mitig. Adapt. Strateg. Glob. Chang.* **2013**, *18*, 1197–1213. [CrossRef]

46. Sobczyk, L.W.; Sobczyk, E.J. Economical and ecological effects of cultivation of basket willow *Salix viminalis*. *IOP Conf. Ser. Earth Environ. Sci.* **2019**, *214*, 012002. [CrossRef]
47. Wagenführ, R.; Wagenführ, A. *Holzatlas*; (E-book); Carl Hanser Fachbuchverlag: München, Germany, 2021.
48. De Neergaard, A.; Porter, J.; Gorissen, A. Distribution of assimilated carbon in plants and rhizosphere soil of basket willow (*Salix viminalis* L.). *Plant Soil* **2002**, *245*, 307–314. [CrossRef]
49. Dzurenda, L.; Geffertova, J.; Hecl, V. Energy Characteristics of Wood-Chips Produced from *Salix viminalis*—Clone ULV. *Drvna Ind.* **2010**, *61*, 27–31.
50. Verma, V.K.; Bram, S.; de Ruyck, J. Small Scale Biomass Systems: Standards, Quality Labeling and Market Driving Factors—An EU Outlook. *Biomass Bioenergy* **2009**, *33*, 1393–1402. [CrossRef]
51. Côté, W.A. Chemical Composition of Wood. In *Principles of Wood Science and Technology*; Springer: Berlin/Heidelberg, Germany, 1968. [CrossRef]
52. EN 13183-1:2002; Moisture Content of a Piece of Sawn Timber—Part 1: Determination by Oven Dry Method. European Committee for Standardization: Brussels, Belgium, 2002.
53. ASTM E871-82:2019; Standard Test Method for Moisture Analysis of Particulate Wood Fuels. ASTM International: West Conshohocken, PA, USA, 2019.
54. DIN 51731: 2013; Testing of Solid Fuels—Compressed Untreated Wood—Requirements and Testing. German Institute for Standardisation: Berlin, Germany, 2013.
55. EN 15103: 2009; Solid Biofuels—Determination of Bulk Density. European Committee for Standardization: Brussels, Belgium, 2009.
56. Mitani, A.; Barboutis, I. Changes Caused by Heat Treatment in Color and Dimensional Stability of Beech (*Fagus sylvatica* L.) Wood. *Drvna Ind.* **2014**, *65*, 225–232. [CrossRef]
57. ISO 1928: 2020; Coal and Coke—Determination of Gross Calorific Value. International Standardization Organization: Geneva, Switzerland, 2020.
58. ASTM E1755-01: 2020; Standard Test Method for Ash in Biomass. ASTM International: West Conshohocken, PA, USA, 2020.
59. ISO 2171: 2007; Cereals, Pulses and By-products—Determination of Ash Yield by Incineration. International Organization for Standardization: Geneva, Switzerland, 2007.
60. Balaban, M.; Uçar, G. Extractives and Structural Components in Wood and Bark of Endemic Oak *Quercus vulcanica* Boiss. *Holzforschung* **2001**, *55*, 478–486. [CrossRef]
61. Maksimuk, Y.; Antonava, Z.; Krouk, V.; Korsakova, A.; Kursevich, V. Prediction of higher heating value (HHV) based on the structural composition for biomass. *Fuel* **2021**, *299*, 120860. [CrossRef]
62. Telmo, C.; Lousada, J. The explained variation by lignin and extractive contents on higher heating value of wood. *Biomass Bioenergy* **2011**, *35*, 1663–1667. [CrossRef]
63. ÖNORM M 7135: 2000; Pellets and Briquettes—Requirements and Test Conditions. Austrian Standards Institute: Vienna, Austria, 2000.
64. Gocławski, J.; Korzeniewska, E.; Sekulska-Nalewajko, J.; Kiełbasa, P.; Dróżdż, T. Method of Biomass Discrimination for Fast Assessment of Calorific Value. *Energies* **2022**, *15*, 2514. [CrossRef]
65. Lamlom, H.; Savidge, R.A. A reassessment of carbon content in wood: Variation within and between 41 North American species. *Biomass Bioenergy* **2003**, *25*, 381–388. [CrossRef]
66. Shulga, G.; Betkers, T.; Brovkina, J.; Aniskevicha, O.; Ozoliņš, J. Relationship between Composition of the Lignin-based Interpolymer Complex and its Structuring Ability. *Environ. Eng. Manag. J.* **2008**, *7*, 397–400. [CrossRef]
67. Bi, T.; Sokhsansanj, S.; Lim, J. *Torrefaction & Densification of Biomass*; Biomass and Bioenergy Research Group, Clean Energy Research Centre, University of British Columbia: Vancouver, BC, Canada, 2012.
68. Brožek, M.; Nováková, A.; Kolářová, M. Quality evaluation of briquettes made from wood waste. *Res. Agric. Eng.* **2012**, *58*, 30–35. [CrossRef]
69. Hu, W.; Yang, X.; Mi, B.; Liang, F.; Zhang, T.; Fei, B.; Jiang, Z.; Liu, Z. Investigating chemical properties and combustion characteristics of torrefied masson pine. *Wood Fiber Sci.* **2017**, *49*, 33–42.
70. Alizadeh, P.; Tabil, L.G.; Adapa, P.K.; Cree, D.; Mupondwa, E.; Emadi, B. Torrefaction and Densification of Wood Sawdust for Bioenergy Applications. *Fuels* **2022**, *3*, 152–175. [CrossRef]
71. Spirchez, C.; Lunguleasa, A.; Antonaru, C. Experiments and modeling of the torrefaction of white wood fuel pellets. *BioResources* **2017**, *12*, 8595–8611. [CrossRef]
72. García-Maraver, A.; Popov, V.; Zamorano, M. A review of European standards for pellet quality. *Renew. Energy* **2011**, *36*, 3537–3540. [CrossRef]

Disclaimer/Publisher's Note: The statements, opinions and data contained in all publications are solely those of the individual author(s) and contributor(s) and not of MDPI and/or the editor(s). MDPI and/or the editor(s) disclaim responsibility for any injury to people or property resulting from any ideas, methods, instructions or products referred to in the content.

Article

Manufacturing and Testing the Panels with a Transverse Texture Obtained from Branches of Norway Spruce (*Picea abies* L. Karst.)

Alin M. Olarescu, Aurel Lunguleasa *, Loredana Radulescu and Cosmin Spirchez

Wood Processing and Design Wooden Product Department, Transilvania University of Brasov, 29 Street Eroilor, 500036 Brasov, Romania
* Correspondence: lunga@unitbv.ro

Abstract: As a result of the imbalances in the forestry market and the increased demand for wood products worldwide, the resource of branches resulting from the exploitation of a forest has attracted special attention from researchers, in order to use these secondary resources judiciously and obtain an added value superior to classic uses. In this context, the current research took into consideration the use of spruce branches to obtain panels with a transverse structure. The work methodology has focused on the process of obtaining panels with a transverse texture and on determining the physical–mechanical properties of the created panels. The results regarding the panel density (determined as a ratio between mass and volume of specimens) showed about 693 kg/m^3, static bending resistance parallel to the face of 5.5 N/mm^2, resistance of adhesion of 5.6 N/mm^2, shear strength parallel to face of 4.1 N/mm^2, and screw pull-out resistance perpendicular to the face of 31.3 N/mm^2, highlighting that the properties were in accordance with the European standards and that the panels obtained were suitable for obtaining furniture products with a special aesthetic aspect. As a general conclusion of the research, it can be stated that spruce branches are a sustainable wood resource with great possibilities to add more value in the form of panels with a traverse texture.

Keywords: spruce; branches' structure; modulus of resistance (MOR); screw withdrawn; adhesion strength; shearing resistance; water immersion

Citation: Olarescu, A.M.; Lunguleasa, A.; Radulescu, L.; Spirchez, C. Manufacturing and Testing the Panels with a Transverse Texture Obtained from Branches of Norway Spruce (*Picea abies* L. Karst.). *Forests* **2023**, *14*, 665. https://doi.org/10.3390/f14040665

Academic Editor: Petar Antov

Received: 13 February 2023
Revised: 13 March 2023
Accepted: 20 March 2023
Published: 23 March 2023

Copyright: © 2023 by the authors. Licensee MDPI, Basel, Switzerland. This article is an open access article distributed under the terms and conditions of the Creative Commons Attribution (CC BY) license (https://creativecommons.org/licenses/by/4.0/).

1. Introduction

The lack or insufficiency of forest resources in recent decades has determined the search for new solutions to replace solid wood in the form of timber or logs. The branches obtained from the exploitation of a forest, be they coniferous or deciduous, has been considered for a long period of time as forest biomass [1], with exclusive uses in combustion and in the technology of composite boards made of chips and wood fibers [2–4]. In the last four decades, tree branches have been studied more and more [5–8] in order to better know their properties and implicitly understand their uses [9]. It was found that the branches have the same structure as the trunk wood [10], with the exception of the reaction wood, particularized in compression wood for softwoods or tension in the case of hardwoods, whose properties are slightly different from the trunk wood.

The large amount of wood from the branches of a tree, estimated at about 10% of the total volume of the trunk, shows that when exploiting a forest area, an impressive volume of wood is available, with physical and mechanical properties close to those of the trunk [11–13]. These branches must be used judiciously and economically efficiently in order to obtain the greatest possible added value, superior to that obtained in the case of use in the field of wood-based composites.

Regarding the design of a new product, the sustainability of the raw material must be taken into account as well as the fact that the products obtained should be cleaner and greener. In recent years, the importance of environmental aspects has increased [14], which is why the product eco-design has become one of the most important conditions for the development of new products. Some of the criteria mentioned in the research of some authors [15–20] are the acquisition of secondary resources such as woody biomass, the choice of materials with a low impact on the environment and the development of products with a high added value. Therefore, through this eco-design (or ecological product design), the design of the new product is brought into line with the environment, both during the product's functionality and after its removal from use.

Wood-based panels are part of the boards category, including particleboard, fiberboard, oriented strand board (OSB), medium-density fiberboard (MDF), plywood, laminated veneer lumber (LVL), glue laminated products (Glulam), cross laminated timber (CLT), etc. That is why the tests for determining the physical–mechanical properties of wood-based panels are the same as those of the previously mentioned boards. Depending on their field of use of EN 13353: 2022, a wetting treatment via immersion was recommended before performing some tests [21] such as gluing resistance, modulus of rupture (MOR), and modulus of elasticity (MOE) in bending strength (EN 300: 2019 for OSB and EN 312: 2010 for particleboard) (Table 1).

Table 1. Type and duration of the prior treatment depending on the environment of use.

Type	Conditions of Use Conform EN 13353: 2022	Type of Prior Treatment
T1	SWP/1 Dry medium (interior)	Immersion in water (20 °C, 24 h)
T2	SWP/2 Wet medium	Boiling, 6 h Cooling in water (20 °C, 1 h)
T3	SWP/3 Exterior medium	Boiling, 4 h Drying in oven (60 °C, 16–20 h) Boiling, 4 h Cooling in water (20 °C, 1 h)

The analysis of previous research in the field of the theme highlights the fact that they studied the properties of wood from branches. Based on these properties, some future uses are predicted. The use of wood from branches in the field of wood-based composite materials is the most common because they are reconstituted engineering products, which even out the structural inhomogeneity of this wood. Starting from these limitations, the current research aimed to highlight the special structure of the cross-sectional area of the wood from the branches by creating a panel that can be used in small furniture. The objectives of this study are to present the optimal technology for manufacturing panels with a transverse texture obtained from spruce (*Picea abies* L.) branches and test these panels. The research started from the hypothesis that the structural inhomogeneity of the wood from the branches will be mitigated by gluing the friezes with a suitable adhesive, and the adequate technology of the panel with a traverse structure will remove any cracks that may appear during the use of the finished product.

2. Materials and Method

Technology of manufacturing and plan of panel testing. The manufacture of panels with a transverse texture from branches includes several phases, such as sorting them in order to choose the appropriate branches, cutting the prisms, forming the blocks of prisms by gluing on the edge, cutting the block into transverse friezes, forming a panel from transverse friezes by applying adhesive on the edge and pressing them, panel processing via formatting and calibration, panel packaging via wrapping, and panel storage (Figure 1).

Due to the small dimensions of the 6-dimensional types of panels made of spruce branches (426 × 335 × 20 mm, noted as P1; 455 × 300 × 20 mm noted as P2, 430 × 314 × 20 mm noted as P3, 510 × 308 × 20 mm noted as P4, 465 × 412 × 20 mm noted as P5, and 420 × 394 × 20 mm noted as P6), the application of the adhesive was performed manually with a trowel. The adhesive specific consumption of 280 g/m^2 of finished panel was used. The panels were glued with two types of cold glue (vinyl and polyurethane Jowapur 687.40 (Rudolf Ostermann GmbH, Bocholt, Germany)).

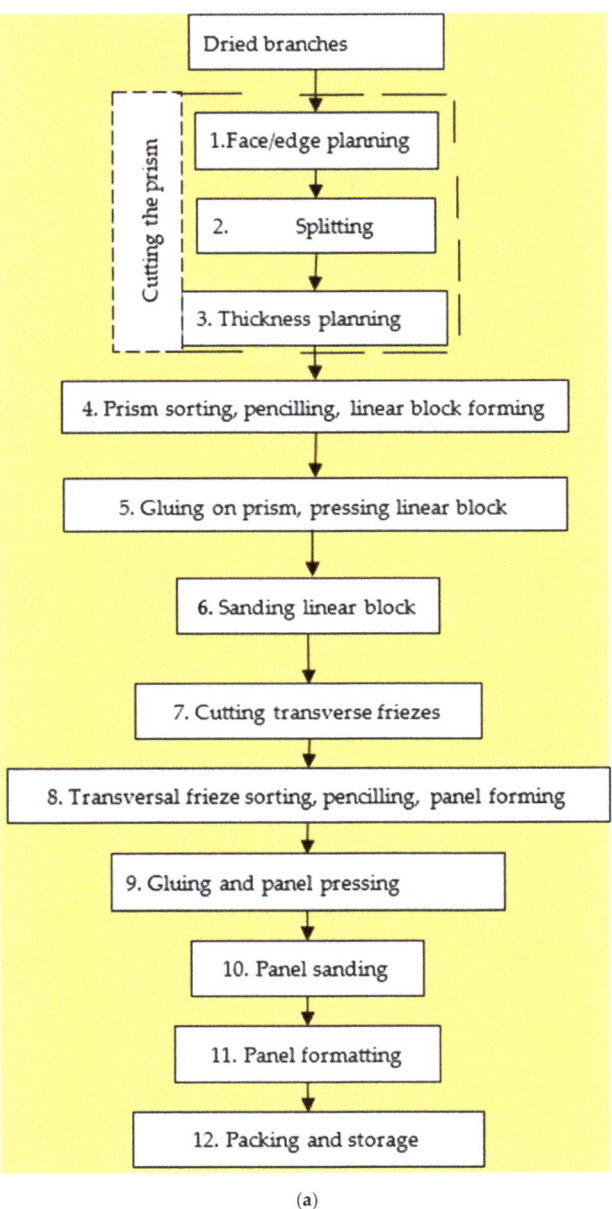

(a)

Figure 1. *Cont.*

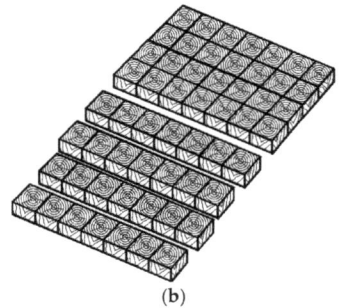

(b)

Figure 1. The manufacturing technological flow of panels with a transverse texture of spruce branches (**a**), and a panel with a transverse texture of spruce branches (**b**).

The preliminary tests carried out for the choice of the adhesive concerned the following mechanical characteristics of the cross-textured panels made of spruce branches: gluing strength, static bending strength, and static bending modulus of elasticity. Regarding the adhesion resistance, the European norm DD CEN/TS 13354:2003 stipulates that before the test [22], the samples should be subjected to a preliminary treatment in cold water immersion t = 20 °C for 24 h, in accordance with the conditions of indoor use [23–29] defined by EN 1995-1-1:2004 and/or EN 335-2:1992. The static bending strength (MOR) and the modulus of elasticity (MOE) were also tested, according to EN 310:1996.

The establishment of the plan of physical–mechanical tests, to which the panels with transverse texture made of spruce branches were subjected, was carried out on the basis of the European standards and the descriptions from the specialized literature [7,12,13,19]. Table 2 shows the physical–mechanical tests.

Table 2. Physical and mechanical tests to which the panels from the branches were subjected.

No.	Name of the Mechanical Test	Justification of the Choice	Norm	Number of Specimens
1.	Dimensions of the panels	It reflects the dimensional accuracy of panel manufacturing	EN 324-1:1993	6
2.	Flatness	The basic technological property of panels	Mantanis et al., 1994 [27]	6
3.	Bonding strength	The basic property of any glued product. It influences most of the mechanical properties	EN 13354:2008	minim 10
4.	Modulus of elasticity and resistance to static bending	Important for horizontal elements in furniture construction (boards, shelves, etc.)	EN 310:1996	6 EN 326-1:1996
5.	Shear resistance in the plane of the plate	Important for vertical or horizontal elements subject to shear (side walls, uprights, crossbars, etc.)	EN 13354:2008	8 EN 326-1:1996
6.	Shear resistance perpendicular to the face plate			
7.	The pull-out resistance of the screws	Screw assemblies are frequently used in furniture construction	EN 13446:2004	6
8.	Determination of the swelling coefficient	It determines the stability of structures in various environmental conditions	Mantanis et al., 1994	14

Dimensions and flatness of the panels. The tests were carried out in the Manufacturing Precision Testing Laboratory in the Wood Industry, Transylvania University Brasov, Brasov, Romania. The testing methodology was in accordance with the requirements of EN 324-1: 1993. The moisture content of the panels at the time of the test was 9.1%, falling within the requirements of EN 13353: 2022, which makes the following specifications regarding moisture content: $8 \pm 2\%$ for use in dry environment, $10 \pm 3\%$ for use in a humid environment, and $12 \pm 3\%$ for use in an outdoor environment.

The measurements were carried out with an OPTOdesQ Measurement Table (Hecth Electronic AG, Stuttgart, Germany) equipped with a magnetic measurement system on the three axes, X, Y, and Z, with a precision of 0.01 mm. Both programming and data acquisition were performed electronically by means of the Hecht OptodesQ software package. The size measurement scheme according to EN 324–1: 1993 is shown in Figure 2.

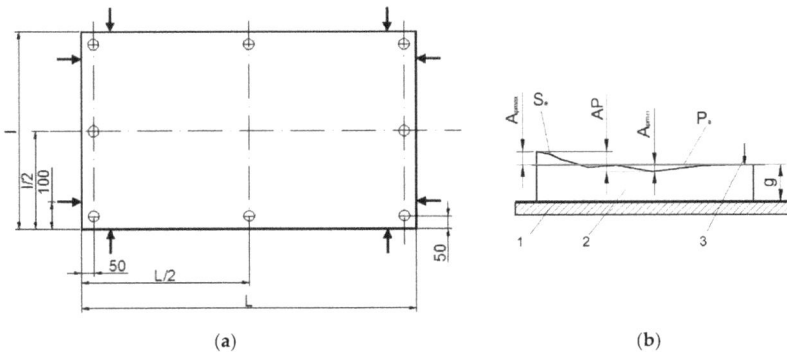

Figure 2. Scheme for measuring panel dimensions according to EN 324-1:1993 (**a**), and measurement of flatness deviation with OPTOdesQ Measurement Table. (**b**): 1—mass of the machine; 2—the measured panel; 3—feeler rod; g—reference thickness; Pa—the adjacent plane; Se—the effective surface; A_{pmin}—minimum (negative) deviation from flatness; A_{pmax}—maximum (positive) deviation from flatness.

The deviation was calculated as the difference between the actual size and the nominal size, expressed in mm with one decimal for the dimensional deviation and three decimals for the flatness deviation. The mean and standard deviation of the results were determined [29], according to EN 326-1 The calculation of the thickness deviation is based on measuring the thickness at the points set by the user and comparing it with the reference thickness g, the machine directly providing the positive or negative differences from it.

The gluing strength of panel with transverse texture. Bonding strength was determined in accordance with the DD CEN/TS 13354:2003 specifications. The tests were carried out on the Universal Testing Machine type ZDM 5t/510 (PPT Group UK Ltd., West Sussex, United Kingdom). The principle of the method consists of pre-treating the specimens according to the service class established according to EN 1995-1-1, and then in determining the maximum compressive shearing force of the bonding surface. Based on the maximum force and the bonding surface, the gluing strength was determined. The method of specimens cutting from the panel, as well as the shape and dimensions of the specimens, are shown in Figure 3. Sixteen specimens were tested.

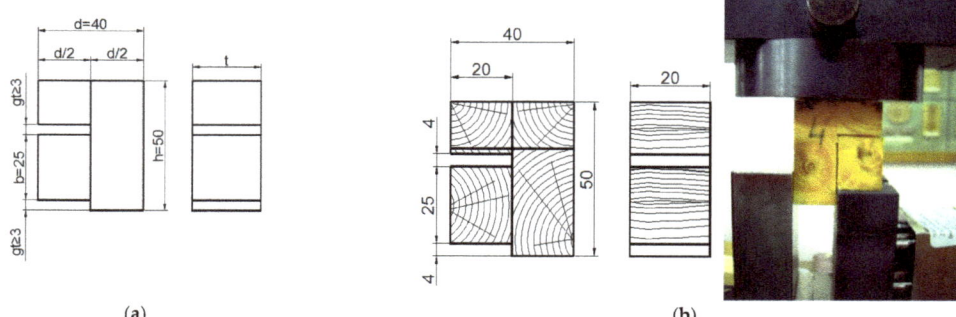

Figure 3. Specimen sizes for determining the gluing strength (**a**), and position of specimen in time of breaking (**b**).

Since the panels were designed for small pieces of furniture, used in indoor conditions, the samples, after cutting, were measured and then immersed in cold water (20 °C) for 24 h. After they were taken out of the water, they were measured at the same points as before the immersion.

The determination of the length and width of the shear area was made by using a Holzman digital caliper with a measuring range of 150 mm and a precision of 0.01 mm and a Holzman Rotating Digital Dial Gage (Holzmann maschinen GmbH, Haslach, Austria) with a measuring range of 0–12.5 mm and precision of 0.01 mm. Constant application of shearing force was made at a rate that produces maximum force (break) in approximately 60 ± 30 s. The shear strength of the gluing was calculated with the following relation Equation (1).

$$f_v = \frac{F}{l \cdot g} [N/mm^2] \qquad (1)$$

where: f_v —the shear strength of the bond, in [N/mm^2], F—the maximum force at which the break occurred, in [N], l—the length of the shear area, in [mm], g—the thickness of the specimen, in [mm].

Determination of resistance (MOR) and elasticity modulus (MOE) in static bending. The determination of the modulus of elasticity and the resistance to static bending was carried out according to EN 310:1996. The elastic modulus and static bending strength were calculated using Equation (2).

$$\sigma_i = \frac{3 \cdot F_{max} \cdot l_1}{2 \cdot b \cdot t^2} \qquad E_m = \frac{l_1^3 \cdot (F_2 - F_1)}{4 \cdot b \cdot t^3 \cdot (a_2 - a_1)} [N/mm^2] \qquad (2)$$

where: σ_i—the static bending resistance, in [N/mm^2], F_{max}—the maximum breaking force, in [N], b—width of the specimen, in mm, t—the thickness of the specimen, in [mm], l_1—the distance between the supports, equal to 20 times the thickness of the specimen, E_m—the modulus of elasticity in static bending, in [N/mm^2], $(F_1 - F_2)$—the force increment on the linear portion of force–deformation ($F_1 = 0.1 \cdot F_{max}$, $F_2 = 0.4 \cdot F_{max}$), in [N], and $(a_2 - a_1)$ represents the increment of the bending arrow corresponding to the force difference $(F_2 - F_1)$, in [mm].

The number of specimens according to EN 326-1:1996 was six valid pieces. According to the requirements of EN 310:1996, the specimens had the following dimensions: a thickness equal to that of the panel from which it comes, a width of 50 ± 1 mm, and the length equal to 20 times the thickness plus 50 mm [25,29]. In order to study the influence of the joining method on the strength of the specimen, two types of specimens were cut, namely: specimens with a central longitudinal line of gluing (Figure 4a) and randomly cut specimens (Figure 4b).

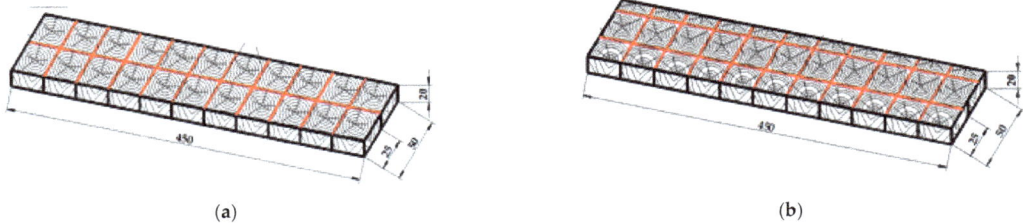

Figure 4. Specimens with a central longitudinal bonding line (**a**), and randomly cut specimens (**b**).

The laboratory machine for testing wood-based boards and panel—model IB × 600—produced by IMAL-PAL Group (San Damaso, Italy) was used for the bending test. When the test was finished, the software of the IB × 600 machine calculated both the static bending strength and the modulus of elasticity.

Determination of shear strength parallel to the panel face. The test was determined in accordance with the requirements of EN 13354:2008. The compressive strength parallel to the faces was calculated as the ratio between the maximum force and the breaking surface of the specimen Equation (3). The scheme of the test is presented in Figure 5.

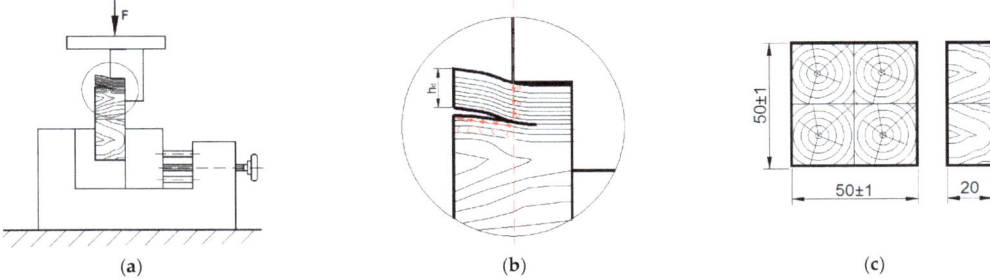

Figure 5. Scheme for determining the shearing resistance in the plane of the plate: (**a**)—outline of the test; F—force of shearing; h_d—shearing height; (**b**)—breaking the samples; (**c**)—dimensions of the samples. Scheme for determining the shearing resistance in the plane of the plate: F—force of shearing; h_d—shearing height.

$$\tau_{f_{ll}} = \frac{F_{max}}{b \cdot l} \quad (3)$$

where: $\tau_{f_{ll}}$—the resistance to compression parallel to the faces, in [N/mm^2], F_{max}—the maximum force, in [N], b—width of sample equal with 50, in mm, l—the breaking length of the specimen equal with 50, in mm^2.

Eight specimens were tested [29], in accordance with the requirements of EN 326-1:1996. The dimensions of the sample were: the thickness equal to that of the panel, width of 50 ± 1 mm, and length of 50 ± 1 mm (Figure 5). A Holzman digital caliper with a measuring range of 150 mm and a precision of 0.01 mm was used to determine the thickness of the specimen. A Holzman Rotating Digital Dial Gage (Holzmann maschinen GmbH, Haslach, Austria) with a measuring range of 0–12.5 mm and a precision of 0.01 mm was used to determine the width and length of the sample area. The tests were carried out on the ZDM 5t/51 Universal Testing Machine. To carry out the test in accordance with the requirements of EN 13354:2008, it was necessary to design and make two test devices. In parallel with the samples obtained from panels with a transverse texture of spruce branches, in order to be able to make a comparison, 6 specimens made of veneered chipboard with a thickness of 20 mm were tested for comparison under the same conditions.

Determination of shear resistance perpendicular to the panel plane. This test was determined in accordance with the requirements of EN 13354:2008. The shear strength parallel to the faces was calculated as the ratio between the maximum force and the breaking surface of the specimen, as in Equation (4). The scheme of determining the shear strength in the plane of the plate is shown in Figure 6.

$$\tau_{f\perp} = \frac{F_{max}}{2 \cdot A} \quad (4)$$

where: $\tau_{f\perp}$—shearing resistance to perpendicular plane, in [N/mm^2], F_{max}—maximum force [N], A—breaking surface of sample, as the product of thickness and 46 mm.

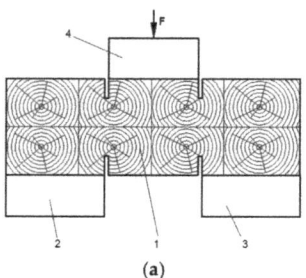
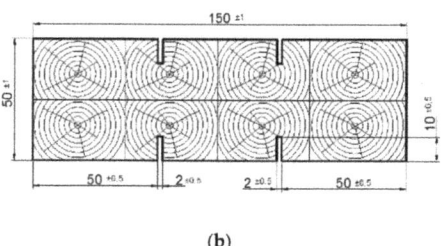

(a)   (b)

Figure 6. Principle scheme of determining the shear strength in the plane of the panel (a), and the dimensions of the sample. (b): 1—sample; 2,3—supports; 4—the force application device.

Eight samples were tested in accordance with the requirements of EN 326-1:1996. A Holzman Rotating Digital Dial Gage with a measurement range of 0–12.5 mm and a precision of 0.01 mm was used to determine the thickness of the specimen. To determine the width and length of the specimen, a Holzman digital caliper with a measuring range of 150 mm and a precision of 0.01 mm was used.

Before the test, for each sample, the thickness of the specimens was measured at two points in the direction of the shear planes, and the arithmetic mean of the two measurements was made. The tests were carried out on the ZDM 5t/51 Universal Testing Machine. To carry out the test in accordance with the requirements of EN 13354:2008, it was necessary to design and make the two supports and the force application device.

Determination of the pull-out resistance of screws. This test was determined in accordance with the requirements of EN 13446:2004 (Figure 7). The resistance to pulling out the screws was calculated as the ratio between the maximum force required to pull out the screw and the area of the screwing surface, as in Equation (5).

$$\tau_s = \frac{F_{max}}{d \cdot l_p} \quad (5)$$

where: τ_s—the resistance to pulling out the screws, in [N/mm^2], F_{max}—the maximum force required to pull out the screws, in [N], d—the diameter of the screw given by the standard, in [mm], and l_p—the screwing length.

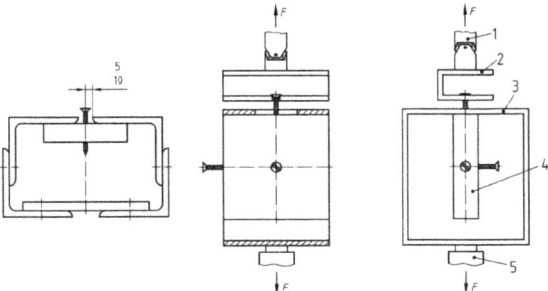

Figure 7. The scheme for determining the resistance to screw pull-out.

Three types of specimens were tested, depending on the structural characteristics of the panel (Figure 8). For each type of specimen, 6 specimens were tested in accordance with the requirements of EN 326-1:1996. The shape and dimensions of the specimens as well as the way of inserting the screws are shown in Figure 8. The screw insertion point can be at the intersection of four joint lines—on a joining line and inside the wood.

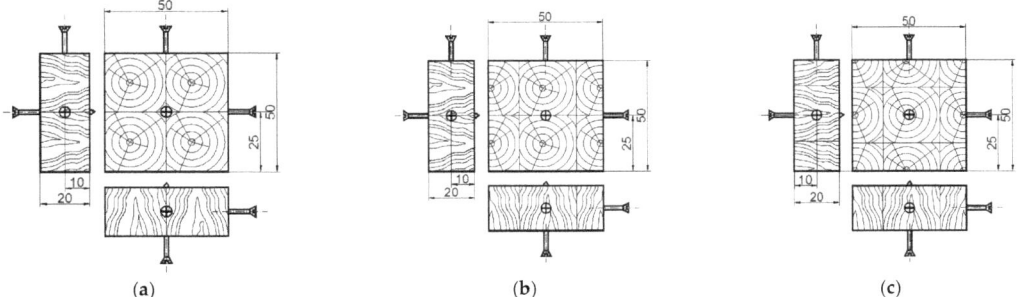

Figure 8. The shape and dimensions of the tested specimens: (**a**)—the screw insertion was made at the intersection of four gluing lines; (**b**)—the screw was inserted on a gluing line; (**c**)—the screw insertion was made in wood.

The tests were carried out on the FMPW—1000 Traction Machine. For this test, ST 4.2 screws according to ISO 1478 were used, with nominal dimensions of 4.2 × 38 mm and a thread pitch of 1.4 mm [30]. The samples were pre-drilled, the hole diameter being 2.7 ± 1 mm in accordance with EN 320:1993. The penetration depth of the screws is regulated by EN 13446:2004 and for the tested specimens the penetration depth was chosen for the faces—pierced [31,32], and, for the edges, 8 times the diameter, or 21.6 mm.

In the specialized literature, the expression of the resistance to the pulling out of the screws was calculated in accordance with the older norms (EN 13446: 2004), and the expression of the result was communicated in N/mm. In order to be able to compare the experimental data, the calculation of the resistance to the pulling out of the screws was made with Equation (6).

$$\tau_{s1} = \frac{F_{max}}{h} \qquad (6)$$

where: τ_{s1}—the pull-out resistance of the screws, in N/mm, F_{max}—the maximum force, in N, and h—the penetration depth, in [mm].

Determination of swelling in thickness. The principle of the test consisted in measuring the dimensions before and after immersing them for 24 h in cold water (t = 20 °C) and determining the swelling coefficient with Equation (7).

$$\beta = \frac{d_{max} - d_{min}}{d_{min}} \cdot 100 [\%] \qquad (7)$$

where: β_d—the lineal swelling coefficient, in [%], d_{max}—the maximum size after immersion, in [mm], and d_{min}—the minimum size before immersion, in [mm].

To determine the swelling coefficient, the same samples were used for determining the bonding strength. The initial moisture content of the specimens was 9.1% (determined using the gravimeter method EN 322:1993) and the density was 687 kg/m^3, determined according to EN 323:1993.

Statistical analysis. All groups of values were processed statistically, obtaining the average of the values, the standard deviation, and the value for a confidence interval of 95% (or the elimination of 5% of the values). This statistical procedure (excluding 5% of the values) was based on a lot of standards and procedures, and has calculated the upper quantile of each group of values according to the arithmetic mean, standard deviation, and Student's *t*-distribution with n + 1 degrees of freedom, where n represents the number of attempts of the respective test. The coefficient of variation was also introduced under the name "Variance", determined with the Microsoft Excel program.

3. Results

3.1. Results Regarding Manufacturing Technology

The specimens from the panels glued with aqueous dispersion polyvinyl acetate (PVC) adhesive after immersion in water became unglued and could no longer be tested [11–13]. From this moment, the use of vinyl adhesive was abandoned. The results obtained by applying the polyurethane adhesive were the following: bonding strength of 5.6 N/mm^2 before immersion and 4.2 N/mm^2 after immersion, compared to 2.5 N/mm^2 as the minimum admissible value, a static bending resistance of 5.5 N/mm^2, compared to 5 N/mm^2 as the minimum admissible value, and the modulus of elasticity of 1027.3 N/mm^2, compared to 600 N/mm^2 as the minimum admissible value. It is observed that the bonding strength was 124.6% higher than the minimum admissible strength provided by EN 13353:2022, the static bending resistance (MOR) was 11.2% higher than the minimum admissible strength provided by EN 13353:2022, and the value of the bending elasticity modulus (MOE) by 71.2% higher than the minimum admissible figure provided by EN 13353: 2022.

3.2. Results Regarding Dimensional and Flatness Deviations

Length, width, and thickness deviations are presented in Figure 9.

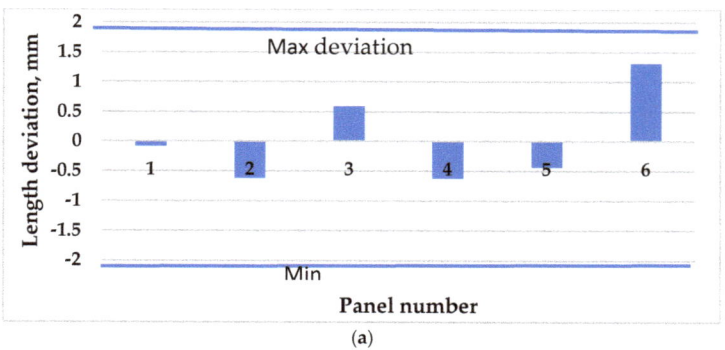

(a)

Figure 9. *Cont.*

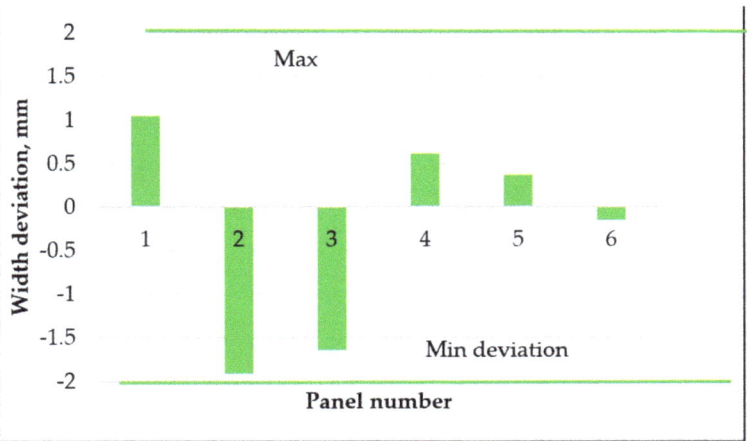

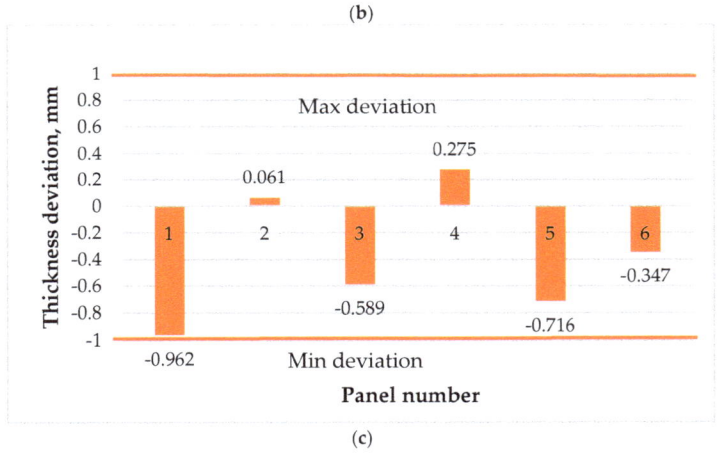

Figure 9. Deviation on panel length (**a**), width (**b**), and thickness (**c**).

In Figure 9, the linear deviations determined according to EN 324-1:1993 and the admissible deviations according to EN 13353: 2022 are compared. According to the data presented, it is found that the dimensional deviations of the panels with a transverse texture made of spruce branches fall within the admissible limits established by EN 324-1. Table 3 shows the deviations from flatness measured with respect to the adjacent plane in the 15 measurement points for each panel.

Table 3. Deviation (D) from flatness of panels with transverse texture obtained from branches.

No.	Panel Number					
	Panel P1	Panel P2	Panel P3	Panel P4	Panel P5	Panel P6
D_{max} mm	−0.2	0.23	0.05	0.16	−0.05	0.14
D_{min} mm	−1.19	−0.09	−0.67	−0.51	−0.61	−0.51
D_{mean} mm	1.19	0.32	0.72	0.67	0.61	0.65
Variance	0.063	0.031	0.051	0.047	0.045	0.046

If both deviations are negative or positive, then the minimum value was taken, and if they are different, their sum was taken. For panels with a length of less than 1000 mm, which are part of the furniture, the admissible deviation from flatness is ±2 mm [12,33], and for boards made of wood chips, the maximum admissible deflection is 1.5–2 mm/m for normally pressed boards and 1.5 mm/m for extruded boards [34]. Eco-panels fit into the requirements stated above (Figure 9, Table 3).

3.3. Bonding Strength Results

The results of the measurements and the interpretation of the results are shown in Table 4.

Table 4. The gluing resistance of panels.

No.	Dimension MC = 9.1%		Dimensions after Immersion in Water		Maxim Force [N]	Bonding Strength [N/mm^2]	Time [s]
	g [mm]	l [mm]	g [mm]	l [mm]			
Mean	19.746	25.023	20.040	25.738	2403	4.88	46.7
SD	0.089	0.120	0.104	0.259	120	1.40	2.33
Except 5%	19.707	24.970	19.995	25.624	2360	4.22	45.1
Variance	0.0045	0.0047	0.0051	0.0101	0.0499	0.0286	0.0498
Admissible					Minim 2.5 N/mm^2 EN 13353 2022		

In order to study the influence of moisture content on the gluing resistance, a set of tests was carried out, on a number of 14 specimens, without being previously treated via immersion in cold water; the results of these tests are shown in Figure 10 and are comparative with the water immersion ones.

Figure 10. Bonding strength of cross-textured panels made of spruce branches.

The gluing resistance of panels with a transverse texture made of spruce branches, determined experimentally according to DD CEN/TS 13354: 2003, falls within the requirements of EN 13353:2022. According to the data from the specialized literature [11,13], the moisture content of the samples almost inversely proportionally influences the bonding strength. In the case of these panels, the gluing resistance after 24-h immersion in cold water (20 °C) was lower by 24.8% than the gluing resistance of the specimens at a 9.1% moisture content. The breakage of the specimens tested at U = 9.1% occurred exclusively in the wood, usually in a different area than the gluing area, and was accompanied by a specific level of noise. For the samples immersed for 24 h in cold water (20 °C), a very high elasticity was observed, and the breaking was performed without noise, usually in the gluing area.

3.4. The Modulus of Elasticity (MOE) and Resistance (MOR) to Static Bending

The results regarding the modulus of elasticity of about 1000 N/mm^2 and resistance to static bending of about 6.1 N/mm^2 are shown in Table 5. The values determined for both types of specimens (T1—the gluing line in the middle of the test pieces, and T2—the gluing lines are arranged randomly on the face of the samples) fell within the requirements of EN 13353: 2022, which regulate the minimum admissible values of a modulus of elasticity over 600 N/mm^2 and resistance to static bending for solid wood panels over 5 N/mm^2. A maximum arrow of 7.58 mm was obtained. The samples were broken, producing a characteristic noise of solid wood. The breaking plan was after the tissues of minimum resistance, namely after the medullary rays and separations on the annual ring in the early wood area, meaning that the gluing strength was greater than that of the wood from the branches.

Table 5. Results of bending strength (MOR, MOE, and maximum arrow).

No.	Density [kg/m^3]		MOR [N/mm^2]		MOE [N/mm^2]		Maximum Arrow [mm]	
	T1	T2	T1	T2	T1	T2	T1	T2
Mean	693.2	692.3	6.09	6.10	1129.97	1006	7.15	7.58
SD [%]	14.16	3.1	1.12	0.41	102.58	44.17	1.24	0.59
Except 5%	682.8	657.7	5.27	5.79	1054.68	956.1	6.24	7.20
Variance	0.020	0.004	0.018	0.067	0.090	0.043	0.023	0.077

Regarding the influence of density on the static bending resistance (MOR), it could be observed that the resistance to static bending increased with increasing density up to a certain value of 705 kg/m^3, then decreases (Figure 11a). The correlation between density and static bending strength is given using a polynomial equation of the second degree, having the density as the parameter and the resistance (MOR) as the dependent variable.

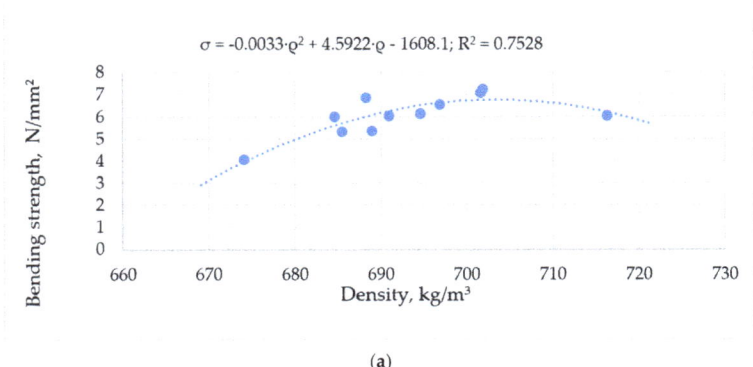

(a)

Figure 11. Cont.

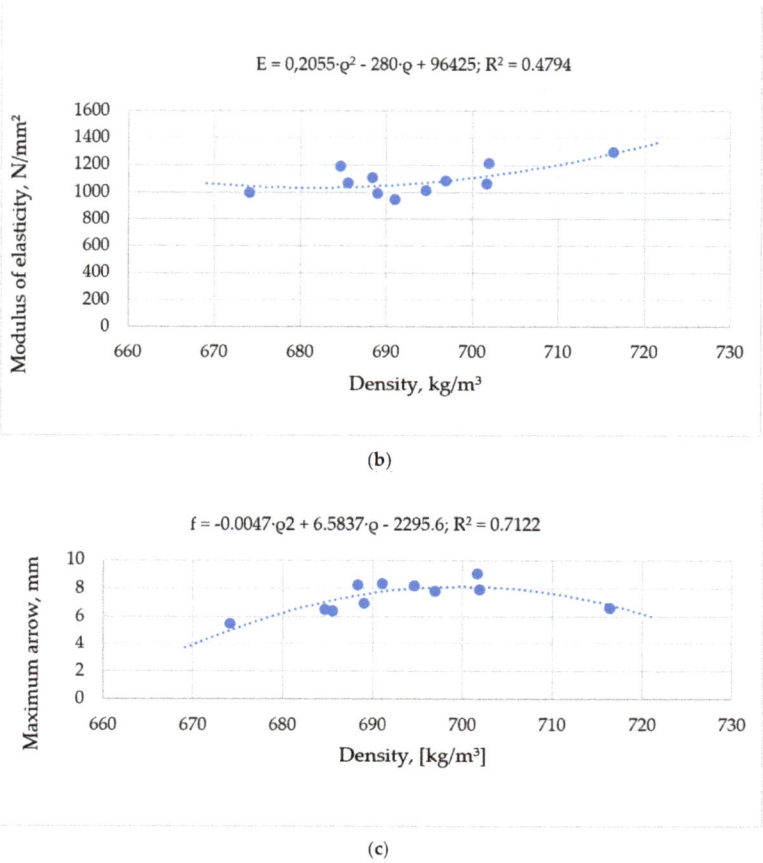

Figure 11. The relationship between density and static bending strength (**a**), density and modulus of elasticity (**b**), and density and maximum arrow (**c**) for panels made of spruce branches.

The correlation between density and the modulus of elasticity (MOE) in static bending is given by a second degree polynomial equation with a concave shape (Figure 11b) and with a minimal value of 685 kg/m³. The correlation between density and the maximum deflection in static bending is given by a second degree polynomial equation with a convex shape (Figure 11c) and with a maximal peak of 700 kg/m³.

3.5. Shear Resistance Parallel to the Panel Plane

The results of the tests on shear resistance parallel to the panel plane are shown in Table 6. There were obtained an average resistance of 4.1 N/mm² and a 46.2 s time of breaking.

Table 6. Shear resistance parallel to panel face (SD—standard deviation).

No.	Length, [mm]	Width, [mm]	Force, [N]	Shear Resistance Parallel to Face, [N/mm²]	Time, [s]
Mean	50.443	50.426	10541.25	4.141	46.212
SD	0.257	0.293	2036.04	0.781	16.472

Shear strength parallel to the panel plane for other materials; for the 20-mm thick veneered chipboard, it was 1.17 N/mm^2, for the 20-mm non-veneered chipboard, it was 1.8 N/mm^2, and for the 16-mm thick high-density fiberboard, it was 1.5 N/mm^2. These values were inferior to cross-structured panels obtained from spruce branches (3.2 N/mm^2). In the case of panels with a cross-section made of spruce branches, shear failure did not occur, as in the case of the chipboard specimens, because the compression with splitting occurred in the upper third of the specimen due to the tangential and normal unit stresses at the surface on the shear plane. The height at which the splitting had occurred varies between 3 and 18 mm, which represents between 6% and 36% of the height of the panel specimen.

3.6. Shear Resistance Perpendicular to the Panel Plane

The results of the tests regarding the shear strength in the plane of the panels are shown in Table 7. There obtained are an average value of resistance of 1.886 N/mm^2 and a breaking time of 33.97 s. It can be seen that the breaking time is shorter than in the previous test, due to the fact that the resistance is also lower.

Table 7. Perpendicular-to-plate shear strength of cross-textured panels (SD—standard deviation).

No.	Thickness 1 [mm]	Thickness 2 [mm]	Mean Thickness	Maximum Force [N]	Shearing Resistance [N/mm^2]	Breaking Time, [s]
Mean	19.788	19.782	19.785	2240	1.886	33.97
SD	0.049	0.051	0.046	488.6	0.411	2.374
Variance	0.002	0.002	0.0023	0.0217	0.0212	0.067

The breakage of the specimens started from the corner of the cuts, advancing in the direction of the medullary rays or in the direction of the annual ring through the early wood zone.

3.7. Results Regarding the Pull-Out Resistance of the Screws

The results obtained after the tests to pull out the screws allowed the formulation of the following conclusions. For screws inserted perpendicular to the plane of the panels, the best resistance to pulling out the screws is given by the situation in which the screw is inserted on a single joint line (type B samples), followed by the situation in which the screw is inserted at the intersection of four joint lines (type A samples), and lastly the situation where the screw was inserted into the wood (type C samples) (Table 8). The pulling out of the screws inserted perpendicular to the plane of the panel was performed silently, with the extraction of the wooden thread formed by the screw, usually without breaking the test piece, and only chips and sawdust in the area of the thread.

Table 8. Screw pull-out resistance parallel to the face panels (SD—standard deviation).

Sample Type	No.	Screw Pull-Out, Parallel to the Panel Face					
		lp, [mm]	d [mm]	Force, [N]	Time, [s]	τ_s [N/mm^2]	τ_{s1} [N/mm]
Type A	Mean	19.707		2624	6.308	31.704	133.158
	SD	0.071		347.4	0.932	4.189	17.596
Type B	Mean	19.675		2701	5.43	32.684	137.275
	SD	0.0302	4.2	261.1	1.445	3.168	13.306
Type C	Mean	19.675		2439	9.595	29.520	123.984
	SD	0.030		617.4	3.066	7.486	31.444
Total	Mean	19.685		2588	7.111	31.303	131.473
	SD	0.047		424.8	2.653	5.143	21.601

For screws inserted parallel to the plane of the panel (on edge), the highest resistance to pulling out the screws is given by the situation where the screw was inserted into the wood but also passes through a joint line (type B samples) followed by the situation where the screw was inserted on four joint lines (type A samples), and lastly the situation where the screw was inserted into the wood (type C samples).

The difference between the pull-out strength of the screws on the different edges was 30.7 N/mm² for edge 1 and type A, 33.1 N/mm² for type C, 22.9 N/mm² for edge 2 type A, and 28.5 N/mm² for type C. For the identical test cases (specimens of type A and C), it was found that the pull-out resistance of the screws from edge 2 was lower than edge 1 by 25.3% in the case of type A and 13.7% in the case of type C specimens (Table 9). This difference was caused by the weakening of the strength of the test piece during the pulling out of the screw from the edge tested firstly.

Table 9. The pulling out resistance of the screws on the edge of panel (SD—standard deviation).

Sample Type	No.	Edge 1				Edge 2	
		τ_s [N/mm²]	τ_{s1} [N/mm]	F [N]	t [s]	τ_s [N/mm²]	τ_{s1} [N/mm]
Type B	Mean	27.111	113.888	2482	5.442	27.358	114.907
	SD	14.942	62.758	474.462	4.01	10.335	43.410
Type C	Mean	33.090	115.817	962.001	9.223	28.549	119.907
	SD	5.989	25.154	474.462	0.992	4.134	17.365
Total	Mean	30.327	119.655	1843.333	7.332	26.300	105.001
	SD	9.926	46.306	665.001	3.072	7.330	30.218

3.8. Coefficient of Swelling on the Thickness and Width of the Panels

The swelling coefficient values were tabulated in Table 10.

Table 10. The swelling coefficient after immersion (SD—standard deviation).

No.	Dimension MC = 9.1%		Dimension after Immersion in Water		Swelling Coefficient on Thickness, [%]	Swelling Coefficient on Width, [%]
	g, [mm]	l, [mm]	g, [mm]	l, [mm]		
Mean	19.74	25.023	20.04	25.738	1.487	2.858
SD [%]	0.089	0.12	0.104	0.259	0.351	1.028
Except 5%	19.707	24.97	19.995	25.624	1.334	2.407
Variance	0.0045	0.0048	0.0052	0.0097	0.0237	0.0357

The value of the swelling coefficient in thickness, i.e., swelling along the fiber, was 1.48%, and the value of the swelling coefficient in width, i.e., tangential swelling, was 2.85%. The coefficients of swelling in thickness after immersion in cold water for 24 h, for the other materials, apart from traverse-structured panels [33–35], were limited to 12% for the chipboard, 12% for the medium-density fiberboard (MDF), 15% for the oriented strand board (OSB), 8% for the plywood, 12% for the softwood as spruce, and only 1.4% and 2.8% (on thickness and width) for the traverse-structured panel from spruce branches.

These values obtained experimentally are due to the high content of compression wood (33–66%), and they support the theory from the specialized literature [12,13], according to which compression wood has a high and irregular longitudinal contraction coefficient and a tangential shrinkage coefficient smaller than normal wood. Thus, the ratio of the two types of swelling was, in the case of the panel with the transverse structure of branches, only 1:2, in relation to about 1:20 for the solid wood [35,36].

4. Discussion

Table 11 summarizes the tests to which the cross-textured panels from spruce branches were subjected, the results obtained, the admissible values, the test methodology, as well as the norm that regulates the admissible value. These values are used for comparing them with the results of authors and limitative values of European standards.

Table 11. Synthesis of the physical–mechanical and technological properties of panels with a transverse texture obtained from spruce branches (SD—standard deviation).

No.	Characteristics	Methodology According to which the Test was Carried Out	Value		SD	UM
1.	Deviation from nominal length	EN 324-1:1993; EN 324-2:1993; EN 13353:2022	Mean Admissible	0.03 ±2	0.001 -	mm
2.	Deviation from the nominal width	EN 324-1:1993; EN 324-2:1993; EN 13353:2022	Mean Admissible	−0.62 ±2	−0.003 -	mm
3.	Deviation from the nominal thickness	EN 324-1:1993; EN 324-2:1993; EN 13353:2022	Mean Admissible	0.47 ±1	0.023 -	mm
4.	Deviation from flatness	[27]	Mean Admissible	1.304 2	0.065 -	mm
5.	Apparent density at MC = 9.1%	EN 323:1993	Mean Admissible	687.05 -	14.1 -	kg/m^3
6.	Internal bond resistance at MC = 9.1%	EN 13354:2008	Mean Admissible	5.61 2.5	0.25 -	N/mm^2
7.	Bonding strength after 24-h immersion	EN 13354:2008; EN 13353:2022	Mean Admissible	4.22 2.5	0.21 -	N/mm^2
8.	Modulus of elasticity (MOE)	EN 310:1996; EN 13353:2022	Mean Admissible	1027.38 600	102.5 -	N/mm^2
9.	Static bending resistance (MOR)	EN 310:1996 EN 13353:2022	Mean Admissible	5.57 5	1.12 -	N/mm^2
10.	Shear resistance parallel to the panel plane	EN 13354:2008	Mean Admissible	4.14 -	0.78 -	N/mm^2
11.	Shear resistance perpendicular	EN 13354:2008	Mean Admissible	1.88 -	0.41 -	N/mm^2
12.	Pull-out resistance of screws perpendicular	EN 13446:2004	Mean Admissible	31.30 -	5.14 -	N/mm^2
13.	Pull-out strength of screws parallel	EN 13446:2004	Mean Admissible	30.82 -	9.9 -	N/mm^2
14.	Coefficient of swelling in thickness	[12,36]	Mean Admissible	1.33 -	0.35 -	%
15.	Coefficient of swelling in width	[12,36]	Mean Admissible	2.40 -	1.00 -	%

All dimensional and flatness deviations fell within the limits provided by the standard [37,38] or expressed by other authors and research in this field. For example, the dimensional deviation in thickness for panel number 2 was 0.061 mm compared to the upper limit of 2 mm according to EN 324-1:1993 and EN 13353:2022, which meant a deviation of the value of 32.7 times.

The apparent density of the panels corresponded to the species used (spruce, with an average wood density of the trunk of 425 kg/m^3 and of 510 kg/m^3 for the branches [12,36], due to the presence of compression wood), but it also increased significantly due to the Jowapur adhesive used and the pressure force used during the formation of the friezes and panels.

The average value for the resistance of the gluing for the panel made of branches of 5.61 N/mm^2 not only corresponded to the limiting value of the European standard (EN 13353: 2022) of 2.5 N/mm^2, but also had an increase of 141.3%. Additionally, these values were in accordance with those expressed by other authors [12,13].

The static bending strength of the traverse-structured panel was 10.71% higher than the minimum admissible value recommended by EN 13353:2022, 19.65% higher than the static bending strength of the chipboard with a 20-mm thickness, and 6.6% lower than the static bending strength of the high-density fiberboard (HDF) [33,34]. Regarding the influence of the cutting method of the samples from the panel on the strength and modulus of elasticity in static bending, the followings were observed.

The shear resistance parallel to the board plane was superior to wood-based boards [36], being higher by 56.53–71.65% than the shear resistance parallel to the board plane of the 20-mm thick chipboard, and with 63.77% higher than the shear strength parallel to the plate plane of the high-density fiberboard. The shear resistance perpendicular to the plane of the board was superior to boards based on fibers and wood chips. This was 78.8% higher than the shear resistance perpendicular to the plane of the chipboard with the 20-mm thickness and the same fiberboard [12].

Pulling out the screws inserted parallel to the plane of the panel was performed with noise, without extracting the wooden thread formed by the screw, and usually caused the specimen to break. When screws were inserted perpendicular to the panels, specimen type A had a resistance of 31.7 N/mm^2 and a breaking time of 6.3 s, specimen type B had 32.6 N/mm^2 resistance and breaking time of 5.4 s, and specimen type C had a resistance of 29.5 N/mm^2 and a breaking time of 9.5 s. The strength and pull-out time of the screws for the three cases of screw insertion parallel to the panel plane were for edge 2 and type A, 22.9 N/mm^2 and 5.5 s, for type B, 27.3 N/mm^2 and 5.4 s, and for type C, 28.5 N/mm^2 and 9 s. Screw strength determined according to the old method had different values for all composite materials. The values were 133.1 N/mm for panel type A on the face, 137.2 N/mm for panel type B on the front, and 123.9 N/mm for panel type C on the front, and were comparable with that of 80 N/mm for solid wood with a density of 600 kg/m^3, of 100 N/mm for solid wood with a density of 650 kg/m^3, and of 150 N/mm for solid wood with a density of 700 kg/m^3 [13,36]. By comparing the data obtained with those from the specialized literature regarding the resistance to pulling out screws [35,36], it was concluded that cross-structured panels fall within the requirements imposed by the European norms [35,36].

Values for screw pull-out resistance for various types of materials [33] are: for screws on the plane of the plate (on the edge) with a chipboard density of 600 kg/m^3, it was 75 N/mm, for a solid wood density of 500 kg/m^3, it was 110 N/mm, and for a solid wood density of 700 kg/m^3, it was 170 N/mm, compared to traverse-structured panels that had 280.1 N/mm for panel type A and edge 1, 114.9 N/mm for panel type B and edge 2, and 119.9 N/mm for panel type C and edge 2.

The pull-out force of the screw inserted perpendicular to the panel was quite different compared to the parallel-inserted ones. Comparative to the resistance of a medium-density fiberboard of 1000 N [33], a different force of 2692 N was obtained for panel type A front, 2701 N for the panel type B front, 2439 N for the type C front panel.

Referring to the coefficient of swelling in thickness and width, the fact that at the initial moisture content of 9.1% after immersion in water for 24 h resulted in a swelling coefficient of 1.33%, the theory according to which the wood from the branches has moisture saturation point of the fiber (FSP) approximative 9% [11–13] was disproved, despite the fact that the wood in the trunk has a fiber saturation point of around 30%. According to this theory, the panels should not have swelled over 9%, because the swelling and shrinking phenomena occur only in the moisture content range under the fiber saturation point (FSP).

The uses of this type of traverse-structured panel could be various, but it is recommended to use it only in indoor conditions. Even in indoor conditions, the designed panels can be used for small furniture products with dimensions up to 1–1.2 m. These types of panels were used for several types of small furniture, such as tables, coffee tables, desks, etc. (Figure 12).

Figure 12. Examples of using traverse structured panels made of spruce branches.

5. Conclusions

The judicious use of spruce branches in the form of wooden panels leads to obtaining a new base of raw material for the furniture industry.

Highlighting the transverse structure of the spruce wood branches, this technology is considered a method to be taken into consideration by the producers of art furniture.

Even if the technology for obtaining these wooden panels with a transverse texture is not yet patented, this remains a superior valorization solution for the branches, and its viability remains to be demonstrated in the coming years.

Following all tests of research, the panels with a transverse texture from spruce branches could be considered one of the best methods for the high valorization of spruce branches.

Author Contributions: Conceptualization, A.L. and A.M.O.; methodology, A.M.O.; software, L.R.; validation, A.M.O., A.L. and C.S.; formal analysis, A.M.O.; investigation, A.M.O.; resources, A.M.O.; data curation, C.S.; writing—original draft preparation, L.R.; writing—review and editing, A.L.; visualization, A.L.; supervision, A.M.O.; project administration, C.S.; funding acquisition, C.S. All authors have read and agreed to the published version of the manuscript.

Funding: This research received no external funding.

Institutional Review Board Statement: Not applicable.

Informed Consent Statement: Not applicable.

Data Availability Statement: No new data were created.

Acknowledgments: In this way, the authors thank the management of the University of Transilvania Brasov for the logistical and financial support to carry out this research.

Conflicts of Interest: The authors declare no conflict of interest.

References

1. Hakkila, P. Utilization of residual forest biomass. In *Utilization of Residual Forest Biomass*; Springer Series in Wood Science; Springer: Berlin/Heidelberg, Germany, 1989. [CrossRef]
2. English, B. Wastes into wood: Composites are a promising new resource. *Environ. Health Perspect.* **1994**, *102*, 168–170. [CrossRef] [PubMed]
3. Jahan-Latibari, A.; Roohnia, M. Potential of utilization of the residues from poplar plantation for particleboard production in Iran. *J. For. Res.* **2010**, *21*, 503–508. [CrossRef]
4. Tichi, A.H. Investigation of the use of old railroad ties (*Fagus orientalis*) and citrus branches (orange tree) in the particleboard industry. *Bioresources* **2021**, *16*, 6984–6992. [CrossRef]
5. Olarescu, A.M.; Lunguleasa, A.; Radulescu, L. Using deciduous branch wood and conifer spindle wood to manufacture panels with transverse structure. *Bioresources* **2022**, *17*, 6445–6463. [CrossRef]
6. Zhao, X.; Guo, P.; Zhang, Z.; Wang, X.; Peng, H.; Wang, M. Wood density and fiber dimensions of root, stem, and branch wood of *Populus ussuriensis* Kom. Trees. *Bioresources* **2018**, *13*, 7026–7036. [CrossRef]
7. Pulido-Rodríguez, E.; López-Camacho, R.; Tórres, J. Traits and trade-offs of wood anatomy between trunks and branches in tropical dry forest species. *Trees* **2020**, *34*, 497–505. [CrossRef]
8. Vurdu, H.; Bensend, D.W. Proportions and types of cells in stems, branches, and roots of European black alder (*Alnus glutinosa* L. Gaertn.). *J. Wood Sci.* **1980**, *13*, 36–40.

9. Luan, S.-J.; Yang, R.-P. Predetermination stem wood quality with the branch wood index of Korean pine. *J. Northeast For. Univ.* **1992**, *3*, 54–61. [CrossRef]
10. Shmulsky, R.; Jones, P.D. *Forest Products and Wood Science: An Introduction*, 7th ed.; Wiley-Blackwell Publishing: Hoboken, NY, USA, 2011.
11. USDA Forest Service. *Wood Handbook, Wood as an Engineering Material*; Forest Products Laboratory. United States Department of Agriculture Forest Service: Madison, WI, USA, 2010.
12. Petrovici, V.; Popa, V. *Chemistry and Chemical Processing of Wood*; Lux Libris Print House: Brasov, Romania, 1997.
13. Nielsen, P. Integration of environmental aspects in product development: A stepwise procedure based on quantitative lifecycle assessment. *J. Clean. Prod.* **2002**, *10*, 247–257. [CrossRef]
14. Zeng, Y. Environment-Based design (EBD): A methodology for transdisciplinary Design. *J. Integr. Des. Process. Sci.* **2015**, *19*, 5–24. [CrossRef]
15. Brezet, H.; van Hemel, C. *Ecodesign—A Promising Approach to Sustainable Production and Consumption*; United Nations Environment Program: Paris, France, 1997.
16. Bras, B. Incorporating environmental issues in product design and realization. *Ind. Environ.* **1997**, *20*, 7–13.
17. Suansa, N.I.; Al-Mefarrej, H.A. Branch wood properties and potential utilization of this variable resource. *Bioresources* **2020**, *15*, 479–491. [CrossRef]
18. Moreira, L.d.S.; Andrade, F.W.C.; Balboni, B.M.; Moutinho, V.H.P. Wood from Forest Residues: Technological Properties and Potential Uses of Branches of Three Species from Brazilian Amazon. *Sustainability* **2022**, *14*, 11176. [CrossRef]
19. Zayed, M.Z.; Wu, A.; Sallam, S. Comparative phytochemical constituents of *Leucaena leucocephala* (Lam.) leaves, fruits, stem barks, and wood branches grown in Egypt using GC-MS method coupled with multivariate statistical approaches. *Bioresources* **2019**, *14*, 996–1013. [CrossRef]
20. EN 13353:2022; Solid Wood Panels (SWP). Requirements. European Committee for Standardization: Brussels, Belgium, 2022.
21. DD CEN/TS 13354:2003; Solid Wood Panels—Bonding Quality—Test Method. European Committee for Standardization: Brussels, Belgium, 2003.
22. EN 1995-1-1:2004+A 1; Eurocode 5: Design of Timber Structures—Part 1-1: General—Common Rules and Rules for Buildings. European Committee for Standardization: Brussels, Belgium, 2004.
23. EN 335-2:2006; Durability of Wood and Wood-Based Products—Definition of Use Classes—Part 2: Application to Solid Wood. European Committee for Standardization: Brussels, Belgium, 2006.
24. EN 310:1996; Wood Based Panels—Determination of Modulus of Elasticity in Bending and of Bending Strength. European Committee for Standardization: Brussels, Belgium, 1996.
25. EN 324-1:1996; Wood Based Panels—Determination of Dimensions of Boards—Part 1: Determination of Thickness, Width and Length. European Committee for Standardization: Brussels, Belgium, 1996.
26. Mantanis, G.I.; Young, R.A.; Rowell, R.M. Swelling of wood. *Wood Sci. Technol.* **1994**, *28*, 119–134. [CrossRef]
27. EN 13354:2008; Wood Based Panels—Small Scale Indicative Test Methods for Certain Mechanical Properties. European Committee for Standardization: Brussels, Belgium, 2008.
28. EN 326-1:1996; Wood Based Panels—Sampling, Cutting and Inspection—Part 1: Sampling and Cutting of Test Pieces an Expression of Test Results. European Committee for Standardization: Brussels, Belgium, 1996.
29. ISO 1478:1999; Tapping Screws Thread. International Organization for Standardization: Geneva, Switzerland, 1999.
30. EN 322:1993; Wood-Based Panels. Determination of Moisture Content. European Committee for Standardization: Brussels, Belgium, 1993.
31. EN 13446:2004; Wood Based Panels—Determination of Withdrawal Capacity of Fasteners. European Committee for Standardization: Brussels, Belgium, 2004.
32. Barbu, M.C. *Wooden Composite Materials*; Lux Libris Print House: Brasov, Romania, 1999.
33. Istrate, V. *Technology of Wooden Agglomerated Products*; Didactical and Pedagogical Print-House: Brasov, Romania, 1983.
34. Giese-Hinz, J.; Jahn, F.; Weller, B. Experimental study of the pull-out resistance of alternative high-strength fasteners for wood-based materials. *Wood Mater. Sci. Eng.* **2018**, *14*, 226–233. [CrossRef]
35. Wood Database 2023. Available online: https://www.wood-database.com/?s=white+spruce (accessed on 5 July 2022).
36. Salem, M.Z.M.; Zayed, M.Z.; Ali, H.M.; El-Kareem, M.S.M.A. Chemical composition, antioxidant and antibacterial activities of extracts from *Schinus molle* wood branch growing. *Egypt. J. Wood Sci.* **2016**, *62*, 548–561. [CrossRef]
37. EN 325:2012; Wood-Based Panels—Determination of Dimensions of Test Pieces. European Committee for Standardization: Brussels, Belgium, 2012.
38. EN 323:1993; Wood-Based Panels—Determination of Density. European Committee for Standardization: Brussels, Belgium, 1993.

Disclaimer/Publisher's Note: The statements, opinions and data contained in all publications are solely those of the individual author(s) and contributor(s) and not of MDPI and/or the editor(s). MDPI and/or the editor(s) disclaim responsibility for any injury to people or property resulting from any ideas, methods, instructions or products referred to in the content.

Article

Physical and Acoustical Properties of Wavy Grain Sycamore Maple (*Acer pseudoplatanus* L.) Used for Musical Instruments

Florin Dinulica [1], Adriana Savin [2,3] and Mariana Domnica Stanciu [3,4,*]

[1] Department of Forest Engineering, Forest Management Planning and Terrestrial Measurements, Transilvania University of Brașov, 500123 Brașov, Romania
[2] Institute of Research and Development for Technical Physics, B-dul Mangeron 47, 700050 Iasi, Romania
[3] Department of Mechanical Engineering, Transilvania University of Brașov, B-dul Eroilor 29, 500036 Brașov, Romania
[4] Russian Academy of Natural Sciences Sivtsev Vrazhek, 29/16, Moscow 119002, Russia
* Correspondence: mariana.stanciu@unitbv.ro

Abstract: The wood used in the construction of musical instruments is carefully selected, being the best quality wood from the point of view of the wood structure. However, depending on the anatomical characteristics of the wood, the resonance of wood is classified into quality classes. For example, sycamore maple wood with curly grains is appreciated by luthiers for its three-dimensional optical effect. This study highlights the statistical correlations between the physical and anatomical characteristics of sycamore maple wood and its acoustic and elastic properties, compared to the types of wood historically used in violins. The methods used were based on the determination of the acoustic properties with the ultrasound method, the color of the wood with the three coordinates in the CIELab system and the statistical processing of the data. The sycamore maple wood samples were divided into anatomical quality classes in accordance with the selection made by the luthiers. The results emphasized the multiple correlations between density, brightness, degree of red, width of annual rings, acoustic and elastic properties, depending on the quality classes. In conclusion, the work provides a valuable database regarding the physical–acoustic and elastic properties of sycamore maple wood.

Keywords: wavy grain sycamore maple wood; anatomical descriptors; acoustic properties; statistical correlation; musical instruments

Citation: Dinulica, F.; Savin, A.; Stanciu, M.D. Physical and Acoustical Properties of Wavy Grain Sycamore Maple (*Acer pseudoplatanus* L.) Used for Musical Instruments. *Forests* **2023**, *14*, 197. https://doi.org/10.3390/f14020197

Academic Editor: Michele Brunetti

Received: 4 December 2022
Revised: 2 January 2023
Accepted: 17 January 2023
Published: 20 January 2023

Copyright: © 2023 by the authors. Licensee MDPI, Basel, Switzerland. This article is an open access article distributed under the terms and conditions of the Creative Commons Attribution (CC BY) license (https://creativecommons.org/licenses/by/4.0/).

1. Introduction

Over time, the practice of luthiers in using maple wood with wavy fibers for the back of violins has been transmitted to the present day. Since the acoustic quality of the violin is the synergistic result of several factors, it cannot be established whether the old luthiers preferred wavy maple wood for its acoustic characteristics or just for its aesthetic appearance. It is certain that researchers [1–3] discovered that the maple wood used by old luthiers, such as Guarnieri and Stradivari, was curly maple wood, but, as a result of aging and preservation treatments applied at that time, the curly maple wood these old luthiers used had very different chemical and acoustic properties when compared to the maple wood used to make modern musical instruments. Procuring this material is quite expensive and difficult because this growth defect is not only genetically produced but is the result of edaphic and climatic factors [4–6]. There are theories according to which the formation of wavy grain depends on the orientation of the cells in the cambium, but the development mechanism has not yet been fully explained. since the environment also contributes to the genetic factors, as shown in [7–9]. According to [10,11], the wavy grain in European sycamore maple wood is considered to be a "natural defect" that is optically observed in the radial direction, in the form of waves in the direction of the grain with respect to the longitudinal axis of the tree, and the pattern being an alternation of light and shadow stripes. North American sycamore and European sycamore maple trees exhibit differences are based on different growth mechanisms, as shown in [10]. The fiber pattern of European sycamore maple wood is natively produced by deviations from straight, regular grain [11–13]. Thus, the most common patterns of sycamore maple wood, resulting from the ability of the wood grain to capture and reflect light, are the following: curly grain, which gives the appearance of wavy waves, because it reflects light differently (Figure 1a); fiddleback, characterized by very frequent undulations, with a small distance between curly grains, used by luthiers in the construction of musical instruments (Figure 1b); flame grain, giving the effect of curls that roll like flames (Figure 1c); quilted grain resembles patchwork patterns (Figure 1d—upper part) and bird's eye, showing small, rounded, bright spots (Figure 1e) [9–12].

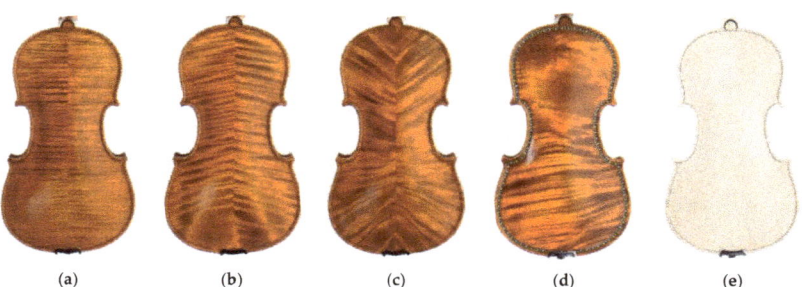

Figure 1. The common figures of sycamore maple wood: (**a**) curly grain; (**b**) fiddleback figure; (**c**) flame grain; (**d**) partial quilted grain; (**e**) bird's eye figure.

Regarding the degree of fiber curl, there are two approaches. The first approach is based on the angle, which highlights the presence of twisted fibers in standing trees, as presented by [9,10], and which is not the subject of our research, the material being collected from trees without apparent defects. The second approach, proposed by Alkadri et al., 2018 [14], is based on the measurement of wavelength, amplitude and grain angle measured in the transition point between peak and valley of the wave. In a previous article, [15], the angle of the fiber was measured according to the method proposed by Alkadri et al., 2018 [14], and an inverse proportional relationship between the wavelength and the curly grain angle was found. Starting from this conclusion, in the present paper we considered that the wavelength was sufficient to express the degree of undulation of the grain.

The current question posed by musical instrument manufacturers is whether having the wavy grain in the wood contributes to the acoustic quality of the musical instruments or has only aesthetic value. In this sense, various studies have been carried out on the physical, mechanical and elastic properties of the wavy-grain sycamore maple wood, but none provided synoptic information regarding the relationship between the pattern of the wood and its elastic and acoustic characteristics [14–18].

The acoustic quality of the musical instrument depends primarily on the acoustic and elastic properties of the material from which it is made; spruce wood for the top plate and wavy maple for the back. These properties are correlated with the physical characteristics of the wood species. Studies on the elastic, acoustic and dynamic properties of resonance wood (spruce and maple) have been the focus of many researchers, providing a rich source of data [9,11–18]. These studies investigated tone wood species from forest basins in Slovenia, the Czech Republic, Austria, Italy, France, Germany, USA.

Thus, related to physical and elastic features of sycamore maple wood, Sedlar et al. [4] reported longitudinal shrinkage, ranging between 0.13% and 0.28%, radial shrinkage, ranging between 3.2 to 5.2%, and tangential shrinkage, ranging between 7.2 to 10.9%, in comparison with straight maple wood reported on by Kollman [12] (longitudinal shrinkage 0.5%; radial 3% and tangential 8%). The density measured for specimens with 12% moisture content, varied from 0.52 to 0.73 g/cm^3 and modulus of elasticity at bending (MOE) recorded values between 6.1–11.3 GPa. Sonderegger et al. in paper [5] found a ratio of the dynamic modulus of elasticity on the three directions L:R:T of 9.8:1.7:1 and of the shear moduli in the shearing planes LT:LR:RT for 4.5:2.9:1. The studies highlighted the mechanical and acoustic properties of resonance wood (spruce and maple), identifying the values of sound propagation speeds in the longitudinal and radial directions, as well as the values of the modulus of elasticity through the following different methods to determine the resonance frequency [15–21]: the intrinsic transfer matrix method used to simulate the propagation of continuous waves or finite impulse in homogeneous, inhomogeneous or multilayered elastic media [16–19], and the non-destructive evaluation method based on ultrasound [22–27].

The research presented in this article aimed to highlight the correlations between the physical appearance of sycamore maple wood and its acoustic and elastic properties, taking into account the way luthiers select the raw material for musical instruments.

Thus, the study started from the hypothesis of the existence of correlations between the physical appearance of sycamore maple wood, including pattern of the wood and its acoustic/elastic properties. For this purpose, fiber undulation pitch of the sycamore maple grain and the color parameters, in terms of quality classes, were entered in digital format and correlated statistically with the following acoustic parameters determined by ultrasonic means: the speed of sound in the palm in the three main directions, the Young's modulus of elasticity, shear moduli and Poisson's ratios. The novelty of the research consisted in verifying the contribution of the wood structure to the acoustic performance of the sycamore maple wood from the Romanian Carpathian Mountains, known for the quality of its resonance wood. Data from the literature does not link the structure of the wood with its acoustic properties, and such studies on wood have not been reported until now.

2. Materials and Methods
2.1. Materials

In order to investigate the physical and mechanical properties, the samples of European sycamore maple wood (*Acer pseusoplatanus* L.) were extracted from types of wood specimens prepared for musical instruments. The specimens were obtained from logs of trees harvested from the Gurghiu area, in the Eastern Carpathian Mountains. The blanks had been naturally dried for a minimum of 3 years (straight grain maple wood used for school instruments) and up to 10 years for maestro instruments (the wavy grain sycamore maple). The specimens used in the tests were cut in the form of a cube, with sides of 40 mm, respecting the three main directions of the wood (radial, tangential, and longitudinal). The physical features of the studied samples are presented in Table 1.

Table 1. The physical features of the sycamore maple wood samples (Legend STDEV—standard deviation).

Physical Features	Grade/Average Value/STDEV							
	A	STDEV	B	STDEV	C	STDEV	D	STDEV
Wood density WD (kg/m^3)	609	3.022	601	4.381	594	8.494	623	16.308
Moisture content (%)	6.8	0.05	7.2	0.1	7.0	0.092	8	0.45

The samples were graded in four classes, denoted as A, B, C and D, in accordance with the classification of wood made by luthiers for the different quality classes of violins. Thus, the A class was considered the best wood maple, due to the wood pattern produced by the wavy grain, class B was also characterized by wavy grain, but rare, class C consisted of specimens with slightly wavy grain and class D had straight grain. The division of maple wood samples into the four quality classes, according to the intensity of the wavy grain, is also used by foresters, according to papers [13,15,23–26]. These categories related directly to the wood surface characterized by the wavy grain structure.

2.2. Methods
2.2.1. Color Measurement

Since the sycamore maple wood is optically sorted based on the macroscopic aspect of the structure, the color parameters were measured according to the CIELab system.

The measurements of color coordinates a* (the color redness/greenness), b* (the color yellowness/blueness) and L* (color brightness or whiteness) were carried out with a Minolta Chroma-Meter CR-400 (Konica Minolta, Tokyo, Japan) [24]. To measure the color parameters, 5 measurements were made on the diagonals of the radial section (from this, 30 measurements resulted, with 6 samples from each class).

2.2.2. Anatomical Features Measurement

Evaluation of the anatomical features of the sycamore maple wood was performed using the WinDENDRO Density image analysis system (Régent Instruments Inc., Québec City, QC, Canada, 2007). Each sample was scanned on its cross section and radial section in order to measure the width of the annual rings, denoted RW, and the fiber undulation pitch of the sycamore maple wood, denoted CWL, according to the method presented in studies [26–30]. In the references [12–17], wavelength was used, but in this study, the authors used the term fiber undulation pitch to differentiate the wavelength known in the acoustic field as the term used for undulation measurement.

Figure 2 shows how the anatomical characteristics of interest were measured for sycamore maple wood. The ring regularity index (RRI) was calculated using the mathematical relation (1), recommended by Dinulică et al., 2015 [29] for wood:

$$RRI = \frac{\max(RW_i) - \min(RW_i)}{\max(RW_i)} \quad (1)$$

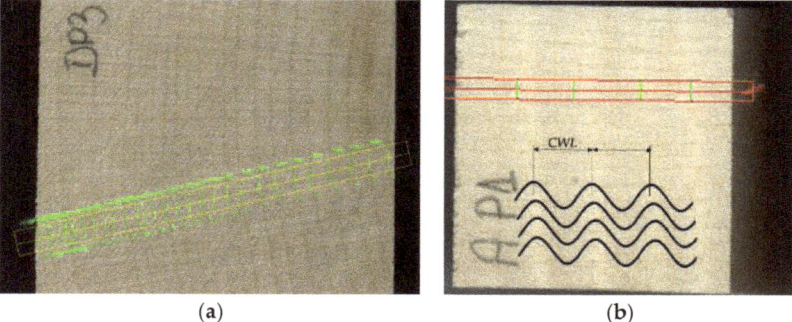

(a)   (b)

Figure 2. Measuring the anatomical characteristics: (**a**) the width of the annual rings on cross section; (**b**) fiber undulation pitch (CWL) on radial–longitudinal section.

2.2.3. Determination of Sound Velocity, Dynamic Elastic Moduli and Shear Moduli by Means of Ultrasonic Devices

The method of ultrasound measurements was presented in previous studies [17,24] and consisted of connecting the ultrasound sensors to the 5073 PR Pulse Receiver–Panametrics equipment [18]. A LeCroy Wave Runner 64Xi digital oscilloscope was used to visualize the signal and measure the propagation time in the wood samples. For each sample, measurements were made at five points on each face of the sample, which corresponded to the three planes of orthotropic symmetry of the wood (longitudinal–radial plane; tangential–longitudinal plane; and radial–tangential plane). The measurements were performed at a temperature of 24 °C and a humidity of 65%.

2.2.4. Statistical Evaluation

A large number of data were obtained, which were first analyzed under the aspect of variability. Using discriminant function analysis (DFA), the possibility of stratifying the values of the acoustic parameters according to the quality class of the sample was verified. DFA provides weights of predictor variables that reflect their ability to distinguish between groups of dependent variables [24,30–32]. The individual ability of acoustic variables to discriminate between quality classes was indicated by the magnitude of the partial Lambda parameter. Thus, the closer the value of the partial lambda parameter to 0, the greater the discrimination power. Normality of distributions was checked with the Shapiro–Wilk test. Relationships between acoustic parameters were verified by testing simple correlation coefficients. The association of the variables involved in the study was explored using Principal Component Analysis (PCA) and the k-means Clustering procedure, and then verified with simple and multiple correlations [29,32].

3. Results and Discussion
3.1. Anatomical Descriptors of the Sycamore Maple Wood

In the sycamore maple wood samples, the anatomical descriptors measured were the width of the annual rings (RW) and the pitch of the fiber undulation (CWL). It was found that RW was approximately 32% greater in the case of sycamore maple with wavy fibers (class A or B), compared to that with straight fibers (class C or D) (Table 2). According to studies [24,28], the higher the value of the regularity index (RRI), the lower the regularity of the rings. According to this hypothesis, it was observed that the RRI in sycamore maple wood was higher in wood with wavy grain than in wood with straight grain (Table 2). Among the ways of expressing the regularity of annual rings, the method of calculation that best reproduced the structural quality of the resonance wood was chosen in accordance with reference [24]. This rendered the regularity as a ratio between the amplitude of the variation (maximum value − minimum value) and the maximum value, a way of expression also used by other researchers. In Table 2, the regularity of annual rings is presented as a percentage value. In previous studies related to the anatomical characteristics of resonance wood in the construction of historical violins, average values of RW, in the range of 1.08–1.29 mm, were identified in historical violins, such as the Stradivarius Edler Voicu (1702), Leeb 1742; Klotz 1747, Babos 1920 and a Stainer Copy [3].

Table 2. Anatomical descriptors of the sycamore wood samples.

Anatomical Descriptors	Grade/Average Value/STDEV							
	A	STDEV	B	STDEV	C	STDEV	D	STDEV
Annual ring width RW (mm)	1.249	0.167	1.218	0.068	0.844	0.119	0.987	0.026
Regularity of annual ring width RRI (%)	76.574	5.838	80.028	5.700	82.932	5.872	83.462	6.045
Fiber undulation pitch CWL (mm)	5.448	0.460	6.2952	0.625	7.4182	1.400	NA	NA

It can be appreciated that the old luthiers chose sycamore wood taking into account these anatomical characteristics, something that was passed down from generation to generation among the luthiers. The use of sycamore maple with wavy fiber was also a tradition of luthiers, a fact proven by the analysis of the wave pitch of the fiber (CWL) both in historical violin boards and in the studied wooden samples. Thus, CWL in the back plates gravitates around the length of 5.7 mm for a Stradivarius Edler Voicu 1702 violin, 4 mm for a Stainer 1716 violin and 6.4 mm for Leeb 1742 violin, as highlighted by [3]. Compared with the values obtained in Table 2, it can be appreciated that the samples from class A described the best class of anatomical quality compared to the wood of heritage violins.

The determinations of the anatomical descriptors highlighted the fact that the width of the annual rings presented a high level of variability that encouraged the stratification of its values. The wavelength showed a moderate level of variability, but the range of values was quite wide (the amplitude of variation was 9.5 mm). The RW of the sycamore maple wood samples was not normally distributed (according to the Shapiro test W = 0.965, $p < 0.0001$), instead the CW was a Gaussian variable (W = 0.977, $p = 0.30$). Differences between samples were highly statistically significant with respect to annual ring width, but not confirmed with respect to CW. It was also found that in sycamore maple wood the tendency was to increase the width of the annual rings (Figure 3a) and to decrease the wavelength with improvement of the quality class (Figure 3b). It can be seen that class D is missing in Figure 2b because the length of the undulation pitch of the fiber in the case of samples from D grade tended to infinity, making it no longer possible to measure the pitch. In class C the range of values was greater. The width of the annual rings in the resonance paltin wood showed a pronounced asymmetry to the left for the size class 0.6–0.9 mm (Figure 3c), and the shape of the wavelength distribution was compatible with the normal law. The distribution module was in the size class 5.5–6.9 mm (Figure 3d). The small deviation revealed the homogeneity within the class and was the result of the way the luthiers selected the samples.

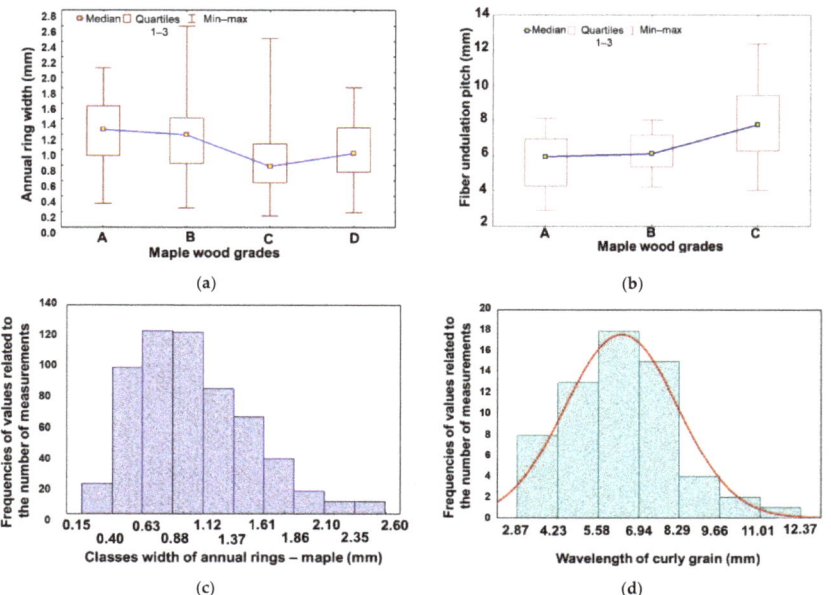

Figure 3. Variation of anatomical descriptors for resonance spruce in relation to the anatomical quality class: (**a**) the variation of the width of the annual rings; (**b**) Variation of the pitch of wavy fiber; (**c**) the distribution of RW; (**d**) distribution of CWL (red line–normal distribution).

3.2. Color of Sycamore Maple Wood Related to Anatomical Quality Classes

The stratification of the color, according to the quality grade, revealed the formation of two groups: one group was formed by classes A and D, and the other group was formed by classes B and C. Samples from classes A and D had darker wood ($L^* < 8.4\%$), with a high concentration of red ($a^* > 3.5$) and yellow in the color composition ($b^* > 17$) (Figure 4). Samples of quality classes B and C, had the highest brightness ($L^* > 84.5\%$). From the perspective of correlations with the other physical descriptors of maple wood, it followed that the density was inversely proportional to the width of the rings and the degree of red, but directly proportional to the wavelength and brightness. The relationship between ring width and sycamore maple wood color did not exceed the threshold of statistical significance. However, there was a tendency to increase the size of the red shade (a^*) and to temper the yellow shade (b^*) as the rings were wider. In conclusion, it could be appreciated that maple wood with curled and dense fiber was darker and had a higher degree of red and yellow in the color composition.

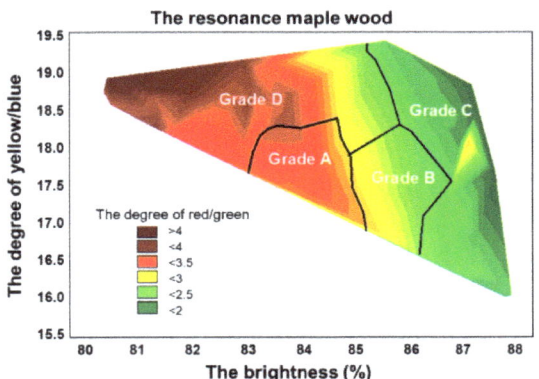

Figure 4. Multiple correlations between color parameters: brightness degree, yellowness degree and anatomical quality grade in sycamore maple wood.

3.3. Acoustic and Elastic Properties of Sycamore Maple Wood

The most important quantities that influence the acoustic quality of a musical instrument are the sound propagation speed, the elastic characteristics, the natural frequency, the respective quality factor, and the damping factor of the wood. Through the ultrasonic method used in this study, the sound propagation speeds in the three main directions of the wood, and its elastic characteristics, were determined, as shown in Table 3.

Table 3. The acoustic and elastic parameters of sycamore maple wood determined by ultrasound method.

Type of Variables	Symbol	Grade of Sycamore Maple Wood Average Values			
		A	B	C	D
Density (kg/m^3)	$\overline{WD}$	610	601	592	624
Sound velocity in wood (m/s)	V_{LL}	4238	3820	3750	3925
	V_{RR}	1773	1896	1866	1808
	V_{TT}	1326	1392	1360	1359
The ratios of sound propagation speeds	V_{LL}/V_{RR}	2.39	2.015	2.009	2.17
	V_{LL}/V_{TT}	3.19	2.74	2.75	2.88
	V_{RR}/V_{TT}	1.33	1.36	1.37	1.33
Young's elasticity modulus (MPa)	E_L	10,968	8775	8359	9626
	E_R	1920	2030	2044	2044
	E_T	1074	1164	1096	1157
Specific longitudinal modulus of elasticity (GPa $*$ g^{-1} $*$ cm^3)	E_L/ρ	17.97	14.60	14.12	15.41
	E_R/ρ	3.14	3.38	3.45	3.27
	E_T/ρ	1.76	1.93	1.85	1.85
Shear Modulus (MPa)	G_{RT}	1259	1259	1259	1259
	G_{LR}	1648	1648	1648	1648
	G_{LT}	1560	1560	1560	1560
Specific shear modulus of elasticity (GPa $*$ g^{-1} $*$ cm^3)	G_{RT}/ρ	2.064	2.74	2.63	2.22
	G_{LR}/ρ	2.24	2.71	2.60	2.42
	G_{LT}/ρ	1.74	1.99	1.90	1.81
Poisson Coefficient	υ_{LT}	0.445	0.423	0.422	0.431
	υ_{LR}	0.394	0.335	0.331	0.365
	υ_{RT}	0.140	0.085	0.072	0.168

The obtained data were consistent with those identified in the specialized literature, even if the analyzed samples were extracted from trees harvested from Romanian forests. Thus, the values determined by Bucur [16,22,33], using the ultrasonic method in the case of curly sycamore maple were: density $\rho = 0.700$ g/cm^3, the velocities $V_{LL} = 4350$ m/s; $V_{RR} = 2590$ m/s; $V_{TT} = 1914$ m/s; and for shear velocity $V_{LT} = 1468$ m/s; $V_{RT} = 812$ m/s; $V_{LR} = 1744$ m/s. For common Sycamore, the values were: $\rho = 0.623$ g/cm^3, the following longitudinal velocities $V_{LL} = 4695$ m/s; $V_{RR} = 2148$ m/s, $V_{TT} = 1878$ m/s; and for shear velocity $V_{LT} = 1148$ m/s; $V_{RT} = 630$ m/s; $V_{LR} = 1354$ m/s. For sycamore, Spycher et al. [13] obtained a density of $\rho = 0.625 \pm 0.022$ g/cm^3, sound velocity $V_{LL} = 3894 \pm 310$ m/s, $V_{RR} = 1662 \pm 97$ m/s, Young's modulus in the longitudinal direction, $E_L = 9.707 \pm 1.13$ GPa and in the radial direction, $E_R = 1.703 \pm 0.236$ GPa. For the control samples, the following values were obtained: $\rho = 0.569 \pm 0.015$ g/cm^3, $V_{LL} = 4180 \pm 143$ m/s, $V_{RR} = 1703 \pm 24$ m/s, $E_L = 9.974 \pm 0.769$ GPa and $E_R = 1.654 \pm 0.066$ GPa. Using the vibrational tests, based on non-contact forced–released vibrations of free-end bars, Bremaud [34] and Carlier et al. [35] reported sycamore with density of $\rho = 0.6$ g/cm^3, sound velocity in longitudinal direction of around $V_{LL} = 3660$ m/s and $V_{RR} = 1560$ m/s. Cretu et al. [17] applied a hybrid method to determine the sound velocity in sycamore maple wood, obtaining $V_{LL} = 4341 \pm 296$ m/s and $V_{RR} = 2001 \pm 58$ m/s.

3.4. Statistical Correlations between Physical and Acoustic/Elastic Features

Since the hypotheses of the study were focused on identifying the links between the appearance and pattern of sycamore maple wood and its acoustic and elastic properties, the experimental data were analyzed statistically, going through several stages of analysis, in terms of the methodology presented in Section 2.2.4. Thus, the discriminant analysis showed that only the wood density had the ability to separate the quality classes of the sycamore maple wood samples. The polarization of the density values according to the physical quality class of the sample can be traced in Figure 5a. It can be noted that there was regression of the size of the density in the first three quality classes. Quality class D samples showed the largest amplitudes of the density magnitude. Quality class A was the most homogeneous in terms of wood density. With the exception of Poisson's ratio v_{RT}, sycamore maple wood was homogeneous in terms of acoustic properties (small coefficients of variation between samples) (Figure 5b). In Figure 5c, it can be seen that the stratification of the sound propagation speed in the longitudinal direction depended on the anatomical quality classes. With the exception of the propagation speed in the tangential direction V_{TT}, the Young's modulus in the radial direction E_R and the shear modulus G_{LT}, the acoustic variables involved in the study presented non-Gaussian distributions (Shapiro–wilk W = 0.85 ÷ 0.98, $p \leq 0.05$). For this reason, significance tests were adopted from the area of non-parametric statistics.

Statistical analysis revealed that the reciprocal links between the acoustic variables in the case of sycamore maple wood were of moderate to strong intensity; the exception was wood density and Poisson coefficient v_{RT}. The size of the wood density was not an indicator for the size of the ultrasound propagation speeds in the three directions, nor for the shear moduli. The closest connection (Spearman correlation coefficient R = +0.976, $p < 0.0001$) was between the shear modulus G_{RT} and the sound propagation velocity in the radial direction V_{RR}. The relationships between the magnitudes of the ultrasound velocity, the Young's moduli and the Poisson coefficients in the three directions were of moderate intensity. Instead, the links between shear moduli G sizes were tighter, as can be seen in the multiple correlations between the acoustic and elastic parameters determined for the sycamore maple wood samples (Figure 6).

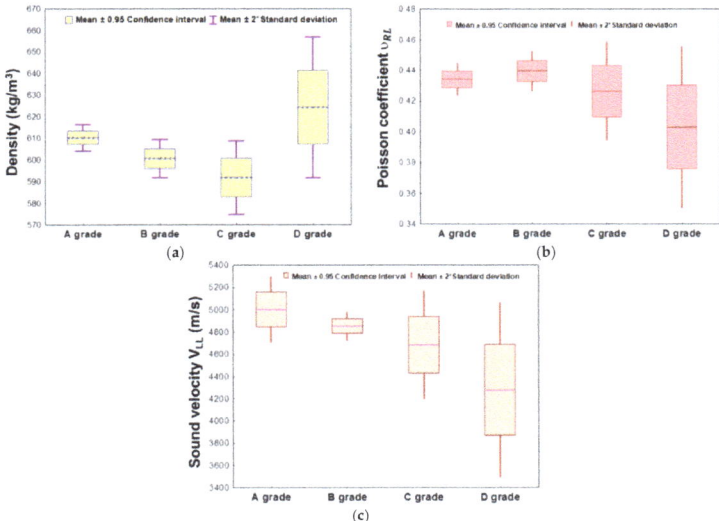

Figure 5. The stratification of sycamore maple wood, according to the specimen grade, in the case of: (**a**) density; (**b**) Poisson coefficient υ_{RL}; (**c**) sound velocity in longitudinal direction V_{LL}.

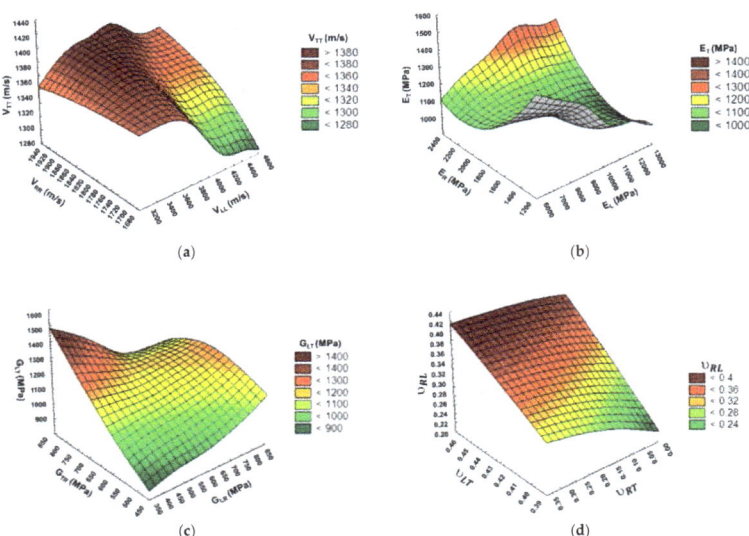

Figure 6. Multiple correlations between some acoustic and elastic parameters determined for the sycamore maple wood samples: (**a**) ultrasound velocity in L, R, and T directions; (**b**) longitudinal elasticity moduli in L, R, and T directions; (**c**) shear moduli G; (**d**) Poisson coefficients.

Using factor analysis, four principal components were extracted, with the first two together explaining 75% of the total variance. The first principal component was defined by the Poisson's ratio υ_{RT} and the quality class of the sample, which varied in contrast with the width of the annual rings and the wood density (Figure 7). The second component was defined by the sound speed in the radial direction V_{RR} and the shear modules G oriented in the RT and LR planes. At the opposite extreme were the longitudinal propagation velocity V_{LL} and Poisson's ratios in the longitudinal direction. So, the anisotropy axes, L and R, were divergent in the magnitude of the physical–acoustic indices. The degree of yellowness (b*) had no relevance in relation to the physical–acoustic properties of the wood. Instead, the brightness varied with the shear moduli G and the radial velocity. The redness (a*) varied in tandem with the modulus of elasticity, velocity of ultrasound propagation and transverse contraction coefficients. The regularity of the annual rings did not seem to influence the acoustic variables, instead the width of the annual rings was closely related to the wood density, as can be seen in Figure 7. Thus, the PCA (Figure 7) highlighted the association tendency of the variables, which was also verified with the k-means clustering analysis. This identified three groups, which separated the Young's modulus E_L from the sound velocity in wood, the Young's moduli E_R and E_T moduli, as well as the shear moduli. The third group would consist of physical indices (quality class, wood structure, wood color, density). So, the physical characteristics dissociated from the acoustic parameters, except for the shear moduli G_{TR} and G_{LR}.

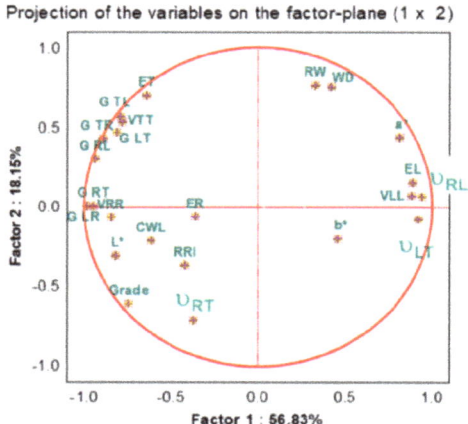

Figure 7. The physical, acoustic, and elastic parameters in the 1,2 planes of the PCA. Legend: Grade, the quality class of sycamore maple wood; CWL, the fiber undulation pitch; RRI, the ring width regularity index; RW, the annual ring width; b*, color yellowness; a*, redness color; L*, brightness color; WD, wood density; EL, Young's modulus along the wood fibers; v_{RT} and v_{LT} Poisson coefficients; V_{LL}, sound velocity in the longitudinal direction of the wood (along the grain); G_{LT} and G_{LR}, shear moduli; V_{TT}, sound velocity in the tangential direction of wood; E_R, Young's modulus in the radial direction of wood.

In Table 4 it can be noted that the variables of the annual rings (width and regularity) were independent of the size of the other variables involved in the study; in particular, the geometry of the rings had nothing to do with the G values in different directions. A certain indication could provide the regularity of the rings in relation to the size of the modulus of elasticity in the radial direction. Thus, sycamore maple wood with rings of relatively constant width presented low values of the modulus of elasticity in the radial direction. Instead, the pitch of the curly fiber (CWL) correlated with the size of the wood density, the shear moduli, the Poisson's coefficients and, especially, the elasticity modulus in the longitudinal direction E_L. The connection can be observed in Figure 7a. The fiber pitch had an unexpectedly large contribution of 63% to variations in the size of elasticity modulus in the longitudinal direction E_L (Figure 7a). Dense-grained sycamore maple wood (i.e., with a small wave pitch) was heavier, conducts ultrasound more easily along the grain, has better elasticity, higher transverse coefficients, and lower shear moduli. The L* and a* components of the color were relevant to the size of the density, the velocities of propagation along the directions, the Poisson coefficients and some shear moduli, as can be seen in Table 4. It can be seen from the bold values, which were significant for $p < 0.05$, in Table 4, that the following elastic and acoustic sizes E_L, υ_{RL}, V_{LL}, G_{LR} and G_{TL} were the properties best described by the physical properties of the wood (in fact, by the CWL and a*).

Table 4. Matrix of simple Spearman correlation coefficients between acoustic parameters and physical characteristics of the sycamore maple wood (Legend: Values marked in bold are significant for $p < 0.05$).

Variables	Coefficients of Simple Correlation *					
	WL	RRI	CWL	L*	a*	b*
WD	0.1905	−0.0534	**−0.6388**	**−0.6096**	**0.6470**	0.1191
V_{LL}	0.2483	−0.0932	**−0.7089**	**−0.4165**	**0.5678**	0.2826
V_{RR}	−0.1378	0.3714	0.3828	**0.5173**	**−0.4791**	−0.3765
V_{TT}	0.1013	0.1172	0.2776	0.2269	−0.1773	−0.2739
E_L	0.2400	−0.1007	**−0.7481**	**−0.5295**	**0.6600**	0.3086
E_R	−0.1691	**0.4390**	0.1248	0.1008	−0.1408	−0.1260
E_T	0.0713	0.0766	0.1207	−0.0408	0.0826	−0.2400
υ_{LT}	0.1272	−0.0940	**−0.6256**	−0.3166	**0.4372**	0.2909
υ_{RL}	0.2579	−0.3039	**−0.6752**	**−0.5097**	**0.6024**	0.3553
υ_{RT}	−0.2487	0.3458	0.1104	0.3443	−0.3052	−0.1295
G_{LT}	0.0243	0.3249	0.2591	0.2296	−0.1809	−0.3144
G_{RL}	−0.0487	0.2917	**0.4840**	0.2913	−0.2869	−0.3634
G_{RT}	−0.1461	0.3729	0.4468	**0.5382**	**−0.5182**	−0.3834
G_{TL}	−0.0082	−0.0165	**0.5748**	0.1165	−0.1556	−0.2643
G_{TR}	−0.0247	0.2083	**0.5624**	0.1409	−0.1869	−0.3992
G_{LR}	−0.2279	0.2655	**0.6264**	**0.4540**	**−0.5166**	−0.4574

Multiple regression allowed the development of predictive models with high explanatory capacity ($R^2 > 0.5$) (Figure 8a). These predictive models, presented in Table 5, allowed estimation of the size of some acoustic and elastic properties, depending on the size of some physical and anatomical characteristics, such as the pitch of the wavy fiber and the yellow color of the sycamore maple wood. A similar method was used by [36] for predictive models in the case of spruce wood. Both the experimental data (Figure 8b) and the model fitted to them (Table 5) showed that, in the case of sycamore wood used for the construction of musical instruments, the magnitude of the velocity of sound propagation along the fiber, for example, decreased with increasing the wave pitch of the fiber.

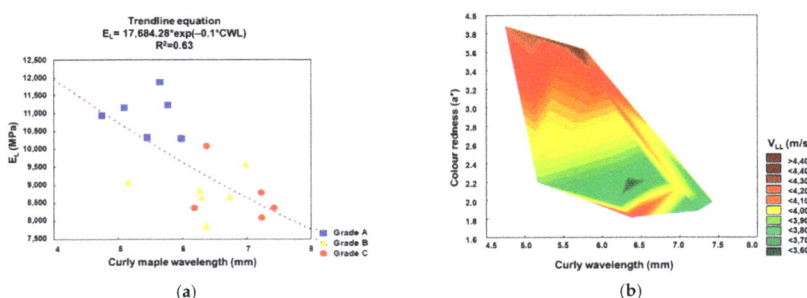

(a) (b)

Figure 8. Correlations between physical, anatomical and elastic parameters of sycamore maple wood: (**a**) regression of modulus of elasticity in longitudinal direction with curly fiber pitch; (**b**) prediction of the multiple correlations between the anatomical, physical and elastic characteristics of maple wood.

Table 5. Wood physical traits as predictors of some wood acoustics and elastic properties.

Dependent Variable	Predictors	Statistical Model	Adjusted R^2	F	p
E_L		$E_L = 6342.12 - 42.87 CWL + 1319.69a^*$	0.7	20.06	<0.001
v_{RL}	CWL, a^*	$v_{RL} = 0.256 + 0.002 CWL + 0.034a^*$	0.6	12.9	<0.001
V_{LL}		$V_{LL} = 3467.26 - 17.74 CWL + 234.71a^*$	0.6	12.95	<0.002

4. Conclusions

The presented study provides both a valuable database regarding the physical and acoustic characteristics of sycamore maple wood from the forests of Romania, as well as the existing statistical correlations between the different measured parameters. Moreover, the application of multiple regression allowed the development of predictive models with high explanatory capacity that allow estimation of the size of some acoustic and elastic properties, depending on the size of some physical and anatomical factors, known or easily determined. The main findings were as follows:

- the link between the values of the physical parameters and the acoustic and elastic properties of sycamore maple wood was confirmed, according to the quality classes used by luthiers for musical instruments;
- the Spearman rank-order correlation coefficient between the wavelength and the width of the rings was −0.367 (p = 0.13) and the correlation between the wavelength and the regularity of the rings was −0.075 (p = 0.79). From here it followed that there was no connection between the geometry of the rings and the curly grain. Therefore, the degree of curly grain can only be justified with the wavelength (if the wavelength is small, which means a pronounced undulation, and if the wavelength is large, it means a low degree of curly grain);
- wood density is the variable that best separates the quality classes of sycamore maple wood samples;
- acoustic properties (sound propagation speed) are not influenced by color parameter, but the redness correlates with elastic properties;
- the closest correlation was observed between the shear modulus G_{RT} and the sound propagation speed in the radial direction V_{RR};
- the anisotropy axes, L and R, are divergent in the size of the physical and acoustic features;
- the pitch of the wavy fiber has a contribution of 63% to the magnitude variations of the longitudinal modulus of elasticity (E_L).

In future studies, quantitative and qualitative analyses of the anatomical elements of sycamore maple wood will be considered according to the quality classes.

Author Contributions: Conceptualization, M.D.S., F.D. and A.S.; methodology, F.D. and A.S.; software, F.D., M.D.S. and A.S.; validation, F.D. and M.D.S.; formal analysis, F.D.; investigation, F.D. and A.S.; resources, M.D.S. and A.S.; data curation, M.D.S.; writing—original draft preparation, M.D.S.; writing—review and editing, F.D. and A.S.; visualization, M.D.S. and F.D.; supervision, A.S. and F.D.; project administration, M.D.S. All authors have read and agreed to the published version of the manuscript.

Funding: This research was supported by a grant from the Ministry of Research, Innovation and Digitization, CNCS/CCCDI—UEFISCDI, project number 61PCE/2022, PN-III-P4-PCE2021-0885, ACADIA—Qualitative, dynamic and acoustic analysis of anisotropic systems with modified interfaces.

Institutional Review Board Statement: Not applicable.

Informed Consent Statement: Not applicable.

Data Availability Statement: Not applicable.

Acknowledgments: We are grateful to the technical staff of Gliga Musical Instruments, Reghin, a Romanian manufacturer of musical string instruments, for supplying the specimens.

Conflicts of Interest: The authors declare no conflict of interest.

References

1. Nagyvary, J.; DiVerdi, J.; Owen, N.; Tolley, H.D. Wood used by Stradivari and Guarneri. *Nature* **2006**, *444*, 565. [CrossRef] [PubMed]
2. Tai, H.C.; Li, G.C.; Huang, S.J.; Jhu, C.-R.; Chung, J.-H.; Wang, B.Y.; Hsu, C.-S.; Brandmair, B.; Chung, D.-T.; Chen, H.M. Chemical distinctions between Stradivari's maple and modern tonewood. *Proc. Natl. Acad. Sci. USA* **2016**, *114*, 27–32. [CrossRef] [PubMed]
3. Stanciu, M.D.; Mihălcică, M.; Dinulică, F.; Nauncef, A.M.; Purdoiu, R.; Lăcătuș, R.; Gliga, G.V. X-ray Imaging and Computed Tomography for the Identification of Geometry and Construction Elements in the Structure of Old Violins. *Materials* **2021**, *14*, 5926. [CrossRef]
4. Sedlar, T.; Šefc, B.; Stojnić, S.; Sinković, T. Wood quality characterization of sycamore maple (*Acer pseudoplatanus* L.) and its utilization in wood products industries. *Croat. J. Eng.* **2021**, *42*, 543–560. [CrossRef]
5. Sonderegger, W.; Martienssen, A.; Nitsche, C.; Ozyhar, T.; Kaliske, M.; Niemz, P. Investigations on the physical and mechanical behavior of sycamore maple (*Acer pseudoplatanus* L.). *Eur. J. Wood Prod.* **2013**, *71*, 91–99. [CrossRef]
6. Krajnc, L.; Čufar, K.; Brus, R. Characteristics and geographical distribution of fiddleback figure in wood of Acer pseudoplatanus L in Slovenia. *Drv. Ind.* **2015**, *66*, 213–220. [CrossRef]
7. Quambusch, M.; Bäucker, C.; Haag, V.; Meier-Dinkel, A. Growth performance and wood structure of wavy grain sycamore maple (*Acer pseudoplatanus* L.) in a progeny trial. *Ann. For. Sci.* **2021**, *78*, 15. [CrossRef]
8. Beals, H.; Davis, T. *Figure in Wood—An Illustrated Review*; Agricultural Experiment Station, Auburn University: Auburn, AL, USA, 1977. Available online: https://aurora.auburn.edu/handle/11200/2414 (accessed on 6 November 2022).
9. Harris, J.M. *Spiral Grain and Wave Phenomena in Wood Formation*; Springer: New York, NY, USA, 1989.
10. Savidge, R.A.; Farrar, J.L. Cellular adjustments in the vascular cambium leading to spiral grain formation in conifers. *Can. J. Bot.* **1984**, *62*, 2872–2879. [CrossRef]
11. Binggeli, P. Invasive Woody Plants. 1999. Available online: https://www.cabidigitallibrary.org/doi/10.1079/cabicompendium.2884#REF-DDB--25 (accessed on 30 December 2022).
12. Kollmann, F.F.P.; Côté, W.A., Jr. Principles of Wood Science and Technology, vol I. In *Solid Wood*; Springer: Berlin/Heidelberg, Germany, 1968.
13. Spycher, M.; Schwarze, F.W.M.R.; Steiger, R. Assessment of resonance wood quality by comparing its physical and histological properties. *Wood Sci. Technol.* **2008**, *42*, 325–342. [CrossRef]
14. Alkadri, A.; Carlier, C.; Wahyundi, I.; Grill, J.; Langbour, P.; Bremaud, I. Relationships between anatomical and vibrational properties of wavy sycamore. *IAWA—Intern. Assoc. Wood Anat. J.* **2018**, *39*, 63–86. Available online: https://hal.archives-ouvertes.fr/hal-01667816 (accessed on 6 November 2022). [CrossRef]
15. Stanciu, M.D.; Coșereanu, C.; Dinulică, F.; Bucur, V. Effect of wood species on vibration modes of violins plates. *Eur. J. Wood Prod.* **2020**, *78*, 785–799. [CrossRef]
16. Bucur, V. Wood species for musical instruments. In *Handbook of Materials for String Musical Instruments*; Springer: Cham, Switzerland, 2016; pp. 283–320. [CrossRef]
17. Crețu, N.; Roșca, I.C.; Stanciu, M.D.; Gliga, V.G.; Cerbu, C. Evaluation of wave velocity in orthotropic media based on intrinsic transfer matrix. *Exp. Mech.* **2022**, *62*, 1595–1602. [CrossRef]
18. Kudela, J.; Kunštar, M. Physical-acoustical characteristics of maple wood with wavy structure. *Ann. Wars. Univ. Life Sci.* **2011**, *75*, 12–18.
19. Yoshikawa, S. Acoustical classification of woods for string instruments. *J. Acoust. Soc. Am.* **2007**, *122*, 568–573. [CrossRef]
20. Stanciu, M.D.; Curtu, I.; Moisan, E.; Man, D.; Savin, A.; Dobrescu, G. Rheological Behaviour of Curly Maple Wood (Acer Pseudoplatanus) Used for Back Side of Violin. *ProLigno* **2015**, *11*, 73–80.
21. Gliga, V.G.; Stanciu, M.D.; Nastac, S.M.; Campean, M. Modal Analysis of Violin Bodies with Back Plates Made of Different Wood Species. *BioResources* **2020**, *15*, 7687–7713. [CrossRef]
22. Bucur, V.; Archer, R.R. Elastic constants for wood by an ultrasonic method. *Wood Sci. Technol.* **1984**, *18*, 255–265. [CrossRef]
23. Fedyukov, V.; Saldaeva, E.; Chernova, M. Different ways of elastic modulus comparative study to predict resonant properties of standing spruce wood. *Wood Res.* **2017**, *62*, 607–614.
24. Dinulică, F.; Stanciu, M.D.; Savin, A. Correlation between Anatomical Grading and Acoustic–Elastic Properties of Resonant SpruceWood Used for Musical Instruments. *Forests* **2021**, *12*, 1122. [CrossRef]
25. Viala, R.; Placet, V.; Cogan, S. Simultaneous non-destructive identification of multiple elastic and damping properties of spruce tonewood to improve grading. *J. Cult. Herit.* **2020**, *42*, 108–116. [CrossRef]
26. Gonçalves, R.; Trinca, A.J. Comparison of elastic constants of wood determined by ultrasonic wave propagation and static compression testing. *Wood Fiber Sci.* **2011**, *43*, 64–75.
27. Fang, Y.; Lin, L.; Feng, H.; Lu, Z.; Emms, G.W. Review of the use of air-coupled ultrasonic technologies for nondestructive testing of wood and wood products. *Comput. Electron. Agric.* **2017**, *137*, 79–87. [CrossRef]

28. Rocaboy, F.; Bucur, V. About the physical properties of wood of twentieth century violins. *J. Acoust. Soc. Am.* **1990**, *1*, 21–28.
29. Dinulica, F.; Albu, C.; Borz, A.S.; Vasilescu, M.M.; Petrițan, C. Specific structural indexes for resonance Norway spruce wood used for violin manufacturing. *Bioresources* **2015**, *10*, 7525–7543. [CrossRef]
30. Pilcher, J.R. Sample preparation, cross-dating and measurement. In *Methods of Dendrochronology*; Cook, E.R., Kairiukstis, L.A., Eds.; Kluwer Academis Publishing: Dordrecht, The Netherlands, 1990; pp. 40–51.
31. Zar, J.H. *Biostatistical Analysis*; Prentice-Hall Inc.: Englewood Cliffs, NJ, USA, 1974.
32. Carlson, D. MANOVA and Discriminant Analysis. In *Quantitative Methods in Archaeology Using R*; Cambridge Manuals in Archaeology; Cambridge University Press: Cambridge, UK, 2017; pp. 244–264. [CrossRef]
33. Bucur, V. Varieties of resonance wood and their elastic constants. *J. Catgut Acoust. Soc.* **1987**, *47*, 42–48.
34. Brémaud, I. Acoustical properties of wood in string instruments soundboards and tuned idiophones: Biological and cultural diversity. *J. Acoust. Soc. Am.* **2012**, *131*, 807–818. [CrossRef]
35. Carlier, C.; Brémaud, I.; Gril, J. Violin making "tonewood": Comparing makers' empirical expertise with wood structural/visual and acoustical properties. In Proceedings of the International Symposium on Musical Acoustics ISMA2014, Le Mans, France, 7–12 July 2014; pp. 325–330.
36. Dinulică, F.; Bucur, V.; Albu, C.T.; Vasilescu, M.M.; Curtu, A.L.; Nicolescu, N.V. Relevant phenotypic descriptors of the resonance Norway spruce standing trees for the acoustical quality of wood for musical instruments. *Eur. J. Forest Res.* **2021**, *140*, 105–125. [CrossRef]

Disclaimer/Publisher's Note: The statements, opinions and data contained in all publications are solely those of the individual author(s) and contributor(s) and not of MDPI and/or the editor(s). MDPI and/or the editor(s) disclaim responsibility for any injury to people or property resulting from any ideas, methods, instructions or products referred to in the content.

Article

The Potential of Smart Factories and Innovative Industry 4.0 Technologies—A Case Study of Different-Sized Companies in the Furniture Industry in Central Europe

Luboš Červený, Roman Sloup * and Tereza Červená

Faculty of Forestry and Wood Sciences, Czech University of Life Sciences Prague, Kamycka 129, 16500 Prague, Czech Republic
* Correspondence: sloup@fld.czu.cz; Tel.: +420-608516302

Abstract: New innovative technologies of Industry 4.0 are the key to the future development of the furniture industry, which is outdated because of its atypical production and small-series production. For applying the novel trends of Industry 4.0 to the furniture sector, the methodical support of managers, the key users of these technologies, is essential. As there is a lack of knowledge regarding implementation of Industry 4.0, this study focuses on the evaluation of the current status of furniture companies in terms of production structure and Industry 4.0 benefits/threats with the aim of proposing methodological solutions for the implementation of this trend across different-sized enterprises. Data are collected using conduct-structured interviews with project managers who describe their own experience with Industry 4.0 implementation in central Europe. All interviews are analyzed using qualitative content analysis. According to the stakeholders, innovative production and non-production technologies are essential for their enterprises. Application of such technologies increases the efficiency of the whole operation by 30%–50% over the five years since the first innovations were introduced, especially in enterprises with atypical production and large enterprises. This study should serve as the tool for adapting the environmental changes and promoting the innovation approaches of the Industry 4.0 strategies on the central European level.

Keywords: forestry and wood sector; furniture technology; smart factory; project managers; furniture industry; Industry 4.0

1. Introduction

The European forestry and wood sector is currently undergoing significant economic changes that are due to the increasing pace of socio-economic and technical changes such as the climate crisis, the reduction in available natural resources, natural disasters, war conflicts and rising energy prices. In addition, consumers are demanding increasingly sophisticated products of high quality, standards and certifications, including support services, to meet their immediate needs.

This sector is unique in its interdependence within the raw material base, which should aim to develop the industries involved, processing capacities and the use of wood as an ecological and renewable raw material for future generations. Wood is of fundamental importance in this case, as it is used by the timber, furniture, paper and energy industries, followed by other downstream sectors such as construction. At the same time, this sector contributes to rural development from an economic, social and environmental point of view. This is also supported by the European Union's (EU's) strategic plan, "Agenda 2030" or "A Clean Planet for All" the new EU strategy for forests by 2030 [1,2], which highlights the need to identify the necessary changes to maximize the use of all energy, material and human resources involved in the value-creation process [3]. Businesses should be able to respond flexibly to these challenges with the latest innovations in their respective fields [4,5] through their value chains. By using both physical and virtual structures, close

collaboration and rapid adaptation can be achieved throughout the project and company life cycle, from production innovation to distribution innovation [6].

One of the tools for adapting to a changing environment is the introduction of information technology, cyber-physical systems and artificial intelligence systems into the production and services in all sectors of the economy [7–9]. The impact of these changes is so significant that we refer to them as the fourth stage of the Industrial Revolution [7]. This phenomenon was presented in Germany at the "Hannover Fair" in Hannover in 2011 as a proposal for a new concept of economic policy for Germany [10–13]. This direction of innovative technologies is characterized by the intelligent vertical and horizontal interconnection of people and machines [7,14–17] and objects and information and communication technology systems [17].

The main prerequisite for the implementation of this strategy is the innovation and modernization of production in the forestry and timber sector, where the main problems are considered to be the high cost of human labor, the outdated and worn-out production equipment and the lack of financial resources for further development. All of this can have a negative impact on the environment or on the sustainability of the whole sector [18].

The importance of technological development in this sector is underscored by the fact that, for example, in 2016 the government of the Czech Republic approved the Industry 4.0 Initiative prepared by the Ministry of Industry and Trade, which aimed to maintain and strengthen the competitiveness in the era of the so-called Fourth Industrial Revolution. In Europe, we can also encounter fully automated furniture plants where the human factor acts only as an additional member. An ecologically minded civilization requires changes in human thinking toward the harmonious coexistence between man and nature [19]. The rational use of environmentally friendly and sustainable renewable materials is a necessary step toward the development of an environmentally friendly industry [20–22].

While the available research shows the great potential of Industry 4.0 for business owners [23], its practical use is limited by a number of factors. One of them is the lack of knowledge and understanding of the potential of Industry 4.0 for different categories of businesses, mostly in atypical manufacturing [6,24].

It is clear that this vision will lead to increased complexity in manufacturing processes in the market at micro and macro levels. In particular, small and medium enterprises (SMEs) in the manufacturing industry are unsure of the financial requirements needed to acquire these new technologies and the overall impact on their business models [6].

Seminar experiences [25] on strategic orientation across enterprises have shown that the actors have serious problems in grasping Industry 4.0 across different concepts and operations. On the one hand, they are not able to link Industry 4.0 with their current corporate strategies; on the other hand, they are not experienced in identifying the state of corporate development and the vision of Industry 4.0. Therefore, they cannot identify specific areas for implementing the attributes of Industry 4.0. It is essential to formulate new models and tools in a way that they provide support with respect to the market needs of the sector and in line with the company's strategy.

For the successful implementation of Industry 4.0 in the furniture practice, it is essential to understand the importance of delivering the benefits of Industry 4.0, especially to project managers who are the key users of these measures. Therefore, the aim of this paper is to analyze what factors may influence the degree of the implementation of Industry 4.0 in the furniture industry depending on the size of the company and the type of production.

Other sub-objectives were:

1. To determine whether the implementation of Industry 4.0 depends on the size and type of production;
2. How the attitudes of project managers toward the introduction of new technologies depend on the size and type of production;
3. How the size of the company depends on the company's strategies;
4. How the risk of not implementing Industry 4.0 depends on the size and type of production.

From the answers obtained, a methodological framework for the implementation of Industry 4.0 will be proposed, which will be applicable to individual sizes of enterprises in the furniture industry and probably in other small-scale industries as well.

Sociodemographic characteristics or other aspects that may influence the subjective assessments of the Industry 4.0 implementation were also investigated. The assessment of attitudes and individual factors allow the identification of problems and obstacles, the solutions to which should then be incorporated in political decision-making and in the design of legislative, subsidy and information systems. These findings can contribute to shaping approaches and applying the objectives of national and European strategies and to the effective promotion of Industry 4.0 in this sector. Participatory methods could serve to increase the need for and consideration of Industry 4.0 in national strategic planning.

1.1. Literature Review

The rapid pace of technological improvements has created a need for rapid adaptation, and the most innovative companies are those that have been able to recognize early on how new digital tools are impacting their business models and what value they can derive from the information generated by their activities.

1.1.1. The Current State of the Furniture Industry

The EU furniture industry, which consists mainly of SMEs, employs around 1 million European workers and produces around a quarter of the world's furniture, representing a market worth EUR 84 billion [26]. According to the classification of economic activities in the EU, furniture manufacturing falls under NACE (Nomenclature statistique des activités économiques dans la Communauté européenne) Division 31. The industry is fundamentally influenced by the customer, social trends, the cost of materials and labor, exports and competitive pressure or the situation on the commodities market. The furniture industry (NACE 31) is currently facing two main challenges. The customer, as a consumer, is looking for a quality product and professional services at the most reasonable prices and the increasing demands for environmental protection and the efficient use of natural resources in the context of the coronavirus pandemic. Smart manufacturing is an inevitable trend to maintain the viability of the NACE 31 sector [9]. Another important fact is that the wood products sector is the least automated industry on a European level—only 0.2% of the production processes in this sector are automated by robots involved in wood processing activities. The sector still performs most tasks manually and is characterized by a low understanding of the potential of automation. The average number of employees in this sector is decreasing at an annual rate of 1.8% [27].

1.1.2. Industry 4.0 in the Procurement Sector

Digital transformation affects business models, production processes and corporate governance. In particular, improvements in information and communication technologies and analytical capabilities are spurring an influx of innovation at all levels of organizations. The opportunities offered by the use of digital technologies in the corporate world are changing the competitive position of companies and the way they interact with employees and customers, improving their market position [28].

The essence of Industry 4.0 is a comprehensive form of factory management that reduces all error factors. In such an environment, there is more efficient production and communication between people, machines and resources according to the principles of social networks [29,30].

The basis of Industry 4.0 is the technological development, advances in information and communication technologies and the internet, that connects the entire value chain [31]. Figure 1 presents the end of the Industry 4.0 concept. In short, the human–cyber–physical system (HCPS) technology enables real-time information acquisition, data analysis, strategic decision making and data processing. It leads to increased efficiency, logistics and a better demand response [32]. It is a completely new philosophy that brings about a societal change that affects a large part of the industry, from technical solutions to work safety to the labor market or the social system. The cornerstones of Industry 4.0 (Figure 1) include the Internet of Things (IoT), Internet of Services (IoS), the HCPS [33,34], artificial intelligence (AI), big data and analytics, cybersecurity, augmented reality (AR), robotics, automation and visualization, cloud storage [10,35] and servitization [36].

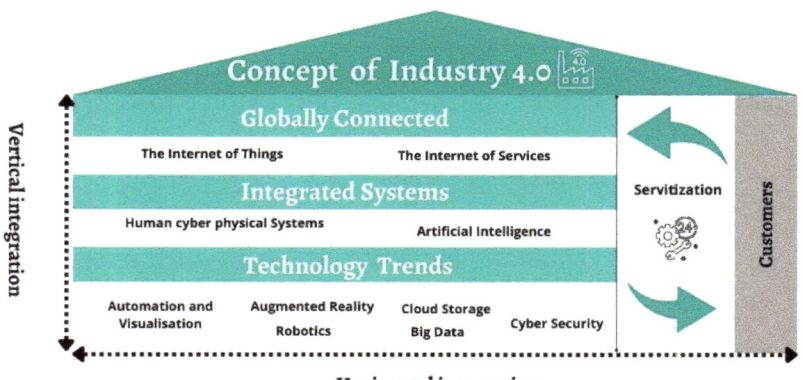

Figure 1. Scheme of Industry 4.0 integrating cornerstones.

The building blocks of Industry 4.0 represent a huge opportunity in terms of sustainability and increased productivity of industrial production and services, as well as the demand for skilled labor [37]. The coordination and integration of strategic and tactical operational decisions across the enterprise supply chain is essential to achieve all the Industry 4.0 objectives. Due to the complex and extensive challenges associated with the production process, strategic decision-making creates constraints in the tactical planning process.

The trade supply chain concept can help in planning decisions in complex industries [38,39]. The supply chain helps the production of raw materials to distribute products in the customers' regions, which helps the development of the micro-region. The forestry and timber supply chain consists of different process stages and different products such as biomass energy paper, semi-finished products for manufacturing and metal materials [40]. Another important level of decision-making in the supply chain is inventory management and planning, which, in an environment of uncertain supply and demand, accounts for 40% of the annual costs in the sector. The management system therefore requires coordinated decision-making on the inventory at all levels and at each facility in the supply chain. Such a supply chain is complex because it involves several units, each responsible for a large number of dependent activities [41].

2. Materials and Methods

The experiences and attitudes of project managers toward Industry 4.0 were investigated in several methodological steps, with an emphasis on a survey conducted through structured interviews and on the possibility of implementing these practices in practice. The interviews were conducted between 1 October 2021 and 1 June 2022.

The study was divided into the following steps:
1. Classification of the basic premises of the innovative technologies of Industry 4.0 and the creation of the structured interviews;
2. Pilot testing of the structured interviews;
3. Implementation of the structured interviews with the respondents;
4. Data processing, analysis and interpretation.

2.1. Step 1. Classification of the Basic Premises of the Innovative Technologies of Industry 4.0 and the Creation of the Structured Interviews

Based on the main assumptions of innovative Industry 4.0 technologies and the basic requirements for project management, the final version of the questions was divided into four thematic sections, namely:

- Attitudes toward the introduction of new technologies in the last two years, including any shortcomings in production;
- Potential for improvement in particular areas of the business;
- The company's strategy for the implementation of Industry 4.0 (obstacles, advantages, readiness for the implementation of innovative technologies and, if applicable, the financial resources used);
- Expected risks if Industry 4.0 is not implemented in the next five years.

The questions were compiled in collaboration with woodworking industry experts, IT and sociologists and served as a basis for the structured interviews with project managers dealing with innovative technologies in the woodworking industry. Specifically, this included companies producing furniture of various sizes and those developing Industry 4.0 software solutions.

Subsequently, enterprises were identified and classified according to the Commission Recommendation of 6 May 2003 concerning the definition of micro, small and medium-sized enterprises [42]. This breakdown was rather crude for the purposes of the research, so a more detailed breakdown of the enterprises (subcategories) was proposed and used in the research (Table 1), influencing the technology adoption in the enterprises according to the number of employees, turnover and type of production.

Table 1. Categories and subcategories affecting technology adoption.

Category	Subcategories
Micro enterprises (up to 10 persons, turnover up to EUR 2 million)	- Up to 3 persons, turnover up to EUR 1 million - Up to 10 persons, turnover up to EUR 2 million
Small businesses (up to 50 persons, turnover up to EUR 10 million)	- Up to 25 persons e, turnover up to EUR 5 million - Up to 50 persons, turnover up to EUR 10 million
Medium-sized enterprises (up to 250 persons, turnover up to EUR 50 million)	- Up to 150 persons, turnover up to EUR 25 million (very untypical production) - Up to 150 persons, turnover up to EUR 25 million (mass production) - Up to 250 persons, turnover up to EUR 50 million
Large enterprises (more than 250 persons, turnover more than EUR 50 million)	- More than 250 persons, a turnover of more than EUR 50 million (built by gradual modernization) - More than 250 persons, a turnover of more than EUR 50 million (built on a "green field")

2.2. Step 2. Pilot Testing of The Structured Interviews

Pilot testing of the structured interviews with a test sample of stakeholders was conducted. The aim of the pilot collection was to test the logic and clarity of the questions. Based on the findings, several questions were subsequently refined.

2.3. Step 3. Implementation of the Structured Interviews with the Respondents

The actual testing (implementation of the structured interviews) was carried out on a total of 31 companies of different sizes in the wood processing industry in central Europe, specifically in Germany, the Czech Republic and Slovakia. The meeting was always pre-arranged in the company so that the intention could be sufficiently explained. The meeting took place either during a personal visit to the company or online via MS Teams. This solution was chosen primarily to allow contact with the respondents without the need for a face-to-face meeting, mainly because of concerns about COVID-19 infections and measures resulting from Government Resolution 1375/2020 concerning restrictions on the free movement of persons [43].

Interviews with the project managers were recorded or transcribed. Subsequently, during a deeper analysis, the data obtained were transcribed and processed into a spreadsheet.

2.4. Step 4. Data Processing, Analysis and Interpretation

In terms of the number of questions answered, key areas and questions were selected for this paper, which were subsequently evaluated. Questions relating to the socio-demographics of the individual respondents and business data required to categories the businesses according to the suggested business sizes were also included. The responses obtained were then analyzed and a synthesis was made; the summary results were interpreted in the form of a table summarizing the main results in points. The tables were divided into four columns. The essence of the first column, "Example of a real company", is the formation of a certain spectrum of examples of companies, perceptible at the same time to a wide range of readers in a nonprofessional environment. The column "Proposed solutions" allows an understanding of the appropriate options for implementing the innovative technologies in each subcategory of enterprises. The column "Benefits of implementing Industry 4.0" provides a realistic view of the benefits of innovation for the enterprise. The column "Risks of non-implementation within 5 years" describes the possible threats that an enterprise may face if it is not implemented within a certain timeframe.

The advantage of this processing is the quick orientation in the results for the uniform categories of the company sizes and the possibility to easily find the solutions proposed to it. The benefit of this whole model is the direct applicability of the results to the furniture industry and the possible applicability of the results to other manufacturing sectors.

3. Results

A total of 94 respondents working in the furniture industry were contacted. Of these, 31 respondents were already involved in the implementation of Industry 4.0. For the purpose of this study, the focus was on data collected from this group of respondents ($n = 31$).

The socio-demographic characteristics of the respondents are shown in Table 2. The length of experience in the field was surveyed, with almost one-third of the respondents having 6–10 years of experience, and about one-fourth having 11–15 years of experience ($n = 10$).

The largest share of responses in terms of the region was dominated by respondents from companies operating in the Czech Republic (54.8%). Another question tracked the highest level of education attained. Overall, the highest number of respondents had a master's degree (41.9%), followed by a secondary education (25.8%), and 58% had a university degree. More than 77% of the respondents had a degree related to the furniture industry.

Table 2. Selected basic characteristics of the respondents.

Data about the Respondent		Is Your Company Involved in Industry 4.0?				Total	
		YES		NO			
		No.	%	No.	%	No.	%
	Total	31	100.0%	63	100.0%	94	100.0%
Period of experience in the field	Less than 1 year	0	0.0%	9	14.3%	9	9.6%
	1–5 years	7	22.6%	12	19.0%	19	20.2%
	6–10 years	10	32.3%	19	30.2%	29	30.9%
	11–15 years	9	29.0%	14	22.2%	23	24.5%
	More than 15 years	5	16.1%	9	14.3%	14	14.9%
The company's market presence	Czech Republic	17	54.8%	27	42.9%	44	46.8%
	Slovakia	5	16.1%	13	20.6%	18	19.1%
	Germany	6	19.4%	14	22.2%	20	21.3%
	Other	3	9.7%	9	14.3%	12	12.8%
Highest education attained	Secondary education—leaving certificate	5	16.1%	20	31.7%	25	26.6%
	Secondary school—high school diploma	8	25.8%	29	46.0%	37	39.4%
	Bachelor's degree	5	16.1%	8	12.7%	13	13.8%
	Master's degree	13	41.9%	6	9.5%	19	20.2%
The link between education and the furniture industry	Yes	24	77.4%	49	77.8%	73	77.7%
	No	7	22.6%	14	22.2%	21	22.3%

Table 3 shows that most of the respondents are in the small business category (38.7%), which corresponds approximately to their market share, as well as to the other size categories reported. The second most represented group are respondents from medium-sized enterprises (32.8%), followed by large enterprises (16.1%), where it is interesting to note that in enterprises with more than 250 employees, two-thirds of these enterprises have already implemented Industry 4.0. On the other hand, only 12.9% of the respondents in micro-enterprises have dealt with Industry 4.0 and only one-third of the respondents have experience with the implementation of Industry 4.0, which is partly due to the lack of interest in innovation and lack of financial resources that could be used for innovation.

Table 3. Number of respondents by size category of the enterprises.

Data about the Respondent	Is Your Company Involved in Industry 4.0?				Total	
	YES		NO			
	No.	%	No.	%	No.	%
Total	31	100.0%	63	100.0%	94	100.0%
Micro-enterprises	**4**	**12.9%**	**7**	**11.1%**	**11**	**11.7%**
Up to 3 persons, turnover up to EUR 1 million	1	3.2%	3	4.8%	4	4.3%
Up to 10 persons, turnover up to EUR 2 million	3	9.7%	4	6.3%	7	7.4%
Small businesses	**12**	**38.7%**	**28**	**44.4%**	**40**	**42.6%**
Up to 25 persons, turnover up to EUR 5 million	5	16.1%	12	19%	17	18.1%
Up to 50 persons, turnover up to EUR 10 million	7	22.6%	16	25.4%	23	24.5%

Table 3. Cont.

Data about the Respondent	Is Your Company Involved in Industry 4.0?				Total	
	YES		NO			
	No.	%	No.	%	No.	%
Medium-sized enterprise	10	32.3%	24	38.1%	34	36.2%
Up to 150 persons, turnover up to EUR 25 million (very atypical production)	5	16.1%	12	19.0%	17	18.1%
Up to 150 persons, turnover up to EUR 25 million (mass production)	3	9.7%	8	12.7%	11	11.7%
Up to 250 employees, turnover up to EUR 50 million	2	6.5%	4	6.3%	6	6.4%
Large Enterprises	5	16.1%	4	6.3%	9	9.6%
More than 250 employees, turnover of more than EUR 50 million (built by gradual modernization)	1	3.2%	2	3.2%	3	3.2%
More than 250 employees, turnover of more than EUR 50 million (greenfield)	4	12.9%	2	3.2%	6	6.4%

3.1. Implementation of Industry 4.0 in Micro Enterprises

This category includes a relatively large number of entities that mostly implement small projects that are not of interest to large entities, thus filling a gap in the market. Micro enterprises are not overly threatened by large companies, focusing mainly on local clients with specific customer requirements. The implementation of Industry 4.0 in micro enterprises is summarized in Table 4.

Table 4. Main results and recommendations (scored) for the micro enterprises and the subcategories ($n = 4$).

Cat.	Subcat.	Example of a Real Company	Proposed Solutions	Benefits of Implementing Industry 4.0	Risks of Non-Implementation within 5 Years
Micro-enterprise (up to 10 persons, turnover up to EUR 2 million)	up to 3 persons, turnover up to 1 million EUR	- Documents, information and tasks are transmitted directly; - Small-scale contracts; - Highly atypical production; - Entities usually consist of one worker; - Outdated machinery; - Mostly hand tools.	- Cloud storage applications, cyber protection; - Establishing a network of business and supply links; - Work with data and 3D data.	- Speeding up the transfer of information; - Efficient use of time; - Increased work efficiency; - Use of SME supply chain services; - The enterprises will contribute indirectly to the development of the whole sector; - Greater efficiency and growth.	- Micro-enterprises fill a gap in the market; - They are able to address atypical customer requirements, which reduces the threat of competitive pressure from large players; - The absence of innovation prevents progressive growth; - The risks in the competition are minimal, the potential on offer is great.
	up to 10 persons, turnover up to 2 million EUR	- Family and individual companies; - Atypical production; - Limited spaces; - High operating costs; - Outdated technical and technological background; - Creation of production documents of both groups mentioned above by "pencil and paper".	- Modernization of machinery; - Innovation of non-productive parts of the company; - For example, cloud storage, advanced data protection and mobility systems; - If this is not under your own control, you need to look for entities providing professional services.		

The respondents report that, in the case of micro enterprises, data and information are transferred directly. Furthermore, these enterprises mainly carry out atypical small-scale production. Therefore, the objective of implementation is not complex automation and digitization. Rather than upgrading bulky machines, it is recommended to create a network of suppliers who provide services in the form of cutting, gluing, milling, etc. Furthermore, the introduction of cloud storage would be useful.

The introduction of modern communication technologies will speed up the transfer of information not only between employees but also between suppliers and customers. The use of supply chain services by SMEs engaged in Industry 4.0 innovation will save time and processes on all sides and contribute to the development of the whole sector. These services will help the SMEs fill the gaps in their production. Micro enterprises are thus indirectly involved in the use of technologies that are oversized for them. At the same time, they can offer better processing and materials that they would not be able to process or details that they would not be able to produce. If individual operators understand this, the benefit will be to process more orders and secure more growth.

3.1.1. Up to 3 Persons, Turnover up to EUR 1 Million

Micro enterprises in this subcategory transfer information and tasks directly. They process small-scale orders with a focus on atypical production. The majority of these entities consist of a single manager and have basic, often outdated machinery. Large woodworking machines and equipment are often partially replaced by hand tools.

Proposed Solutions

Cloud storage, along with cyber protection, are essential points for innovation in the smallest type of enterprise. Creating and leveraging a network of business–supplier relationships providing services in the form of cutting, gluing, milling, etc. is essential. It is thanks to cloud storage that the efficient sharing of production documents is made possible. A de facto option for a micro enterprise is to adopt the design software of the supplier company from which it receives services in the form of materials. By processing the 3D data by the contracting company itself, it is possible to create compatible documents that are shared with the interested company, on the basis of which it executes the services. This 3D data can be further used in communication with the customer, assembly line, etc.

Benefits of Implementing Industry 4.0—For Both Subcategories of Large Enterprises

The respondents agreed that the implementation of Industry 4.0 will mainly speed up the transfer of information between employees and the supply chain. Cloud storage variants enable data sharing not only within the internal environment but also in the aforementioned cooperation with other entities. The effective use of modern technologies brings time savings, an increased work efficiency and an increased competitive advantage.

The use of supply chain services by SMEs engaged in Industry 4.0 innovations saves time and processes on all sides and contributes to the development of the whole sector. By using services, micro enterprises can indirectly participate in the development of the whole sector. They also reduce the price of raw materials and obtain a product that is processed with high quality on professional machines. This also contributes to processing more orders and ensuring greater growth.

Risk of Non-Implementation within 5 Years—For Both Subcategories of Large Enterprises

Micro enterprises fill a gap in the market with their spectrum and flexibility. As they are able to address sophisticated and highly atypical customer requirements, they are not threatened by competitive pressure from large players. The absence of innovation will prevent effective development and a progressive competitive ascent.

The risks in the competition are minimal, but the potential it offers is great.

3.1.2. Up to 10 Persons, Turnover up to EUR 2 Million

These are mostly family and individual businesses, businesses with up to 10 employees engaged in atypical production, usually with limited space and high operating costs. The machines are often outdated and lack modern software to allow compatibility with other equipment. The production documents of both these groups of micro enterprises are usually produced in an inefficient 'pencil and paper' way.

Proposed Solutions

The modernization of the machinery and the global innovation of non-production parts of the company are essential. Cloud storage, advanced data protection and mobility systems and engineering software are recommended. If the enterprise does not have expert in-house employees, it is necessary to seek entities providing IT services and other necessary services.

Micro enterprises should consider whether it is more efficient for them to produce furniture in-house or to have a range of cut, edged and milled semifinished products produced in collaboration with higher-end specialist operations that provide services as part of the trading and supply process. This potential is not sufficiently exploited or understood by businesses, and the production of a micro or small business can increase by multiples of the existing capabilities. As medium-sized enterprises become more technically and technologically equipped, these services become more affordable. By using services, the micro enterprise indirectly participates in development without fully addressing Industry 4.0.

3.2. Implementation of Industry 4.0 in Small Businesses

A small business with a long-term production spectrum is likely to feel the need to modernize and digitize its operations. If it understands the benefits on offer, the business will move toward greater efficiency. According to the respondents, the whole process of modernization is quite lengthy, bringing major changes in both the production and non-production parts that must be taken into account by the enterprises as summarized in Table 5, requiring modernization. It should be stressed that the process of technological change must be clearly elaborated upon with predetermined objectives and, above all, it must be carried out in steps that must always be perfectly executed.

3.2.1. Small Businesses, Sub-Category 25 Persons, Turnover up to 5 Million EUR

According to the characteristics of the respondents, these are companies that are undergoing a transformation from a small family-owned entity to a company with a wider range of contracts. Information concerning the production is distributed directly among the employees without the use of modern communication technologies. The production operation of cutting is carried out using conventional saws. The absence of modern banding machines is common in a company of this size. Outdated computer numerically controlled (CNC) centers are used where drilling and partial manual machining takes place. Processing of raw lumber is common in this category of company, which enters production as the main raw material in combination with the processing of agglomerated wood-based material.

Proposed Solutions

The development requires the creation and digitalization of the design and programming department, stock management, modernization of machines, and the division of labor of individual workers. The creation of work cells and departments that need to be gradually modernized and developed is required. Recording the time and material consumed directly at the workplace in a digital interface will streamline the entire business process. Introduction of an internal company network and distribution of data from designers or programmers to machine operators (shared storage with real-time data distribution). The design department becomes the heart of the furniture company, along with the networks that ensure the transfer of information.

Table 5. Main results and recommendations (scored) for small businesses and the subcategories (*n* = 12).

Cat.	Subcat.	Example of a Real Company	Proposed Solutions	Benefits of Implementing Industry 4.0	Risks of Non-Implementation within 5 Years
Small businesses (up to 50 persons, turnover up to 10 million EUR)	up to 25 persons, turnover up to 5 million EUR	- Transformation from a small family entity to a broad-based one; - Companies often do not have departments such as production design and programming; - The impossibility of integrating modern technologies; - Verbal data distribution.	- Establishment and digitalization of the design and programming department, warehouse management, modernization of machines, division of labor of individual employees; - Creation of work cells and departments; - Digital time and material records; - Implementation of an internal corporate network and data distribution.	- Real-time display of current data; - Data availability; - Increase in production capacity; - Optimization of production costs; - Automatic readers for registration and data collection, label scanning; - The steps are provided by self-organizing structures.	- Unmodernized entities are unable to offer the required level of cooperation, are expensive and inefficient; - Inefficient employees, lack of competence in Industry 4.0; - Loss of customers, reduction in design quality.
	up to 50 persons, turnover up to 10 million EUR	- Corporate governance, a narrow circle of managers; - Great efforts in management and communication; - High diversity of technical and technological background and production processes; - A combination of modern and obsolete machines; - Combination of different input commodities; - Companies feel the need to modernize, they don't have the know-how; - The need to modernize spatial and technological facilities.	- Introduction of complex digitalization of the company; - Design department, the digital twin of the product; - Modernization of equipment; - Cooperation with specialized IT companies - Sophisticated management software, total personnel and project management; - Creation of a position managing company development and employee training; - Choice and specialization of processed commodity (solid sawn timber, wood-based panel material).	- Elimination of repetitive operations; - The design department generates all data automatically, digitally; - Applicable to all types of production; - Specializing production in solid or plate material will streamline technical and technological processes; - Freeing up space for underutilized technologies; - Trade and supply links; - Increased internal transparency.	- In the absence of employees, there is a risk of paralysis of the enterprise; - Without sophisticated software, there is a risk of information loss and business collapse; - Failure to determine the direction of production will result in high operating and development costs; - The spatial layout will limit the growth of the company.

Benefits of Implementing Industry 4.0

Among the benefits of implementing Industry 4.0, the respondents cited the data collection and distribution to cloud storage or internal servers. Viewing real-time, up-to-date data outside the company allows for better governance and operational management. Data from production and non-production departments are used for future company growth and process optimization. Modern machines allow one to handle increased production capacities.

These devices include automatic readers for registration and data collection, as well as for the scanning of labels, contributing to an automated process in the company. These and other processes reduce the dependency on key persons and reduce the risk of downtime for these important persons in the running of the business. Thanks to the innovations, the company can offer its customers and business partners an interesting collaboration in terms of price and quality.

Data from the production and non-production departments that are recorded by machines and workers are collected in cloud storage. If the company cannot handle the data now, the data can serve in the future development of the company.

Risk of Non-Implementation within 5 Years

Businesses that do not invest in the development and retraining of workers are unable to offer the required level of cooperation and are expensive and inefficient. Untrained employees in Industry 4.0 often cannot work with basic data or design software. This discourages and terminates the possibility of collaboration with mature companies and reduces success in competitive development.

3.2.2. Up to 50 Persons, Turnover up to EUR 10 Million

A business of this size requires a great deal of effort in management and communication. At this level, non-modernized businesses are led by one or two key people, and their sudden absence often affects the entire existence of the company. Enterprises have a wide variety of technological equipment and production processes depending on the processing of input commodities (e.g., solid sawn timber and agglomerated board materials). We encountered plants that work with outdated machines or software, as well as companies that install state-of-the-art self-cutting centers, or combinations of both. Companies feel the need to modernize, but they do not have sufficient know-how and do not understand the potential of Industry 4.0. The requirement is directed toward modernization of the equipment, space and technological facilities.

Proposed Solutions

Comprehensive digitization of the preproduction part, i.e., all the data entering production from the design department, will streamline the entire production flow. Companies are advised to create a cyber twin of the product. A digital copy of the product will enable the data transfer and modification and simulation of the individual production phases. The company should consider a complete reconstruction of the software and hardware infrastructure of the enterprise, aiming at the complete interconnection of all sections of the company.

Establishing cooperation with specialized IT companies ensures the operation of the entire company. These specialist companies have tailored software solutions that connect and automate individual departments, reducing repetitive human resource activities. The software in place enables data to be flipped from accounting and design programs and to be seen in a global view by all people and project management.

Respondents suggest the creation of a job position ensuring the development of the company and the training of employees. They further state that where the job position has been created already, there has been an effective shift in the overall development of the company.

Benefits of Implementing Industry 4.0

The company's innovation in digitalization will eliminate repetitive operations that burden workers in all positions of the company. The design department generates all the digital data needed to implement a complex production flow by modeling the so-called "live model", automatically generating production documentation, drawings, material orders, CNC machining programs, parts carrying information about the material, edges, production dimensions, production process and more. The technology can be applied to batch plants, as well as to entities engaged in serial and highly atypical production, which is inherent in a company of this size.

According to the respondents, a change in access to the raw material input into the production flow is essential. Specialization of the production in the processing of raw sawn timber or board material is necessary to make all the processes more efficient. An example is the elimination of single-purpose machines for processing raw sawn timber.

The companies surveyed use partnerships with other companies to replace the in-house technologies for processing solid sawn timber that have been discontinued. These can, for example, supply solid wood semifinished products, thus creating business supply links.

The creation of a development department and a position responsible for development will ensure the effective implementation, training of staff and communication with all entities involved in the development.

Risk of Non-Implementation within 5 Years

Failure to implement these technologies leads to the exploitation of key employees whose potential is not effectively used. In their absence, the company is at risk of paralysis.

In the absence of sophisticated software for data management, personnel management and business administration, there is a risk of information loss and business collapse. Companies that do not set the direction of production will face the high costs of producing and developing two different products. An unresolved spatial layout and the absence of modern technology will limit the company; thus, the enterprise will start to lose the possibility of effective growth.

Companies engaged in atypical and highly atypical production must evaluate their direction and the future development of their order volumes. The digitization of the company and the introduction of innovative technologies in the non-manufacturing departments of these companies will always pay off.

The company should evaluate its production direction. The processing of the commodities of solid sawn timber and agglomerated materials requires additional technological specifics, machinery and spatial layout, while the division of development, production and machinery increases the costs of technology and company facilities. Respondents state: "The machinery originally used on a monthly basis should be replaced by the services of external suppliers, thus creating room for the modernization and innovation of their own facilities."

3.3. Implementation of Industry 4.0 in Medium-Sized Enterprises

In the case of introducing innovations in a medium-sized enterprise, the respondents agreed that the classic joinery workshop is losing its typical form (Table 6). Small woodworking machines are replaced by nesting milling centers, CNC machining centers with automatic part stacking, continuous lines connected by conveyors, continuous painting lines and other equipment. This realistic view of modern joinery production differs from the reality of today's vocational training for the various branches of joinery.

One of the major issues of concern of the respondents is the lack of preparedness of furniture manufacturing graduates from secondary schools and colleges. It often takes a year for them to understand the whole issue of digitalized production and modern technologies, while the employer expects the graduate to be immediately involved in production. This is unrealistic in relation to the classical joinery education, where modern joinery production is separated from classical production by the acquired competences. This practically makes it a separate discipline, which would be better identified in the teaching and when these requirements are included in the teaching.

Respondents also report that innovative technologies, both manufacturing and non-manufacturing, are essential to their business while increasing the efficiency of the overall operation. Specifically, respondents engaged in atypical production report efficiency gains ranging from 30%–40%. In enterprises that use batch production, labor efficiency gains of up to 50% over a 5-year period are observed. Efficiency gains are based primarily on reducing communication flows, errors and repetitive operations at all levels of the enterprise.

Table 6. Main results and recommendations (scored) for medium-sized enterprises and their subcategories (n = 10).

Cat.	Subcat.	Example of a Real Company	Proposed Solutions	Benefits of Implementing Industry 4.0	Risks of Non-Implementation within 5 Years
Medium-sized enterprise (up to 250 persons)	up to 150 persons, turnover up to 25 million EUR (very atypical production)	- Highly operational decision-making; - The management is usually handled by not very sophisticated software or by the classic personal assignment of work; - Long-term planning here is very challenging, highly operational management; - There are companies with advanced technical and technological backgrounds, but also companies with major shortcomings.	- Introduction of complex digitalization at the level of product parts; - Production planning, persons management using advanced complex software; - The life cycle of a project is planned by the software depending on its 3D data; - In production, the human member is always indispensable; - High demands on cloud storage and cyber protection.	- Global sophisticated software providing complex operations, data management, control and planning of all projects and persons in the company's flow; - The design department automatically generates all the documents for the production flow from the cyber twin; - Using QR codes, efficient data collection and record keeping; - Increasing labor productivity.	- The absence of a sophisticated software solution brings threats, data loss, cyber-attacks; - Increased demands on staff; - Constant pressure, stress and high responsibility; - Increased incidence of mental illness among workers; - Reduced fungibility, failure to meet deadlines, threat.
	up to 150 persons, turnover up to 25 million EUR (mass production)	- Multiple repetitive operations; - Long-term planning of production capacities, materials, human resources and more; - Business built on long-term contracts with customers; - Adapted facilities for the production of a given product; - Technical and technological equipment is more advanced than in atypical production companies.	- Digitalization of the entire communication flow management and technical and technological equipment; - Upgrading to the Intelligent Factory; - Shifting human resources from manufacturing to non-manufacturing positions; - Creating intelligent work cells of people, machines and software; - Optimization and automation and robotization of operations.	- Creation of new positions, filling by existing employees; - The management and operation of the company is provided by sophisticated software; - A society more resilient to individual failure; - More efficient production flow, flexible product changes, reduced costs, increased competitiveness;	- Inability to respond to price and quality competition; - Pressure of competition from countries with low production costs or using the attributes of the modern smart factory; - Impossibility of meeting delivery dates and prices; - Human factor failures often threaten the entire production flow of a company.
	up to 250 employees, turnover up to 50 million euros	- Modern high-capacity plants; - Experience with some automation and digitization; - High level of machinery, its constant modernization; - Companies are already addressing Industry 4.0 and its attributes; - Automation, robotization necessary for the operation of production.	- Using all the building blocks of Industry 4.0; - The smart factory and its cyber twin; - Big data collection and evaluation, use of IOT, IOS, HCPS; - Data monitoring and evaluation, for automated decision-making systems; - Created teams of people, software, machines.	- It ensures the running of the business; - Long-term planning and self-management; - Process automation; - Continuity and efficient operations of an intelligent enterprise.	- In the future, it will jeopardize the entire operation of the company and its infrastructure; - Without the use of the innovative technologies offered, it will not be possible to sufficiently and comprehensively manage traffic of this magnitude.

3.3.1. Up to 150 Persons, Turnover up to EUR 25 Million (Very Atypical Production)

Businesses with atypical production face highly operational decision-making choices. Long-term planning is very challenging, and any material shortages or production errors will fundamentally disrupt the production flow and lead times, bringing the need for operational management. The division of labor, task control, project handover and communication across the company is usually handled by not very sophisticated software or by the classic personal allocation of work.

There are companies on the market with an advanced technical and technological background with modern machines, sawmills with chaotic storage systems and CNC nesting centers, but also companies that have major shortcomings in this regard, which reduce their production efficiency.

Proposed Solutions

Implementing the comprehensive digitalization at the product part level, production planning along with people management using an advanced complex software that provides fast real-time information transfer with the ability to view it from anywhere is required.

Stakeholders propose the implementation of sophisticated software to ensure the project life cycle. From the acceptance of the order, the project passes through the company, packs all the data (technical and economic) and is planned by the software itself with respect to its 3D data and the project life cycle, including the possibility of showing the customers themselves in the approval process and involving them in the production cycle.

The stakeholders emphasize: "In a company of this size, engaged in the modernization of enterprises, the human being as a member of production is indispensable. All machinery must be adapted to the needs of the final product and production flow." Here, man acts as a machine operator, an operator of equipment, and performs specific jobs that are difficult for machines to grasp.

High demands are placed on cloud storage and cyber protection, which must not be underestimated.

Benefits of Implementing Industry 4.0

Global sophisticated software providing complex operations, data management and control and planning of all projects and people in the company's flow, including production planning and real-time records of all orders and products, is required.

The design department creates a cyber twin of the product, which is a digital copy of the final product. In the process of data generation, cutting plans, data for CNC centers, electronic manufacturing processes, material ordering, creating a plan of expected production times and schedules of real production are automatically created. Subsequently, performance records of the machine and people, the material consumed and other data valuable to the company are collected in the manufacturing process.

Using QR codes or barcodes, the system is able to carry production documents directly from the design department and display them anywhere in the production or assembly.

There is a reduction in the number of technical and economic workers and an increase in labor productivity.

Risk of Non-Implementation within 5 Years

Without a sophisticated software solution, a company exposes itself to many dangers such as material shortages in production, data loss or cyberattacks.

Missing data records on corporate or external networks place increased demands on the existing communication flow of individual employees.

The management of a company, often by one person, and its possible reduced substitutability or failure to meet dispatch deadlines in the context of production outages and the need for high operational management are a threat in the long term. Constant pressure, stress and high responsibility are more likely to cause psychological illnesses among workers.

3.3.2. Up to 150 Persons, Turnover up to EUR 25 Million (Mass Production)

In companies engaged in batch and repetitive production, the production volume, material requirements, human resources and more can be planned more effectively over the long term. Companies often have long-term product supply contracts with their customers and may also have customized facilities for the production of specific types of products. The costs associated with producing a single part, minimizing waste, and efficient preparation strongly influence the efficiency of the business. In this subcategory of the firm, manufacturing becomes the heart of the enterprise.

The machinery tends to be more advanced than in atypical manufacturing companies. We can find, here, partial or full automation, modern banding centers, continuous paint booths and others.

Proposed Solutions

Multiple repetitive operations in serial production, digitized control of the entire communication flow, together with the modernization of plants and technical and technological equipment, create the ideal basis for a modern automated intelligent factory.

As part of the automation of operations, the need for manpower in production, which is carried out by machining, is decreased, with human resources being transferred to non-production areas. The formation of work teams takes place depending on the production cells. Teams are formed by human resources, machines or software addressing the optimization and automation of the necessary tasks.

Benefits of Implementing Industry 4.0

According to the respondents, innovative companies show an increased demand for skilled human labor in non-manufacturing parts. New jobs are being created so that employees do not have to worry about their jobs; they just have to adapt to the workload, which often becomes less physically demanding. Intelligent work teams in which humans and machines work together are more efficient and flexible, communicating with each other using display devices, and they are able to task each other.

The management and operation of the company is handled by information storage software, making the company more resilient to individual failure. There is the long-term planning of production, materials, logistics, etc. As a result, these innovations bring the possibility of better production flow, eliminating labor shortages and reducing costs.

Risk of Non-Implementation within 5 Years

The respondents point to the company's lack of development, reducing its ability to respond to price and quality competition and threatening the business in the long term. In the future, these enterprises will face more competition from markets with lower production costs, for example from eastern Europe or developing countries.

Delivery dates and product prices are decisive. Without innovation, the company will be severely compromised by more advanced and ready competitors using the attributes of the modern smart factory. There are often repetitive manufacturing operations of hundreds of units that can almost always be replaced by machines in these enterprises. Human failure often threatens the entire production flow of a company.

3.3.3. Up to 250 Employees, Turnover up to EUR 50 Million

Modern advanced production plants in serial or repetitive production that already have experience with some automation and digitalization are necessary. The machinery is at a high level, constantly being renewed and modernized. Companies are already addressing the question of the individual attributes of Industry 4.0. They are using cloud storage, cyber protection, partially or comprehensively collecting digital information, etc.

Companies engaged in mass production also have elements of automation and robotics that are effectively applied and essential to the operation of the company.

Proposed Solutions

Using all the building blocks of Industry 4.0, specifically, we can talk about the implementation of the smart factory and the digital twin of the enterprise for the long-term production and human resource planning.

The company continuously monitors and collects data that serve as the basis for automated decision-making systems.

Benefits of Implementing Industry 4.0

The attribute of full Industry 4.0 integration ensures the operation of the business, long-term planning and self-management. This ensures the creation of intelligent production blocks, creates teams of people, software and machines and maximizes process efficiency.

Example: a customer creates a product in the e-shop in pre-prepared libraries (cyber twin of the product), the design can be consulted by an in-house architect, ordered and then paid for. In the next step, automated processes in the IoS interface, the automatic ordering of materials, the creation of production documentation, production planning, the start of a partially or fully automated production process (all repetitive operations performed by several people are eliminated) all take place. The item passes through a production flow where 3D data is displayed, and the efficient control and planning of the entire production process takes place. In the production flow, big data are collected by the IoT, stored on cloud repositories where they are evaluated and distributed to the machines and equipment in the IoS interface. Thanks to HCPS, machines and people work together to create efficient structures.

Risk of Non-Implementation within 5 Years

Underestimating the importance of the global perspectives offered by Industry 4.0 can jeopardize the entire operation of a company and its infrastructure in the future. The opportunities and innovations that the market offers must be exploited, monitored and continuously integrated into the company's strategy. Otherwise, data, contracts and quality employees may be lost.

Without the use of the innovative technologies offered, it will not be possible to sufficiently and comprehensively manage traffic of this magnitude.

The digitalization and complexity of the project flow from the confirmed order to assembly with the overall display of documentation and data collection of products or individual product parts represents a major shift and streamlining of process management. Corporate stakeholders recommend incorporating software solutions that allow the evaluation and simulation of the flow of orders through the company and incorporate the above aspects. The company that has implemented this measure reports that, in the three years since implementation, production has increased by up to 50%, with an investment of hundreds of thousands of euro. In production, it is about managing people and projects, displaying production data, reading barcodes, collecting data and all the information that is used to evaluate the information that makes it possible to manage individual processes.

3.4. Implementation of Industry 4.0 in Large Entesrprises

Intelligent large-scale furniture manufacturing plants (Table 7) are a harbinger of the future that all industries are heading toward. Respondents agreed that the introduction of innovation increases labor productivity and reduces the need for direct production staff, which is increasingly difficult to find in the market. These enterprises are also subject to the general requirements set for medium-sized enterprises, e.g., the need for a training response. Here, it is possible to fully integrate all the attributes of Industry 4.0 and implement sophisticated solutions as in the previous groups. Enterprises usually have sufficient know-how, employing a specialist to deal with various subsidy titles. The company also has a specialist focused on development and innovation.

Table 7. Main results and recommendations (scored) for large enterprises and their subcategories ($n = 5$).

Cat.	Subcat.	Example of a Real Company	Proposed Solutions	Benefits of Implementing Industry 4.0	Risks of Non-Implementation within 5 Years
Large company (more than 250 employees, turnover more than €50 million)	built by gradual modernization	- Limited by the spatial layout of the businesses; - The mechanical and technological progress here is much worse than that of a "greenfield" company.	-Application of all the building blocks of Industry 4.0; - Application of CE principles; - Waste utilization, energy sources; - Supporting the next generation of the workforce.	- Full automation and digitalization of the company; - Connecting the company, increasing efficiency and competitiveness; - The amount of innovation and efficient layout is influenced by the spatial layout of the enterprise; - Efficient use of waste.	- Underestimation of certain attributes during the design, threatens the efficiency of production, lack of employees, etc.; - Lack of stability in competition; - The threat is lack of staff, high energy costs, large amounts of funds tied up in warehouses, loss of data and more; - Failure to address the energy situation and partial self-sufficiency threatens the business with high costs.
	built on a "green field"	- Enough finance, know-how; - Implementation on a "green field" with minimal layout restrictions; - The attributes of Industry 4.0 can be effectively implemented; - Efficient layout of operations and machines; - Fully automated large capacity plants; - Serial production.		- Construction on the "green field" is not limited by the layout of the production hall; - The most efficient production lines ensuring automatic movement of parts, their machining, assembly and packaging; - Finding new ways to increase productivity.	

3.4.1. Built by Gradual Modernization

Large enterprises, created by the gradual growth of their capabilities and capacities, often take several decades to take shape. Companies of this kind are often limited by the space available in their production facilities. Halls are often renovated, as old buildings are limited by support columns, low ceilings or narrow halls. Mechanical and technological progress is far inferior to that of a greenfield company. Companies of this type have sufficient experience and know-how to innovate and develop the business. The efficiency of the whole company is increased by approximately 50% compared to the original operation.

Proposed Solutions—For Both Subcategories of Large Enterprises

According to the stakeholder, the application of all the building blocks of Industry 4.0 is necessary. The use of circular economy (CE) principles in waste management with the possibility of using a cogeneration unit and waste treatment is required. With the gradually increasing demands for cleaner production, these steps will be necessary in the future. Supporting a new generation of workers by working with secondary and higher education institutions and retraining workers is necessary.

Benefits of Implementing Industry 4.0

The modern automated factory and digital interconnection of the company will bring very significant efficiency gains. It all depends on the range of products, the seriality and the layout of the company's premises. The layouts allow for an efficient arrangement of woodworking machines along with the fully automatic movement within the overall production flow. Minimizing waste and using it efficiently will help reduce energy costs.

Risk of Non-Implementation within 5 Years—For Both Subcategories of Large Enterprises

A company of this size must take all the attributes of Industry 4.0 into account when designing or reconstructing its operations. If it underestimates some of the attributes, it risks failure, inefficient production, a lack of employees, etc. In a competitive struggle, an insufficiently innovative company will be at risk. In the near future, the enterprise may be replaced by a company with a clear vision and goal. Interviewees of large enterprises point to the threats of a lack of skilled workers competent in Industry 4.0, high energy costs, large amounts of funds tied up in warehouses, loss of data and others. The lack of preparation in partial energy self-sufficiency, for example, by waste treatment or solar panels, threatens the enterprise in times of high energy fluctuations.

3.4.2. Built on a "Green Field"

Businesses that have sufficient finance and know-how will build a factory on a "greenfield" site that is not constrained by the layout of already existing buildings and structures. In this variant, all the attributes of Industry 4.0 can be effectively planned, integrated, fulfilled and controlled. The given layout of the operations and machines can be adapted directly to the targeted production. Here, we can talk about fully automated large-scale units engaged in mass production.

Benefits of Implementing Industry 4.0

If a company plans to build a greenfield company, it is not limited by the layout of the production halls. This is where the smart factory concept with all the attributes of Industry 4.0 comes into play. The goal of the investors is to design the most efficient production line, ensuring both the automatic movement of manufactured parts and their machining, surface treatment, assembly and packaging. Companies are constantly looking for new ways to increase productivity. The waste generated can be a major burden, but its recovery in energy production reduces its threat and the costs associated with running a business. Decision-making processes within the CE will bring substantial efficiencies to the entire company.

According to stakeholders of large companies, the amount of investment depends on the level of added technological value and the complexity of the processes. The return on the entire investment should be approximately five years and the investment can range from hundreds of thousands to millions of euro. However, the added value of the entire production flow usually increases by more than 50% within five years in large enterprises compared to non-innovative enterprises.

3.5. Practical Implementation of Industry 4.0 in the Furniture Industry

From the experience of the respondents who are involved in the overall implementation of Industry 4.0, a model of the complete implementation of Industry 4.0 was established on the basis of structured interviews.

Rising costs and competitive pressures require a high degree of automation, intelligence and flexibility. Industry 4.0 is a solution capable of coordinating the flow of information between all departments in a company through technologies and networks that facilitate communication between the process participants (machines, people and devices), making their daily tasks easier and eliminating repetitive processes.

A smart factory (Figure 2) is a data-driven enterprise where intelligent devices can perform computation, communication or precision control. The day-to-day operation of a smart factory relies heavily on the maturity of the software (3D printing, cloud computing platform, manufacturing execution system, virtual reality, smart logistics, etc.) and technology equipment (robotic, self-cutting centers, glue centers, CNC machine tools, etc.).

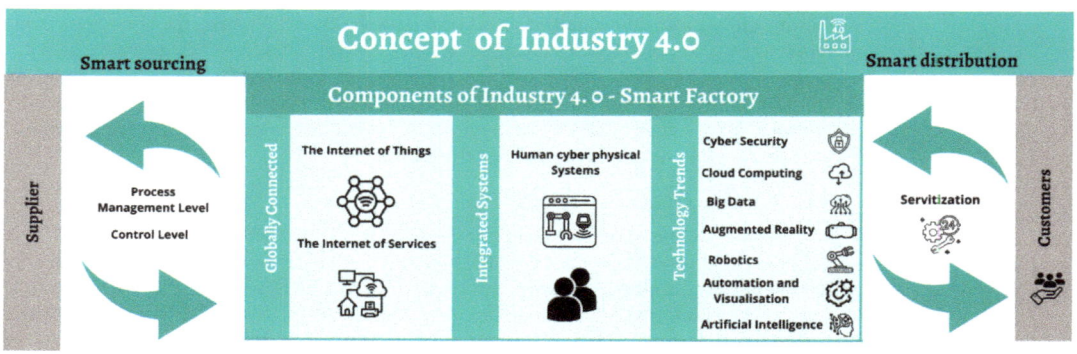

Figure 2. The new concept of Industry 4.0 connecting the smart factory closely with customers and suppliers.

In terms of project management, the entire furniture production process depends on the size of the company and the specifics of the production. The furniture manufacturing company itself should be divided into departments (sections) through which individual projects move toward an efficient goal. The section must have clearly defined rules, structure and job descriptions for each employee, so that each employee knows the scope, goal and extent of their work.

The entire smart grid is divided into two parts. The transfer of information in the company is undertaken by machine-to-machine (M2M) applications that change the dialog between man and machine. The second part is the exchange of information in the factory, which takes place through the Internet of Things. The IoT provides big data collection and system communication with transfer to cloud storage. During the actual collection, on the network or in cloud storage, the data are evaluated through the Internet of Services, which forms a global internet connection between systems and services. Furthermore, decentralized information is distributed on the basis of which the systems can make decisions. Thanks to these technologies, HCPS, driven by computer algorithms from physical sources, provides self-organizing structures and enables interaction between the different participants in the process within artificial intelligence systems. With increasing digitalization, the risk of cyberattacks is also increased, making cybersecurity an integral part of furniture operations.

To organize a smart factory, it is necessary to have an overview of all the materials involved in the production of furniture. It is necessary to know, at all times, what needs and requirements production has and how to satisfy these requirements in the most efficient way. Automatic inventory management together with software classifies all materials (raw materials, semi-finished and finished products) and provides all this information via lists to the responsible personnel, as well as manages the logistics and outputs of the final product. It is also linked to the calculation of purchase requirements and automatically generates orders to suppliers, respecting the production and storage criteria used in the company. Through servitization and trade management, a complete service is provided to customers and suppliers, including the quality control of the incoming material and waste processing. The vision for the future is the indirect involvement of the customer in the production process, e.g., through an intelligent e-shop, display of current production statuses, etc.

Different groups of attributes can be intertwined within the maturity of an enterprise and the following tables indicate the basic distribution of their capabilities. The technologies used should always be one level more advanced than the company's current need. All innovative processes moving toward Industry 4.0 technologies streamline and optimize the production time, reduce any potential errors and allow physically demanding and sometimes dangerous tasks to be transferred from man to machine. A human team member solves a complex problem and an AI team member handles less complex activities or more repetitive operations. This creates a competitive advantage, ensuring the optimal use of the resources, as well as environmental responsibility.

4. Discussion

The results reveal several insights and best practices regarding the effective implementation of Industry 4.0. In this study, we focused on furniture companies that are involved in Industry 4.0 and used guided interviews with project managers to find out the current status and benefits of Industry 4.0.

In the following, the most important similarities are discussed and compared with the current state of research using the person, organization and technology model.

The analysis of the interviewed enterprises and the guided interviews revealed that there is a relatively low level of readiness of SMEs for the specific use of Industry 4.0 compared to large enterprises engaged in mass production. This finding can be explained by the fact that large enterprises have a much higher availability of resources for the use of technology, have the know-how and understand the importance of innovative technologies. In their size and manufacturing focus, they are practically indispensable and have the space to focus on strategically oriented activities. This finding is in line with previous studies dealing with the implementation of Industry 4.0 [6]. According to the findings of the respondents, in the case of a large enterprise (more than 250 employees, turnover more than EUR 50 million), Industry 4.0 is an essential component. If an entity has sufficient funding, know-how and builds a greenfield factory, all the attributes of Industry 4.0 can be met more easily than for a company that is gradually modernizing and is limited by the different spatial layouts of its facilities. Large and medium-sized mass-production companies can apply all the attributes of Industry 4.0 to their portfolio, thereby increasing their production and capacity while reducing their costs. Interviewees point out that "Software solutions, machinery and companies supplying innovative turnkey Industry 4.0 solutions are mainly specialized for large enterprises with mass or repetitive production". This is confirmed by the article analyzing furniture companies. The authors point to the lack of supply chain and hardware and software services for SMEs [36].

One of the many practical outcomes and practical impacts will be to connect SMEs in the implementation of Industry 4.0 solutions, which will bring competitive advantages to these businesses such as lower costs, higher yields and the rational use of green and sustainable renewable resources without the need for their own high investments. The practical experience of the respondents shows that as a business grows from a micro-enterprise disposition to a large enterprise, it undergoes major development, innovation and internal changes.

According to the findings, it is essential that the transformation of operations leads to simple yet flexible industrial robotic cells. A large part of the wood manufacturing industry needs to revamp its production systems and develop new manufacturing technologies or software solutions [18]. An important factor for the success of the application of the individual parts of a smart factory is to start with the limited implementation of individual cells rather than a comprehensive reconstruction. It is important to proceed only when the work process is understood by the workforce, i.e., human resources [27]. Some entities considering the application of robotic and automated lines consider modern technology as a tool to solve most of their problems. Open communication within the company and education are necessary to avoid misunderstandings and possible failures [17,19,27]. This is confirmed by the respondents, who point to the need for strategic planning of the entire implementation with clear objectives for each section of the company. Repeated communication and clarification of the intentions must also take place with employees, for whom it is important to identify with the company's objectives and the implementation of Industry 4.0. They must come to understand that automation and robotics help them to increase efficiency and replace particularly demanding and monotonous work. Through the process of servitization and customer involvement in the production process to address environmental issues, it can either directly generate economic benefits or indirectly, through environmental or operational performance, increase customer satisfaction. This route can also maximize the volume of all downstream production and supply processes [44,45]. Unfortunately, the very meaning of servitization is not properly understood by all Industry 4.0 companies. Even if they perform certain services in this regard (e.g., design department, e-shop, etc.), there is no targeted integration and value creation [32].

Stakeholders interviewed unanimously point to two of the most significant barriers to the implementation of Industry 4.0 in the wood-based materials processing sector (furniture manufacturing). The first factor is small-scale production, which requires very flexible mechanical cells in the manufacturing plant itself. This is accompanied by high programming requirements reducing profitability and efficiency. Second, the machining of solid wood requires a very sophisticated approach, as machining parameters must be constantly adapted to the characteristics and processes of the wood being processed. The processing of raw timber requires the use of single-purpose machines (e.g., dimensioning saws, trimming saws, horizontal milling machines and presses), which use up space and staff capacity. The large amount of waste, energy and human labor consumption thus becomes a burden not only for the company itself but also for the environment [46,47].

The next most frequently cited barrier to respondents taking any action was the problem of investing time and money in development. This also applies to other sectors such as forestry or textiles [48–50]. Before deciding to invest in Industry 4.0, business owners should see examples that these practices are profitable and provide many other benefits. Current business representatives consider innovative solutions to be difficult to manage financially and unprofitable. Implementing an innovative solution is undoubtedly more difficult than continuing with the traditional way of working. However, the implementation of Industry 4.0 can be adapted to the capabilities and needs of a given company.

According to the respondents, gaps in the business supply chains on both the buyer and supplier side are central to the implementation of Industry 4.0, with the biggest problem being the introduction and co-operation with various entities that do not use modern technologies and sharing and communication with these partner entities is difficult. Nowadays, the concepts and tools of business management are rapidly expanding. Top management is, therefore, focused on implementing the entire Industry 4.0, including the business and supply chain [35]. A well-integrated and -managed supply chain is considered a powerful strategic and logistical "weapon" that is difficult to imitate and provides a long-term competitive advantage [38,39].

All of the managers interviewed in the case of large and medium-sized enterprises see a general problem in the fact that service providers offering a combination of hardware and software are mostly specialized for large companies. There is practically no company that offers a small enterprise services in a comprehensive solution of technical and technological equipment and sales of certain "know how" addressing the complete services of a production and non-production nature. Here, we come up against a fundamental fact, which is the lack of information about possible solutions in the whole furniture manufacturing sector. Another problem is the incompatibility between the machines providing the different manufacturing operations and the software controlling or managing them, and the intercommunication between different software machines from different manufacturers. There is a great opportunity for companies to offer business analysis and implement a solution that mediates communication between the warehouse, design software, human resource management software and accounting software and across other operations of the company [7]. The aim is to streamline the communication flow, display the necessary data and eliminate repetitive operations applicable in wood-processing companies, which is also confirmed by Jasinska [51]. The interviewed enterprises engaged in development often employ a specialist who programs the product. Many companies still do not see the potential that lies, for example, in cooperation with specialist outsourcing entities [27]. It is important to note that, for some processes, it is more efficient to use the services of specialized IT entities than to carry out the process with in-house IT staff [52], such as data management and cyber protection, software interconnection, etc. When implementing, it is advisable to proceed in parts; it is recommended to complete one part first and then work on the next part.

One of the main principles of Industry 4.0 is the creation of operational production plans and a reduction in the physical inventories. Given the current situation and the status of the supply chain, this is now very problematic and a revival would be advisable. Across the industry, supplies of materials needed for production are delayed or stopped altogether as a result of the COVID-19 pandemic [53,54], the Russian–Ukrainian war, high inflation and across-the-board price increases for all products and services. The absence of a wide range of materials, fittings and electronics supplied from Asian countries is a current issue, where transport times play a major role. This problem highlights the unpreparedness of society and industry in the broader spectrum of all the supply and production flows, forcing companies to build up stocks of materials and components in which large amounts of money are tied up. With the gradual increase in transport prices along with the green deal, producers of scarce goods produced outside the EU will be forced to move production to European countries as well. By reducing transport times and ensuring the self-sufficiency of European countries, more rapid deliveries would also be achieved. At the same time, however, there are specific differences in the conditions of individual countries in the furniture industry [55].

Furniture companies should use the opportunity associated with green energy promotion to improve furniture waste recycling mechanisms to minimize energy consumption [21]. One of the main factors that causes resource overload is the global system based on linear flow of materials and energy, which causes the depletion of natural resources and the generation of large amounts of waste [56]. The appropriate solution is a circular model based on a circular economy (CE), which allows for sustainable development. Another option is to use waste as a source of energy in the company, for example, for cogeneration units producing electricity that will then be used in production, which will significantly reduce the production costs of the company and can have a significant effect, especially in the current energy crisis. This is confirmed by some companies that are already producing electricity from their residue.

Stakeholders cite staff shortages, both in production and technical management positions, as the most common risk of non-implementation. In these cases, the enterprise is often managed by one person and its possible reduced substitutability or failure to meet shipment deadlines in the case of a production failure and the need for high operational management are threats in the long term. This risk is most evident in medium-sized and large enterprises, where Industry 4.0 technologies make production more efficient, and the number of technical and economic staff is reduced. All four industrial revolutions have had a major impact on work in terms of education. According to the interviewees, there is currently a large deficit of competent and skilled workers (workforce) in the European labor market with experience in the field and competence in innovative technologies, as confirmed by a study from Spain [57,58]. Along with the requirement to retrain employees, the workforce needs to meet sophisticated production conditions and acquire soft skills [59,60]. The automation and industrialization of plants will also have an impact on human resources in the sense of Human Resources 4.0. In the long term, jobs lost will be replaced by jobs that meet the needs of the future market [10,61]. An aspect that raises concerns about the implementation of Industry 4.0 is the current education system setup. With the advent of digitization and robotics, production processes as such are changing fundamentally and the education system should respond immediately [62]. Outdated joinery plants have nothing in common with modern production. School graduates are usually fresh out of apprenticeship and do not master digitized processes, which significantly hampers the development of companies. Respondents strongly perceive this problem and cite cooperation between schools and technically advanced companies as a solution. Here, pupils would undergo an apprenticeship that would help raise awareness of a wide range of opportunities, for example, as in the automotive industry. Some of the enterprises interviewed are already implementing this concept and are positive that it is of great importance for the students' education. This subsequently applies to the enterprises for jobs as graduates of the schools involved.

5. Conclusions

The new innovative technologies of Industry 4.0 are key to the future development of the furniture sector. Methodical support for managers who are the key users of these technologies is essential in applying the new Industry 4.0 trends to the furniture sector. In this study, we were focused on furniture companies that are involved in Industry 4.0 and used guided interviews with project managers to find out the current status and benefits of Industry 4.0. The study proposed a methodological framework in each of the key areas for implementing Industry 4.0 in companies, broken down by the company size. Furthermore, recommendations for future research in this sector emerged from this study. In particular, the focus was on new trends in furniture manufacturing and the effective implementation of Industry 4.0 within different company size categories According to the findings of the respondents, it is necessary to involve small companies in the process of implementing Industry 4.0 in this sector in addition to large companies where Industry 4.0 is commonplace, as they represent a significant potential for implementing Industry 4.0, especially in terms of supply chains. The different approaches to the implementation and technologies of Industry 4.0 need to be simplified, especially for SMEs. They are mainly limited by a lack of financial resources, knowledge and business organization. Therefore, it is necessary to use only partial attributes that small enterprises can realistically use. In conclusion, all the interviewed managers agreed that innovative technologies of production and non-production types are essential for their companies, where their applications have undergone rapid shifts, increasing the efficiency of the entire operation in the range of 30%–50%, reducing the communication flow, error rate and repetitive operations at all levels of the enterprise and ensuring the efficient use of renewable resources in line with the Sustainable Development Goals (SDGs).

Further follow-up research can help to clarify other aspects of Industry 4.0 implementation. Using different samples of enterprises with respect to, for example, the nationality of the wood and forest industry sector, can elucidate any nuances that exist. In addition to this study, a survey among different stakeholder groups, especially among supply chain representatives of enterprises, will also be conducted. It is also appropriate to focus the Industry 4.0 communication strategy on supply chains and SMEs, which represent a significant potential for the success and development of Industry 4.0 in the furniture sector.

To summarize the results of this study, it can serve as a basis for addressing strategic decision-making in project management in the application of Industry 4.0 in enterprises; it can also be successfully applied in other sectors as the principles for each category of enterprise will be very similar. To succeed in the competition in the long term, enterprises must continue to evolve.

Author Contributions: Conceptualization, L.Č., R.S., T.Č.; methodology, R.S.; validation, L.Č., T.Č.; formal analysis, L.Č., R.S.; investigation, T.Č.; sources, L.Č.; writing—drafting, L.Č., R.S., T.Č.; writing—revision and editing, R.S., T.Č.; visualization, T.Č.; supervision, R.S.; project administration, R.S.; fundraising, R.S. All authors have read and agreed to the published version of the manuscript.

Funding: This research was funded by the non-project research funds of the Faculty of Forestry and Wood Technology of the Czech University of Life Sciences, Prague.

Acknowledgments: The authors wish to thank all the stakeholders who took part in this research and made it possible. The authors also thank Harvey Cook for proofreading and editing the article.

Conflicts of Interest: The authors declare no conflict of interest.

Abbreviations

AI	Artificial intelligence
AR	Augmented reality
HCPS	Human–cyber–physical system
IoT	Internet of Things
IoS	Internet of Services
CE	Circular economy
CNC	Computer numerical control
EU	European Union
SMEs	Small and medium enterprises
SDGs	Sustainable Development Goals
NACE	Nomenclature statistique des activités économiques dans la Communauté européenne

References

1. Gordeeva, E.; Weber, N.; Wolfslehner, B. The New EU Forest Strategy for 2030—An Analysis of Major Interests. *Forests* **2022**, *13*, 1503. [CrossRef]
2. 2Rana, A.; Rawat, A.S.; Afifi, A.; Singh, R.; Rashid, M.; Gehlot, A.; Akram, S.V.; Alshamrani, S.S. A Long-Range Internet of Things-Based Advanced Vehicle Pollution Monitoring System with Node Authentication and Blockchain. *Appl. Sci.* **2022**, *12*, 7547. [CrossRef]
3. Michal, J.; Březina, D.; Šafařík, D.; Babuka, R. Sustainable Development Model of Performance of Woodworking Enterprises in the Czech Republic. *Forests* **2021**, *12*, 672. [CrossRef]
4. Rametsteiner, E.; Weiss, G. Assessing Policies from a Systems Perspecitve—Experiences with Applied Innovation Systems Analysis and Implications for Policy Evaluation. *For. Policy Econ.* **2006**, *8*, 564–576. [CrossRef]
5. Pudivítrová, L.; Jarský, V. Inovační Aktivity v Lesním Hospodářství České Republiky. *Zprávy Lesn. Výzkumu* **2011**, *56*, 320–328.
6. Erol, S.; Schumacher, A.; Sihn, W. Strategic Guidance towards Industry 4.0—A Three-Stage Process Model. In Proceedings of the International Conference on Competitive Manufacturing, Stellenbosch, South Africa, 27–29 January 2016; ResearchGate: Stellenbosch, South Africa, 2016.
7. Brettel, M.; Friederichsen, N.; Keller, M.; Rosenberg, M. How Virtualization, Decentralization and Network Building Change the Manufacturing Landscape: An Industry 4.0 Perspective. *Int. J. Inf. Commun. Eng.* **2014**, *8*, 37–44. [CrossRef]
8. Wang, S.; Wan, J.; Li, D.; Zhang, C. Implementing Smart Factory of Industrie 4.0: An Outlook. *Int. J. Distrib. Sens. Netw.* **2016**, *12*, 1–10. [CrossRef]
9. Wang, Y.; Ma, H.S.; Yang, J.H.; Wang, K.S. Industry 4.0: A Way from Mass Customization to Mass Personalization Production. *Adv. Manuf.* **2017**, *5*, 311–320. [CrossRef]
10. Azman, N.A.; Ahmad, N. Technological Capability in Industry 4.0: A Literature Review for Small and Medium Manufacturers Challenges. *J. Crit. Rev.* **2020**, *7*, 1429–1438. [CrossRef]
11. Posada, J.; Toro, C.; Barandiaran, I.; Oyarzun, D.; Stricker, D.; De Amicis, R.; Pinto, E.B.; Eisert, P.; Döllner, J.; Vallarino, I. Visual Computing as a Key Enabling Technology for Industrie 4.0 and Industrial Internet. *IEEE Comput. Graph. Appl.* **2015**, *35*, 26–40. [CrossRef]
12. Sivathanu, B.; Pillai, R. Smart HR 4.0—How Industry 4.0 Is Disrupting HR. *Hum. Resour. Manag. Int. Dig.* **2018**, *26*, 7–11. [CrossRef]
13. Seymour, T.; Hussein, S. The History Of Project Management. *Int. J. Manag. Inf. Syst.* **2014**, *18*, 233–240. [CrossRef]
14. Shehadeh, M.A.; Schroeder, S.; Richert, A.; Jeschke, S. Hybrid Teams of Industry 4.0: A Work Place Considering Robots as Key Players. In Proceedings of the 2017 IEEE International Conference on Systems, Man, and Cybernetics, SMC 2017, Banff, AB, Canada, 5–8 October 2017; IEEE: Banff, AB, Canada, 30 October 2017; pp. 1208–1213.
15. Strandhagen, J.W.; Alfnes, E.; Strandhagen, J.O.; Vallandingham, L.R. The Fit of Industry 4.0 Applications in Manufacturing Logistics: A Multiple Case Study. *Adv. Manuf.* **2017**, *5*, 344–358. [CrossRef]
16. Stock, T.; Seliger, G. Opportunities of Sustainable Manufacturing in Industry 4.0. *Procedia CIRP* **2016**, *40*, 536–541. [CrossRef]
17. Veile, J.W.; Kiel, D.; Müller, J.M.; Voigt, K.-I. Lessons Learned from Industry 4.0 Implementation in the German Manufacturing Industry. *J. Manuf. Technol. Manag.* **2019**, *31*, 977–997. [CrossRef]
18. Loučanová, E.; Paluš, H.; Dzian, M. A Course of Innovations in Wood Processing Industry within the Forestry-Wood Chain in Slovakia: A Q Methodology Study to Identify Future Orientation in the Sector. *Forests* **2017**, *8*, 210. [CrossRef]
19. Zhou, Y.; Xu, W.; Pan, Y.; Wang, F.; Hu, X.; Lu, Y.; Jiang, M. Deep Eutectic-like Solvents: Promising Green Media for Biomass Treatment and Preparation of Nanomaterials. *BioResources* **2022**, *17*, 5485–5509. [CrossRef]
20. Sellitto, M.A.; Camfield, C.G.; Buzuku, S. Green Innovation and Competitive Advantages in a Furniture Industrial Cluster: A Survey and Structural Model. *Sustain. Prod. Consum.* **2020**, *23*, 94–104. [CrossRef]
21. Zhu, J.; Niu, J. Green Material Characteristics Applied to Office Desk Furniture. *BioResources* **2022**, *17*, 2228–2242. [CrossRef]
22. Bressanelli, G.; Perona, M.; Saccani, N. Challenges in Supply Chain Redesign for the Circular Economy: A Literature Review and a Multiple Case Study. *Int. J. Prod. Res.* **2019**, *57*, 7395–7422. [CrossRef]

23. Santos, C.; Mehrsai, A.; Barros, A.C.; Araújo, M.; Ares, E. Towards Industry 4.0: An Overview of European Strategic Roadmaps. *Procedia Manuf.* **2017**, *13*, 972–979. [CrossRef]
24. Schumacher, A.; Erol, S.; Sihn, W. A Maturity Model for Assessing Industry 4.0 Readiness and Maturity of Manufacturing Enterprises. *Procedia CIRP* **2016**, *52*, 161–166. [CrossRef]
25. Erol, S.; Jäger, A.; Hold, P.; Ott, K.; Sihn, W. Tangible Industry 4.0: A Scenario-Based Approach to Learning for the Future of Production. *Procedia CIRP* **2016**, *54*, 13–18. [CrossRef]
26. Silvius, G.; Ismayilova, A.; Sales-Vivó, V.; Costi, M. Exploring Barriers for Circularity in the EU Furniture Industry. *Sustainability* **2021**, *13*, 11072. [CrossRef]
27. Landscheidt, S.; Kans, M. Method for Assessing the Total Cost of Ownership of Industrial Robots. *Procedia CIRP* **2016**, *57*, 746–751. [CrossRef]
28. Castelo-Branco, I.; Cruz-Jesus, F.; Oliveira, T. Assessing Industry 4.0 Readiness in Manufacturing: Evidence for the European Union. *Comput. Ind.* **2019**, *107*, 22–32. [CrossRef]
29. Mabkhot, M.M.; Al-Ahmari, A.M.; Salah, B.; Alkhalefah, H. Requirements of the Smart Factory System: A Survey and Perspective. *Machines* **2018**, *6*, 23. [CrossRef]
30. Wiśniewska-Sałek, A. Sustainable Development in Accordance With the Concept of Industry 4.0 on the Example of the Furniture Industry. *MATEC Web Conf.* **2018**, *8*, 37–44. [CrossRef]
31. Akbari, M.; Hopkins, J.L. Digital Technologies as Enablers of Supply Chain Sustainability in an Emerging Economy. *Oper. Manag. Res.* **2022**, *15*, 689–710. [CrossRef]
32. Huxtable, J.; Schaefer, D. On Servitization of the Manufacturing Industry in the UK. *Procedia CIRP* **2016**, *52*, 46–51. [CrossRef]
33. Saniuk, S.; Grabowska, S.; Straka, M. Identification of Social and Economic Expectations: Contextual Reasons for the Transformation Process of Industry 4.0 into the Industry 5.0 Concept. *Sustainability* **2022**, *14*, 1391. [CrossRef]
34. Zhou, J.; Zhou, Y.; Wang, B.; Zang, J. Human–Cyber-Physical Systems (HCPSs) in the Context of New-Generation Intelligent Manufacturing. *Engineering* **2019**, *5*, 624–636. [CrossRef]
35. Dhiaf, M.M.; Atayah, O.F.; Nasrallah, N.; Frederico, G.F. Thirteen Years of Operations Management Research (OMR) Journal: A Bibliometric Analysis and Future Research Directions. *Oper. Manag. Res.* **2021**, *14*, 235–255. [CrossRef]
36. Červený, L.; Sloup, R.; Červená, T.; Riedl, M.; Palátová, P. Industry 4.0 as an Opportunity and Challenge for the Furniture Industry—A Case Study. *Sustainability* **2022**, *14*, 13325. [CrossRef]
37. Flynn, J.; Dance, S.; Schaefer, D. Industry 4.0 and Its Potential Impact on Employment Demographics in the UK. *Adv. Transdiscipl. Eng.* **2017**, *6*, 239–244. [CrossRef]
38. Swain, M.; Zimon, D.; Singh, R.; Hashmi, M.F.; Rashid, M.; Hakak, S. LoRa-LBO: An Experimental Analysis of LoRa Link Budget Optimization in Custom Build IoT Test Bed for Agriculture 4.0. *Agronomy* **2021**, *11*, 820. [CrossRef]
39. Baghizadeh, K.; Zimon, D.; Jum'a, L. Modeling and Optimization Sustainable Forest Supply Chain Considering Discount in Transportation System and Supplier Selection under Uncertainty. *Forests* **2021**, *12*, 964. [CrossRef]
40. Mirabella, N.; Castellani, V.; Sala, S. LCA for Assessing Environmental Benefit of Eco-Design Strategies and Forest Wood Short Supply Chain: A Furniture Case Study. *Int. J. Life Cycle Assess.* **2014**, *19*, 1536–1550. [CrossRef]
41. Fu, R.; Qiang, Q.; Ke, K.; Huang, Z. Closed-Loop Supply Chain Network with Interaction of Forward and Reverse Logistics. *Sustain. Prod. Consum.* **2021**, *27*, 737–752. [CrossRef]
42. EUR-Lex—32003H0361—EN Commission Recommendation of 6 May 2003 Concerning the Definition of Micro, Small and Medium-Sized Enterprises. Available online: http://data.europa.eu/eli/reco/2003/361/oj (accessed on 18 October 2022).
43. Government of CR Resolution of the Government of the Czech Republic. Available online: https://www.vlada.cz/assets/media-centrum/aktualne/volny-pohyb-1375.pdf (accessed on 5 December 2022).
44. Azevedo, S.G.; Carvalho, H.; Cruz Machado, V. The Influence of Green Practices on Supply Chain Performance: A Case Study Approach. *Transp. Res. Part E Logist. Transp. Rev.* **2011**, *47*, 850–871. [CrossRef]
45. Zhu, Q.; Sarkis, J.; Lai, K. Institutional-Based Antecedents and Performance Outcomes of Internal and External Green Supply Chain Management Practices. *J. Purch. Supply Manag.* **2013**, *19*, 106–117. [CrossRef]
46. Lopes de Sousa Jabbour, A.B.; Jabbour, C.J.C.; Godinho Filho, M.; Roubaud, D. Industry 4.0 and the Circular Economy: A Proposed Research Agenda and Original Roadmap for Sustainable Operations. *Ann. Oper. Res.* **2018**, *270*, 273–286. [CrossRef]
47. Zhu, J.; Wang, X. Research on Enabling Technologies and Development Path of Intelligent Manufacturing of Wooden Furniture. *J. For. Eng.* **2021**, *6*, 177–183. [CrossRef]
48. Ahmad, S.; Miskon, S.; Alabdan, R.; Tlili, I. Towards Sustainable Textile and Apparel Industry: Exploring the Role of Business Intelligence Systems in the Era of Industry 4.0. *Sustainability* **2020**, *12*, 2632. [CrossRef]
49. Feng, Y.; Audy, J.-F. Forestry 4.0: A Framework for the Forest Supply Chain toward Industry 4.0. *Gestão Produção* **2020**, *27*, e5677. [CrossRef]
50. Müller, F.; Jaeger, D.; Hanewinkel, M. Digitization in Wood Supply—A Review on How Industry 4.0 Will Change the Forest Value Chain. *Comput. Electron. Agric.* **2019**, *162*, 206–218. [CrossRef]
51. Jasińska, K.; Szala, Ł. Potential Directions for Improving Production Processes in Industry 4.0 Conditions Based on a Polish Furniture Enterprise—A Case Study. *Inform. Ekon.* **2021**, *2021*, 27–46. [CrossRef]

52. Renda, A.; Zavatta, R.; Tracogna, A.; Tomasell, A.R.; Busse, M.; Wieczorkiewicz, J.; Mustilli, F.; Simonelli, F.; Luchetta, G.; Pelkmans, J.; et al. The EU Furniture Market Situation and a Possible Furniture Products Initiative. Available online: https://www.ceps.eu/ceps-publications/eu-furniture-market-situation-and-possible-furniture-products-initiative/ (accessed on 8 August 2022).
53. Dongfang, W.; Ponce, P.; Yu, Z.; Ponce, K.; Tanveer, M. The Future of Industry 4.0 and the Circular Economy in Chinese Supply Chain: In the Era of Post-COVID-19 Pandemic. *Oper. Manag. Res.* **2022**, *15*, 342–356. [CrossRef]
54. Yu, Z.; Razzaq, A.; Rehman, A.; Shah, A.; Jameel, K.; Mor, R.S. Disruption in Global Supply Chain and Socio-Economic Shocks: A Lesson from COVID-19 for Sustainable Production and Consumption. *Oper. Manag. Res.* **2022**, *15*, 233–248. [CrossRef]
55. Florio, M.; Peracchi, F.; Sckokai, P. Market Organization and Propagation of Shocks: The Furniture Industry in Germany and Italy. *Small Bus. Econ.* **1998**, *11*, 169–182. [CrossRef]
56. Masi, D.; Kumar, V.; Garza-Reyes, J.A.; Godsell, J. Towards a More Circular Economy: Exploring the Awareness, Practices, and Barriers from a Focal Firm Perspective. *Prod. Plan. Control* **2018**, *29*, 539–550. [CrossRef]
57. Romero Gázquez, J.L.; Bueno Delgado, M.V.; Ortega Gras, J.J.; Garrido Lova, J.; Gómez Gómez, M.V.; Zbiec, M. Lack of Skills, Knowledge and Competences in Higher Education about Industry 4.0 in the Manufacturing Sector. *Rev. Iberoam. Educ. A Distancia* **2020**, *24*, 285–313. [CrossRef]
58. MPSV Kompetence 4.0—Mapování Budoucích Kompetencí Jako Součást Systémových Opatření pro Vymezení Požadavků Trhu Práce. Available online: https://www.mpsv.cz/kompetence (accessed on 3 December 2022).
59. Pinzone, M.; Fantini, P.; Perini, S.; Garavaglia, S.; Taisch, M.; Miragliotta, G. Jobs and Skills in Industry 4.0: An Exploratory Research. In Proceedings of the Advances in Production Management Systems. The Path to Intelligent, Collaborative and Sustainable Manufacturing, Hamburg, Germany, 3–7 September 2017; IFIP: Hamburg, Germany, 2017; pp. 282–288.
60. Cotet, G.B.; Balgiu, B.A.; Zaleschi, V. Assessment Procedure for the Soft Skills Requested by Industry 4.0. In *MATEC Web of Conferences*; Bondrea, I., Simion, C., Inţă, M., Eds.; EDP Sciences: Bucharest, Romania, 9 August 2017; Volume 121, pp. 1–8.
61. Maisiri, W.; Darwish, H.; van Dyk, L. An Investigation of Industry 4.0 Skills Requirements. *S. Afr. J. Ind. Eng.* **2019**, *30*, 90–105. [CrossRef]
62. Dosi, G.; Moschella, D.; Pugliese, E.; Tamagni, F. Productivity, Market Selection, and Corporate Growth: Comparative Evidence across US and Europe. *Small Bus. Econ.* **2015**, *45*, 643–672. [CrossRef]

Article

Investigation on *Phoenix dactylifera*/*Calotropis procera* Fibre-Reinforced Epoxy Hybrid Composites

Mohammad Hassan Mazaherifar [1], Hamid Zarea Hosseinabadi [1], Camelia Coșereanu [2,*], Camelia Cerbu [3,*], Maria Cristina Timar [2] and Sergiu Valeriu Georgescu [2]

[1] Department of Wood and Paper Science and Technology, Faculty of Natural Resources, University of Tehran, Karaj 31585-4314, Iran
[2] Faculty of Furniture Design and Wood Engineering, Transilvania University of Brasov, B-dul Eroilor nr. 29, 500036 Brasov, Romania
[3] Department of Mechanical Engineering, Faculty of Mechanical Engineering, Transilvania University of Brasov, B-dul Eroilor nr. 29, 500036 Brasov, Romania
* Correspondence: cboieriu@unitbv.ro (C.C.); cerbu@unitbv.ro (C.C.)

Abstract: This paper presents the investigations conducted on three types of fibre-reinforced epoxy-resin hybrid composites with different structures, manufactured using midrib long fibres of date palm (*Phoenix dactylifera* L.) and *Calotropis procera* fibres. The two types of fibres were formed into flat sheets, without adding other chemicals or resins, and employed as reinforcing layers in the structure of the multi-layered laminate composites. Three-layer and five-layer epoxy-reinforced laminates were manufactured from the sheets of date-palm fibres and *Calotropis* sheets bonded with laminar epoxy resin. Water resistance investigation and mechanical testing under tensile, bending and impact loads were conducted in the research in order to evaluate and compare the performance of the resulting composites. Emphasis was put on the effect of various factors, such as the type of reinforcement material and the number of plies in the laminate on the mechanical behavior of the composites. The interpretation of those results was supported by the stereo-microscopic investigation of the adhesion between the layers of the composites, and the vertical density profile (VDP), which showed the repartition of the density on the composite thickness depending on the layer material. The results of the mechanical performance of the composites showed lower values of tensile strength, tensile modulus of elasticity and impact resistance and an increase of water absorption (WA) and thickness swelling (TS) for the five-layer composites compared to the three-layer composites. Contrarily, the addition of *Calotropis* fibres improved the flexural strength and the flexural modulus of elasticity. The alkali treatment of the *Calotropis* fibres improved the mechanical performance of the composites compared to the ones made with untreated fibres, because of an apparent increase in cellulose content and free hydroxyl groups revealed by FTIR spectra.

Keywords: date palm; long fibres; *Calotropis* fibres; epoxy resin; laminate composites; mechanical testing

Citation: Mazaherifar, M.H.; Hosseinabadi, H.Z.; Coșereanu, C.; Cerbu, C.; Timar, M.C.; Georgescu, S.V. Investigation on *Phoenix dactylifera*/*Calotropis procera* Fibre-Reinforced Epoxy Hybrid Composites. *Forests* 2022, 13, 2098. https://doi.org/10.3390/f13122098

Academic Editor: Petar Antov

Received: 8 November 2022
Accepted: 5 December 2022
Published: 8 December 2022

Publisher's Note: MDPI stays neutral with regard to jurisdictional claims in published maps and institutional affiliations.

Copyright: © 2022 by the authors. Licensee MDPI, Basel, Switzerland. This article is an open access article distributed under the terms and conditions of the Creative Commons Attribution (CC BY) license (https://creativecommons.org/licenses/by/4.0/).

1. Introduction

Composite materials are used as alternatives to conventional materials in industries such as automotive, aerospace and buildings because of their improved characteristics including higher mechanical strengths and reduced specific weight. The increasing demand for sustainable and renewable materials brought to the attention of specialists the possibility of using natural lingo-cellulosic fibres in several applications, correlated to their light weight and high strength. On the other hand, the employment of various natural lingo-cellulosic fibres instead of wood fibres reduces the forest trees' exploitation rate. One of the renewable resources with high potential for use is date palm (*Phoenix dactylifera* L.), which is widely cultivated in the Middle East and North Africa for its fruit crops, covering important areas in these regions [1,2]. Saudi Arabia, Algeria, Iran, Iraq, and Egypt have the highest palm ranks in the world [3] and the area under cultivation indicates a percentage of 21% of the world's date-palm groves belonging to Iran. The world's total number of date palms is more than 120 million [4]. Approximately 11.8 million hectares are under cultivation for date-palm trees, distributed in 94 countries. Overall, the date-palm waste, generated from seasonal pruning and trimming, can be estimated at an average of 35 kg per tree [5]. Considering the 120 million date-palm trees, and 35 kg of waste per tree, 4,200,000 tons of natural fibres can be consumed in various fields, a fact which helps to prevent cutting forest trees around the world. Consequently, these residues, which are renewable, have good potential to be used in the composite production industry [6], such as for the reinforcement of polymer composites for automotive or maritime industries, for construction as geotextiles, and the reinforcement of asphalt concrete or gypsum plaster [2,7].

Long fibres extracted from the midrib of the date palm and spanned into yarns and further alkaline treated have physical, chemical, morphological and mechanical properties comparable to those of other natural fibres, such as sisal, hemp, and flax [7]. A tensile strength of date-palm midribs of 11.4 Mpa, higher than for bamboo or sisal fibres, was reported in the literature [8]. Date-palm fibres can be obtained from annual pruning by-products such as spadix stems, midribs, and leaflets [9], but the longest ones can be obtained only from midribs, which contain them in their natural matrix [7].

Calotropis procera is a small perennial tree of the *Apocynaceae* family and it is native to Africa, the Arabian Peninsula, Western Asia, the Indian subcontinent and India, and was also introduced in South Africa, Australia, Latin America and the United States because of its economic benefits [10–14]. Commonly known as the giant milkweed, apple of sodom, or calotrope [13], this plant grows slowly, and it is drought-resistant and evergreen with softwood, having thick branches that may grow up to 6 m. The fruits split at maturity to release numerous seeds, around 350–500 seeds per fruit [15], with bundles of white silk or Papus fibres having several applications in the chemical industry, building industry and medicine [16–18]. Apart from these applications, Papus fibre is a natural and renewable material, composed of 64.0 wt% of cellulose, 19.5 wt% hemicellulose, and 9.7 wt% lignin, being a source of ligno-cellulosic fibres with low density and high strength [13,19]. These fibres are lightweight hollow tubes [20–22] with thin walls and low density, but high strength and hydrophobic properties, being very suitable not only as insulation materials but also as natural fibres for reinforced composites [20]. An alkaline treatment applied to the *Calotropis* fibres extracted from the stem can improve the mechanical properties when they are used in reinforced epoxy polymer composites. This was explained as a result of the increase of the cellulose content and decrease of the fibre density following alkaline treatment, resulting in effective bonding at the fibre–matrix interfaces [23].

As reinforcements in composites, fibres have to have high strength, high stiffness, and low density, whilst the matrix requires good shear properties. Carbon fibres, glass fibres, and aramid fulfill these requirements and are preferred as reinforcement materials of the advanced composites [24–26], and they are being used in the automotive industry for automobile bodies, civil engineering applications for strengthening walls, or in maritime applications for ship hulls [27]. Natural fibres are studied by many researchers as alternative materials for synthetic fibres in the industry of composite materials, due to their low cost and lightweight properties, bringing important advantages when used in polymer matrix composites. A study on 2D woven kenaf fibre-reinforced acrylonitrile-butadiene-styrene (ABS) has shown that the alkali treatment of the kenaf woven fabric led to the increase of the adhesion between fibres and ABS, with a beneficial effect on the tensile strength of the composite [28]. Generally, the drawback of the natural fibres is their low resistance to water, which is why it is preferred to use them for applications in the indoor environment. In this context, date-palm fibres represent an agricultural waste suitable as a reinforcement for polymeric composites used in the automotive industry as an interior component [29]. Alkali treatment applied to natural fibres increased the compatibility between the fibres, and polymers in fibre-reinforced polymer composites, improving the mechanical performance of the resultant structures [30–32]. This effect was explained by wax removing and the fact that the fibres tend to be densely packed due to the removal of hemicellulose [32].

In the present study, three types of multi-layered laminate composites with different structures were developed and manufactured, employing long date-palm (*Phoenix dactylifera* L.) fibres extracted from the midribs and *Calotropis* white silk fibres as reinforcing materials, and epoxy resin as the matrix. The envisaged applications of these composites are in the automotive, aerospace and construction industries, as potential alternatives to the synthetic fibre-reinforced composites, which are more expensive and require complex manufacturing technologies. The advantage of using these natural fibres is the availability of a large amount of waste, their low cost and low CO_2 footprint. The research conducted on the developed composites focused on the effect of some factors, such as the type of reinforcing material, the alkali treatment of *Calotropis* silk, and the number of plies in the laminate on the mechanical properties of the composites. For a more comprehensive study, a stereo-microscopic analysis of the fibres and of the adhesion between layers, an investigation of the vertical density profile (VDP), and an FTIR analysis of *Calotropis* fibres before and after alkaline treatment and resistance to water were conducted in order to assist in the interpretation of the mechanical tests results.

2. Materials and Methods
2.1. Materials
2.1.1. Date Palm Midrib Long Fibre Extraction

Based on the anatomic features of date-palm parts, just the midrib consisted of long fibres positioned in the natural matrix [7]. The entire process of the fibre extraction following the procedure used by other researchers [7] is presented in Figure 1.

Figure 1. An illustration of the process of date-palm midrib long fibre extraction.

To provide the raw materials needed for extraction of the date-palm fibres, date-palm midribs were obtained from the date-palm groves of Hormozgan province (Rudan city), Iran. The middle parts of the midribs were used to extract the fibres. After cutting the midribs at the length of 400 mm, their moisture content was increased by immersing the samples in water for seven days and changing the water at least once every 24 h.

After seven days of immersion in water, the samples were cut manually into thinner strands with a knife or blade, and the outer skin of the midrib was removed from each strand, so to allow a better penetration of sodium hydroxide into the core. The fibres close to the outer skin of the midribs are stiffer fibres than the fibres of the middle parts, so when the samples were converted to strands, these fibres were removed. Then, the strands were immersed in an alkaline solution of sodium hydroxide 1.5% (ratio of 1:20 as strand: NaOH solution) and cooked in a Bain Marie device at a temperature of 95 °C for three hours. The NaOH solution was used for delignification. Then, the samples were squished by manual compression with rollers. During this process, the lignin and extracts that were softened during the cooking process were squished, and the fibres were separated. The fibres were afterwards separated with a metal comb with a very small distance between the teeth, and we performed this manual combing operation until all the waste was removed from the fibres. After extracting the fibres, in the last step, a neurulation of alkaline fibres with acetic acid (concentration of 5%) by immersion for 2–3 min was performed. First, the fibres were washed with water so that they were completely clean, and this operation was repeated after the immersion into the acid, so finally, the color of the fibres was brighter. The washing steps with water were conducted to produce cleanliness and further delignification.

The extracted date-palm fibres were longitudinally oriented and arranged to form flat sheets for stratified composite layers. The date-palm midrib long fibres were put in water until they converted to flat sheet form, then they were longitudinally oriented and pressed by hand and left for 24 h to dry under normal environmental conditions (temperature of 20 °C and 55% relative humidity of the air). The date-palm sheets formed this way had sizes (length × width) of about 350 mm × 350 mm.

The midrib long fibres of date palm were used in the composition of three-layer laminate (D), with adjacent layers having perpendicularly oriented fibres, in the basic structure of the five-layer composites DC and DTC (Table 1).

Table 1. The structure of experimental composites.

Code	Date Palm Fibres	*Calotropis* Fibres
D	3 layers	-
DC	3 layers	2 layers (untreated)
DTC	3 layers	2 layers (treated)

2.1.2. Calotropis Fibres Extraction and Treatment

To provide the fibres needed for *Calotropis* sheets, the mature fruits (Figure 2a) were collected from Sistan and Baluchestan province (Iran), and the white silk (Papus fibres) were separated from the seeds (Figure 2b).

The *Calotropis* sheets with sizes (length × width) of 350 mm × 350 mm were formed by arranging the fibres as seen in Figure 2c. The sheet that formed was sprinkled with water and left for drying for 24 h in an indoor environment, at a temperature of 20 °C and 55% air relative humidity. Without being pressed, the fibres connected to one another and formed a continuous sheet weighing approximately 3 g. These sheets were used to manufacture the five-layer composite (DC), as shown in Table 1.

The alkali treatment of *Calotropis* sheets has the role of removing unwanted substances such as oil or wax from the fibres' surfaces, providing improved mechanical properties due to the increased amount of cellulose [23]. First, the fibres were immersed for 30 min in a 5% solution of sodium hydroxide (NaOH) prepared with distilled water, in order to remove non-cellulosic impurities from the fibres.

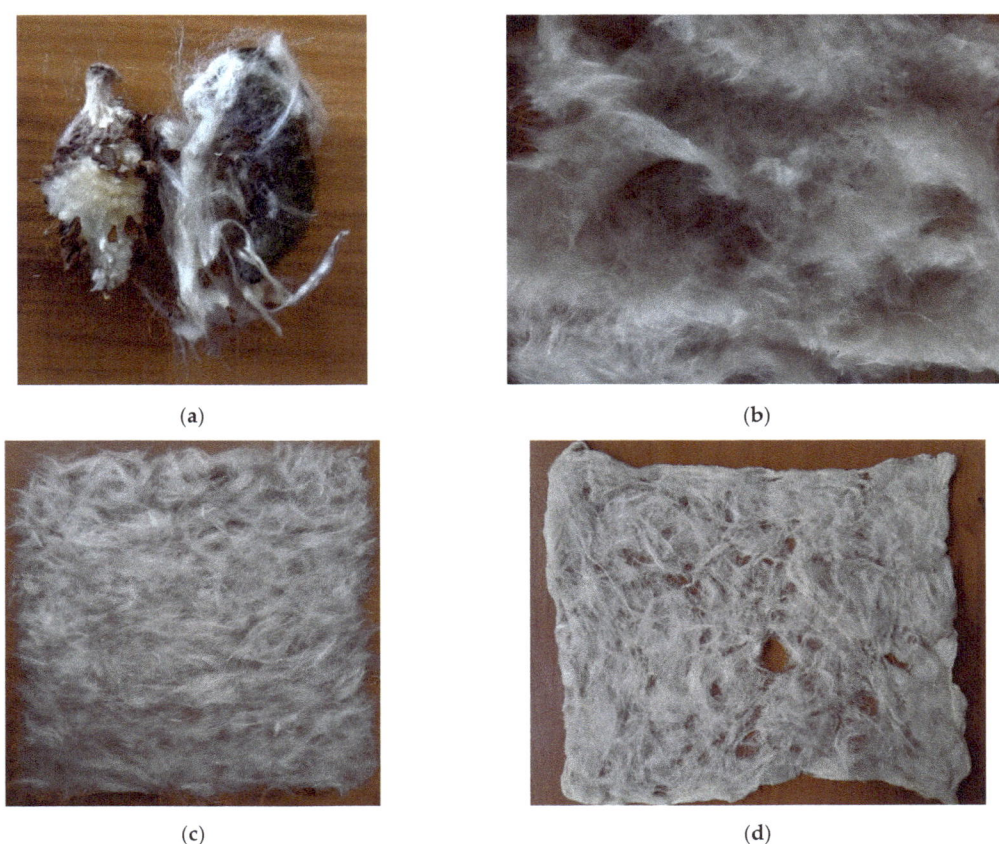

Figure 2. *Calotropis procera* fibres: (**a**) mature fruits; (**b**) extracted fibres arranged to form a thin sheet for composite layer; (**c**) untreated *Calotropis* sheet; (**b**) treated *Calotropis* sheet.

In the next step, the fibres were washed with distilled water to remove the sodium hydroxide solution, and after that, the fibres were immersed in dilute hydrochloric acid (HCl) solution for 1 min. The fibres were washed again with distilled water, and after drying, they were arranged to form a uniform rectangular layer of approximately 350 mm × 350 mm for composite manufacturing (Figure 2d). These treated *Calotropis* sheets were used in the manufacturing of five-layer composites (DTC) (Table 1).

2.2. Composites Manufacturing

Multi-layered fibre-reinforced epoxy composites were manufactured. The date-palm fibres and *Calotropis* fibres were formed into entangled sheets as previously presented. The Laminar BK epoxy resin, manufactured by Vosschemie GmbH company located in Uetersen, Germany, was used as a binder between the layers of fibres.

Three types of composites (Table 1) were manufactured under laboratory conditions: a three-layer composite made only from date-palm fibre layers (two longitudinally oriented layers for the faces and one transversally oriented for the core) (composite D) and five-layer composites made from three layers of date-palm fibre sheets and two layers of *Calotropis* sheets (one variant with untreated *Calotropis* fibres coded as DC and one variant with treated ones, coded as DTC). The moisture content of date-palm fibres was 3% before manufacturing the composites.

Epoxy BK is a two-component epoxy resin, is solvent-free, clear and transparent, and can be used as a laminating resin. The mixing ratio of the two components (A/B parts weight) is 100/60 and the hardness index of the resin is Shore D-80. The densities of the two components are 1.15 g/cm^3 for component A and 1 g/cm^3 for component B. The working temperature of the surface on which BK epoxy resin was applied was about 20 °C, and the working time for applying it in the structure of one composite panel was up to 30 min. According to the technical sheet, the BK epoxy resin reaches its final strength after 3–5 days.

The resin was applied by brush on each individual layer for fibres impregnation (Figure 3a). Each layer of the composite was firstly weighed with an accuracy of 0.01 g, and the epoxy resin amount was calculated by multiplying the weight of each layer as follows: five times for date-palm fibres and seven times for *Calotropis* fibres. The core layer of the date-palm composite was placed to ensure that the direction of fibres was perpendicular to those of the face layers (Figure 3b). White silicone baking paper was used at the top and bottom of the sandwich composite in order to avoid adhesion to the top and bottom supports (Figure 3c). The formed sandwich composite covered on both faces by silicone baking paper was placed between two blockboard panels, and a weight of 20 kg was placed on the top panel for cold pressing the composite layers at a pressure of 0.019 bar for three days. The composites were then conditioned for six days at a temperature of 20 °C and 55% air relative humidity before testing. Three replicates of each composite type (D, DC, DTC) were manufactured. After six days, the panels were sized at 300 mm × 300 mm. The final composite panels after sizing presented a compact appearance and a rigid structure (Figure 3e). Testing samples were then cut according to the specific standards requirements for measuring vertical density profile, water absorption, flexural strength and flexural modulus of elasticity, tensile strength and tensile modulus of elasticity, and impact resistance. The number of samples and the methods of testing were in accordance with the corresponding technical specifications, as further detailed in the respective sub-sections.

Figure 3. Composite manufacturing: (**a**) impregnation of the first layer with BK epoxy resin; (**b**) arrangement of middle date-palm layer; (**c**) composite covered by white silicone baking paper; (**d**) composite placed between two block board panels and pressed by two calibrated weights of 10 kg; (**e**) final composite panel.

2.3. Vertical Density Profile (VDP)

The vertical density profile (VDP) was investigated using the X-ray density profile analyzer DPX300 (IMAL, San Damaso, Italy). Five square-shaped specimens cut from each composite panel type were tested. The specimens had dimensions (length × width) of 50 mm × 50 mm, and the density profile was measured along the entire thickness of the sample. The specimens were first weighed with an EU-C-LCD precision scale (Gibertini Elettronica, Novate Milanese, Italy) and their dimensions were measured by the density profile analyzer.

2.4. Microscopic Investigation

The stereo-microscope NIKON SMZ 18-LOT2 (Nikon Corporation, Tokyo, Japan), was employed in the microscopic investigation and measurement of the fibres used for the composites, and also for better visualizing the structures of the manufactured composites and the adhesion between the component layers. In order to perform the measurements of the fibres' diameters, images with 240× magnification were taken. For the composites, the microscopy was performed on the edge of the structures with 22.5× magnification, and the constituted layers and the adhesion area between them were highlighted.

2.5. FTIR Analysis

Fourier transform infrared spectroscopy (FTIR) was employed to investigate the chemical features of the untreated *Calotropis* fibres, considered as control, as well as the chemical changes brought about by the alkaline treatment applied to three samples of *Calotropis* fibres. An ALPHA Bruker spectrometer (produced by Bruker Optik GmbH, Ettlingen, Germany) equipped with an ATR (attenuated total reflection) module was used to record the FTIR spectra in the range 4000–400 cm^{-1} at a resolution of 4 cm^{-1} and 24 scans/spectrum. Three individual spectra were recorded for each sample. The recorded spectra were further processed employing OPUS software for baseline correction and smoothing, and an average spectrum was calculated for each type of sample. After the average spectra were normalized (Max-Min normalization), they were compared to reveal chemical changes occurring as a result of the treatment applied. The characteristic absorption bands were assigned according to the references in the literature.

2.6. Water Immersion

Samples with dimensions (length × width) of 50 mm × 50 mm were cut from the composite panels in order to determine the thickness swelling and water absorption by immersion in water according to the SR EN 317: 1996 standard [33]. The test samples (five replicates for each type of composite) were immersed in a water bath at a temperature of 20 °C for 24 h. The sizes of the samples were measured by an electronic caliper with an accuracy of 0.01 mm. The samples were weighed before starting the test and after 24 h of immersion into the water using an electronic scale with the accuracy of 0.01 g. The thickness of the samples was measured every time at the diagonal cross point. The values were recorded for structures D and DC, investigating whether the layer of *Calotropis* fibres has an influence or not on the water absorption (WA) and on the thickness swelling (TS).

2.7. Mechanical Performance

The testing method and the number, shape and dimensions of the specimens used for each mechanical test performed were according to the corresponding European standards. Flexural strength and flexural modulus of elasticity values were determined according to the EN 310:1993 standard [34] and the equipment used for the test was the Zwick/Roell Z010 universal testing machine (Ulm, Germany).

Tensile tests were carried out on the universal testing machine LFV50-HM, 980 (Walter and Bai, Switzerland), which is digitally controlled. The maximum force provided by the machine is 200 kN. The tensile specimens were cut from the composite panels and their dimensions were established according to the European standard EN ISO 527-4 [35]. The loading speed was 1.5 mm/min according to the same standard. The data recorded every 0.1 s were: tensile force F, elongation Δl of the tensile specimen, and time t. The tensile modulus of elasticity and tensile strength were determined according to the standard EN ISO 527-4 [35]. The tensile modulus of elasticity was determined on the linear portion of the stress–strain curve recorded for each specimen.

The impact strength was investigated by the Charpy test, carried out on the pendulum impact tester HIT50P manufactured by Zwick/Roell (Ulm, Germany). The maximum capacity of the impact pendulum HIT50P with digital controlling is 50 J for the impact energy. The impact test set-up and the dimensions of 100 mm × 10 mm for the rectangular specimens were in accordance with the European standard ISO 179-1 [36]. The thickness of the specimens is provided by the thickness of the panel from which the specimens were cut. The dimensions of the cross-sections were accurately measured before testing for each specimen by the electronic caliper with the accuracy of 0.01 mm. In the Charpy impact test, the impact hammer hits the middle of the specimen, which is simply supported at both ends, and the elastic failure energy W is displayed digitally by the pendulum impact tester HIT50P, as in other experimental research works [37,38]. The impact strength or resilience denoted with K is computed as the ratio between the elastic failure energy W and the area A of the cross-section for each specimen by using Equation (1):

$$K = W/A \qquad (1)$$

2.8. Statistical Analysis

The statistical analysis employed the determination of standard deviation in Microsoft Excel for a confidence interval of 95% and a significance level of 0.05 ($p < 0.05$). Single-factor Analysis of Variance (ANOVA) was performed with the Microsoft® Excel package for analyzing the way in which the average values of tensile and flexural strengths, modulus of elasticity (for bending and tensile loads), impact resistance, water absorption (WA), and thickness swelling (TS) were significantly affected by the participation of both treated and untreated *Calotropis* fibres in the structure of the composites and also by the direction of the fibres during the test (longitudinal and transversal). The direction of the fibres was considered to be the direction that the fibres in the composite face. Furthermore, Tukey's test was utilized to specify the significant differences created by the treatments.

3. Results and Discussion

3.1. Vertical Density Profile (VDP)

In Figure 4a–c, vertical density profile examples for each type of manufactured composites are presented.

The measured thicknesses for the three types of composite laminates varied as follows: between 6 mm and 6.77 for the D structure, between 6.43 mm and 7.4 mm for the DC structure, and between 6.82 mm and 7.42 mm for the DTC structure.

As seen in Figure 4a, minimum densities were recorded for positions of approximately 2.2 mm and 4.3 mm on the thickness of the composite, measured from the face. These minimum points on the graph correspond to the epoxy resin layers on the thickness, for which the calculated density of the mixture of the two components was 1094 kg/m^3. The maximum recorded densities were 1174.1 kg/m^3 for the left peak and 1208.7 kg/m^3 for the right peak, positioned at the thickness of 2.35 mm and 4.55 mm from the face of the composite and corresponding to the date-palm fibres layers (core and face layer). The vertical density profile clearly shows the structure of the composite, as follows: the three zones with low density variations and containing the maximum peaks belong to the date-palm layers (with thicknesses of approximately 2 mm), whilst the two minimum density points indicate the position of the epoxy resin layers. The recorded ratio between the minimum density and the average density was 0.99.

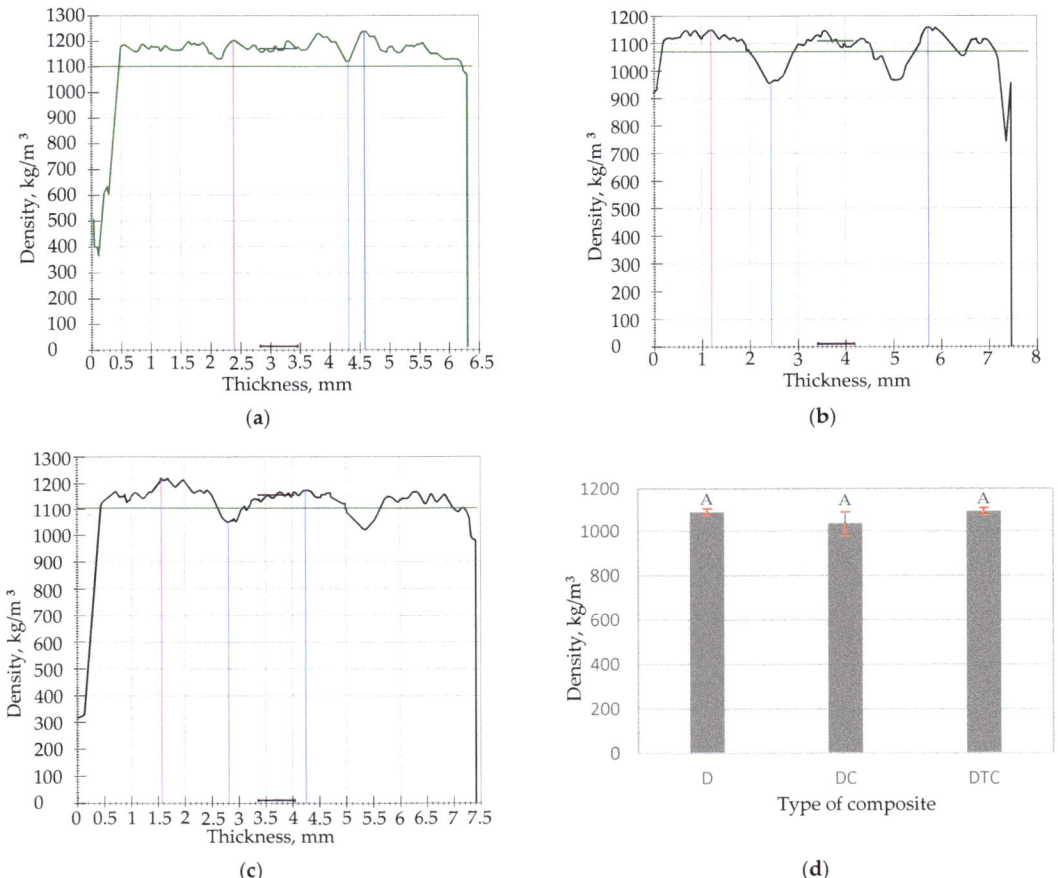

Figure 4. Vertical density profiles of the tested composites: (**a**) D; (**b**) DC; (**c**) DTC; (**d**) average values of measured densities.

The VDP of the DC structure shown in Figure 4b indicates higher differences between minimum points and maximum peaks of the graph. The minimum density recorded for this structure was 956.3 kg/m^3, located at the thickness of 2.45 mm from the outer face, indicating the presence of the *Calotropis* fibres layer, which had a density of approximately 0.025 kg/m^3 before pressing into the composite structure. The recorded ratio between the minimum density and the average density was 0.89, which was lower for the D structure. The VDP for the DC structure shows the layer's position along the thickness of the composite, with small variations of the densities corresponding to the date-palm layers, similar to the findings of VDP for the D structure.

In the case of the DTC composite, when treated *Calotropis* fibres were used, the VDP graph (Figure 4c) shows smaller differences between maximum density peaks and minimum points than in the case of using untreated *Calotropis* fibres (structure DC). However, the presence of the treated *Calotropis* layers is also noticed through the visible difference of densities between the maximum and minimum values. Instead, the maximum density peak is higher than for structure D, recording a value of 1220 kg/m^3. The ratio between the minimum density and the average density was 0.96 in this case, closer to the value calculated for the D structure.

The results show that the treatment applied to the *Calotropis* fibres affects the density of the composite, lowering the differences in densities between the *Calotropis* fibre layers and date-palm layers. This observation is also proved by the graph from Figure 4d, which shows a lower average density value for structure DC compared to the other two. The findings of the present investigation are in line with the results obtained by other researchers [2], who determined a density of 1227.27 kg/m^3 for date-palm fibres, considering it to be lighter compared to cotton, jute, flax or sisal, or around 1200 kg/m^3 for date-palm and date-palm/bamboo fibre-reinforced epoxy hybrid composites [28]. So, one of the advantages brought by the proposed laminate composites in this study is their light weight compared with other potential similar structures reinforced with natural or synthetic fibres, such as Kevlar, carbon and glass epoxy composites, for which the densities are greater than 1230 kg/m^3, as was found in the literature [25].

3.2. Microscopic Investigation

The microscopy of the date-palm fibres, and of the *Calotropis* fibres (both untreated and treated) with 240× magnification, are shown in Figure 5a–c.

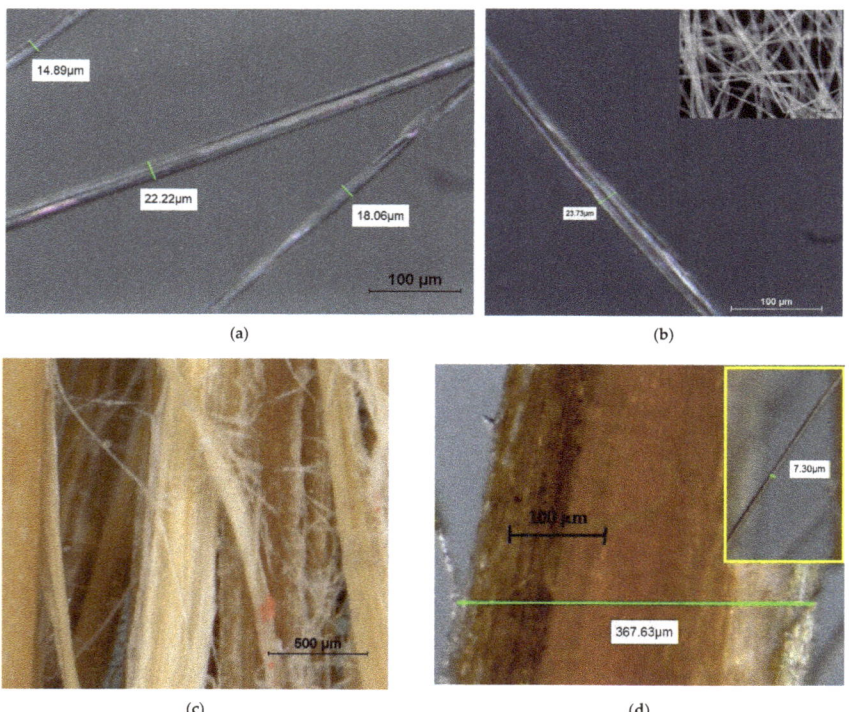

Figure 5. Microscopic investigation of the fibres: (**a**) untreated *Calotropis* fibres (240× magnification); (**b**) treated *Calotropis* fibre (240× magnification); (**c**) date-palm fibres (60× magnification); (**d**) measured bundle of date-palm fibres and the measured diameter of the single fibre in the detail image (240× magnification).

The measurements of the fibres' diameters reached values between 0.22 mm and 0.37 mm for the date palm, between 0.015 mm and 0.022 mm for the untreated *Calotropis*, and between 0.023 mm and 0.031 mm for the treated *Calotropis*. The measurements of untreated *Calotropis* are in line with the measurements recorded by other researchers based on SEM micrographs [20]. They also noticed that treating *Calotropis* with NaOH for 5 min at room temperature produces winding fibres. Other researchers observed that the treatment with NaOH significantly increased the diameter of the fibres [22], a fact revealed by microscopic investigation from the present research. As a first conclusion to the microscopic investigation, the *Calotropis* fibres have diameter sizes ten times smaller than those of the date-palm fibres. As seen in Figure 5a,b, the *Calotropis* fibre is transparent, and no visible transformation was noticed when comparing the treated one with the untreated one, except the increased diameter.

As regards the date-palm fibres, as they appear in Figure 5c,d, they are cylindrical bundles composed of single fibres measuring less than 10 µm in diameter.

The microscopic images of D, DC, and DTC are presented in Figure 6. The images shown in Figure 6a–c were taken with 22.5× magnification, and the detail presented in Figure 6d was 60× magnified. The images were taken on the edge of the composite specimen, so that the three-layer in the case of the D composite and the five-layer composites for both DC and DTC were visible.

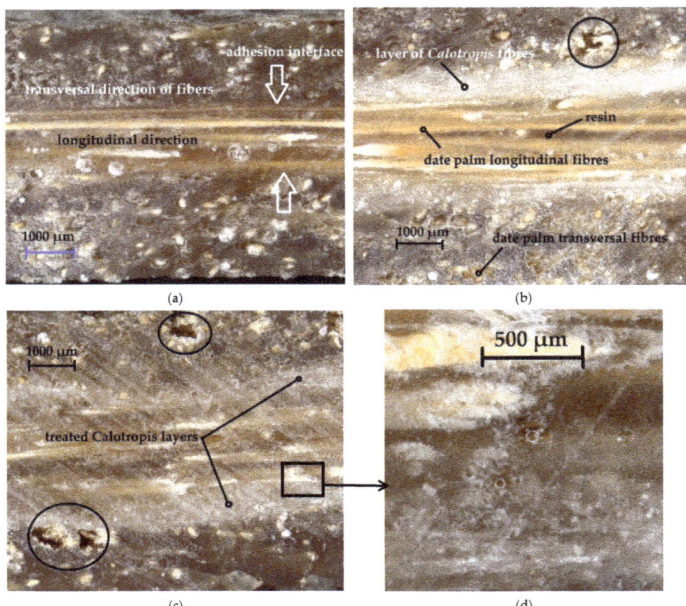

Figure 6. Microscopic investigation of the composite structures: (**a**) D (22.5× magnification); (**b**) DC (22.5× magnification); (**c**) DTC (22.5× magnification); (**d**) detail of the interface between date-palm layer and treated *Calotropis* layer (60× magnification) marked at (**c**) with black rectangle.

All three structures were placed under the microscope so as to point out the longitudinal section of the date-palm fibres for the core and their crosscut section for the faces.

In Figure 6a, representing the edge of the three-layer composite (D), the compact structure and the good adhesion interface between the layers can be observed. For the second composite (Figure 6b) made from three layers of date-palm fibres and two layers of untreated *Calotropis* sheets, the longitudinal and crosscut sections of the date-palm fibres are highlighted. The dark brown color from this image represents the hardened epoxy resin. The layer of untreated *Calotropis* fibres is white colored and very well visible in Figure 6b. Less visible is the layer of treated *Calotropis* fibres in Figure 6c, corresponding to the composite DTC. An explanation could be the yellowish color of this layer due to the chemical treatment applied, but also a possible better impregnation with epoxy resin. The circled areas highlighted in Figure 6a,b represent the gaps (voids) that occurred in the composite structures, especially in the vicinity of the *Calotropis* layers. They are caused by the insufficient impregnation with resin in certain areas, as a result of the hand layup technology used for impregnation, which could not ensure uniformity on the whole surface.

The existence of the voids at the interface area located between date-palm fibres and epoxy resin matrix was unveiled also by other researchers in their study [29]. In comparison, bamboo fibres displayed a better consolidation in the matrix resin and a greater contribution to the mechanical strength. Mixing date-palm fibres with bamboo fibres in an epoxy resin matrix could be the subject of further research, having as an objective the improvement of the mechanical performance of such composite materials.

3.3. FTIR Analysis

For the untreated *Calotropis* fibre (Control), the spectra (Figure 7) presented high absorptions bands at around 3330 cm^{-1}, 2910–2800 cm^{-1}, 1730 cm^{-1}, and several peaks in the fingerprint region (presented in Figure 7), especially in the ranges 1640–1500 cm^{-1} and 1370–900 cm^{-1}, indicating the presence of the main constituents of lignocellulosic natural materials such as cellulose, hemicelluloses, and lignin, in good accordance with the literature [39]. The high and broad absorption with a maximum at around 3330 cm^{-1} is assignable to H-bonded hydroxyl (-OH) groups in alcohols (cellulose, hemicelluloses) and phenols (lignin), while absorption at 2910 cm^{-1} is assigned to C-H stretching vibrations in methylene (-CH$_2$-) and methyl (-CH$_3$) groups in cellulose, hemicelluloses, and also possibly in waxes or oils, if present. The high absorption at 1730 cm^{-1} (unconjugated carbonyl) can be mainly associated with the acetyl groups in hemicelluloses and possibly also with other esters, such as waxes or oils, which might be present on the fibres [40]. Another absorption band is that at 1604 cm^{-1} (aromatic ring) with a shoulder at about 1645 cm^{-1}, which might be assigned to conjugated carbonyl bonds or aromatic ketones in the structure of lignin, but also to O-H bending vibration for absorbed water [39,41]. The absorption at 1370 cm^{-1} is assigned to cellulose and hemicelluloses, while that at 898 cm^{-1} (C-H vibration) is characteristic of crystalline cellulose. The strong absorption at 1240 cm^{-1} might be assigned to the acetyl groups in hemicelluloses: 1255 cm^{-1} according to [41], 1230 cm^{-1} assignable to -C-C plus C-O plus C=O stretch for acetyl in xylan according to [41], and also to the syringyl ring in the structure of lignin, 1235 cm^{-1} syringyl ring in lignin, and C-O stretch in lignin and xylan, according to [42,43]. The absorption at 1507 cm^{-1} is the most characteristic absorption of lignin (aromatic skeletal vibration).

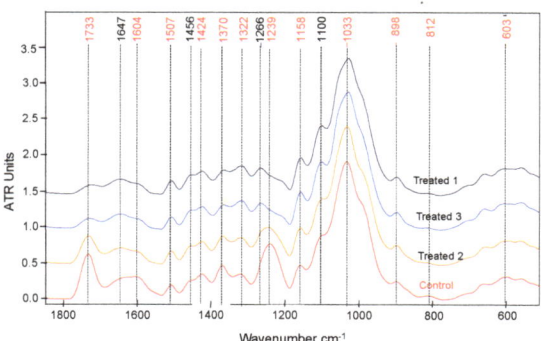

Figure 7. FTIR analysis of the treated and untreated *Calotropis* fibres (comparative spectra in fingerprint region 1800 cm^{-1}–600 cm^{-1}).

For the treated *Calotropis* fibre (sample 2), the spectrum was quite similar to the control sample, with some differences in the intensity of some characteristic absorptions, as follows: slight increase of -OH absorption band at 3300 cm^{-1}; a slight decrease of absorptions at 1733 cm^{-1} and 1370 cm^{-1}, corresponding to the decrease of hemicelluloses content; the increase of absorption at 1645 cm^{-1}, which becomes evident whilst the former absorption peak at 1604 cm^{-1} decreased to a shoulder, possibly indicating some changes in the structure of aromatic compounds (lignin) and/or increased content of absorbed water due to an increased hygroscopicity. The absorption at 1239 cm^{-1} decreased similarly to the band at 1730 cm^{-1}, most likely pointing out degradation of hemicelluloses, or saponification (hydrolysis in an alkaline medium) with elimination of the acetyl groups.

The spectra for the treated samples 1 and 3 were similar to one another while both showing more substantial changes when compared to the spectrum for control *Calotropis* fibres than was the case for treated sample 2. The main changes were: increase of the -OH groups absorption band at 3300 cm^{-1}; more significant decrease of absorption at 1733 cm^{-1}; clear evidence of the absorption at 1645 cm^{-1}, which includes as a shoulder the absorption at 1604 cm^{-1}; the disappearance of the absorption peak at 1239 cm^{-1} or its shifting to 1265 cm^{-1}, assignable to guaiacyl ring plus C=O stretch in guaiacyl lignin [40]. This might indicate two types of processes: (1) decrease of acetyl groups in xylan (hemicelluloses) by alkaline hydrolysis or decrease of hemicelluloses content (correlated with the decrease of 1733 cm^{-1}) and (2) de-metoxylation of lignin, respectively transforming some syringil rings in the structure of lignin into guayacil rings. Additionally, the small absorption peak at 899 cm^{-1} seems to be slightly increased, indicating a relative increase of (crystalline) cellulose content due to degradation and removal of other components, mainly hemicelluloses, as the most characteristic lignin absorption band at 1507 cm^{-1} seems to be little affected by the applied treatment.

FTIR spectra indicated that the alkaline treatment applied in this research brought about a de-acetylation and decrease of hemicelluloses content by saponification and alkaline hydrolysis and possibly some chemical changes in the structure of lignin, such as some demethoxylation of the syringil rings. Accordingly, the FTIR spectra highlighted an apparent increase in cellulose content and free hydroxyl groups. These findings are in good accordance with similar reported research. For instance, in [40] it was found that the pre-treatment of *Calotropis* fibres with sodium hydroxide resulted in an increase of cellulose content from 64.47% to 69.93% and its crystallinity index from 36% to 39.8%, while hemicelluloses content decreased from 9.64% to 6.72%, lignin decreased from 13.56% to 11.25% and wax content also decreased from 1.93% to 1.12%. All these changes had a positive influence on the characteristics of the selected fibres, which are important for their utilization as reinforcing materials in polymer eco-friendly composites.

3.4. Water Immersion

The recorded values for the water absorption (WA) and thickness swelling (TS) are presented in Figure 8 for the D and DC composites investigated.

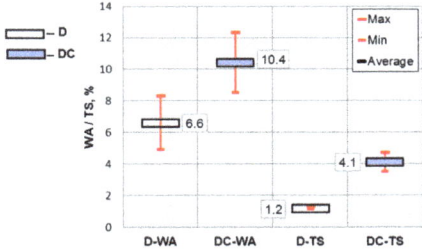

Figure 8. Water absorption (WA) and thickness swelling (TS) after 24 h of immersion of D and DC composites into water.

The presence of the *Calotropis* fibres in the structure of the composite reduces the water resistance, increasing the values recorded for water absorption (WA) and thickness swelling (TS) in the conditions of immersing the samples in water for 24 h. As seen in Figure 8, the recorded average value of WA for the three-layer composite made of date-palm fibres (D) was 6.6% and higher with 57.6%, up to a value of 10.4% for the composite with two additional layers of untreated *Calotropis* fibres (DC). The thickness swelling (TS) increased for composite DC with 3.42%. The higher recorded values of WA and TS can be explained by the hydrophilic character of treated and untreated *Calotropis* fibres, highlighted by the high content of hydroxyl groups that increased following alkaline treatment, revealed by FTIR in this research and other studies [17,22]. One of these studies [22] showed that the natural wax coating of these fibres is partially removed when an alkaline treatment is applied, thus improving their absorption property. This property is in favor of its applicability as an absorbent but becomes a disadvantage when considering the water behavior of *Calotropis* fibre-reinforced composites. Therefore, for the potential applications of the composites developed in this research in humid environments, or in water contact (for boats for example), the use of *Calotropis* fibres in the structure is not recommended. The performance of *Calotropis procera* fibres as reinforcements in an epoxy matrix was investigated by several researchers [44] using fibres extracted from the branch of the tree. For 30 weight % fibres in the composite structure, the WA value was 7 %, lower than the results obtained for the five-layer laminates proposed in this study. Date palm/bamboo fibre-reinforced epoxy composites immersed in water for eight days [29] showed a fast increase in the first 24 h, with the TS and WA recorded values around 6%, similar to structure D investigated in this paper. Another study [2] showed that the date-palm fibres have a lower porous structure and implicitly a lower absorption capacity compared with sisal, wheat straw, hemp and kenaf fibres, providing the important advantage of using date palm in the place of other natural fibres for composite manufacturing.

3.5. Mechanical Performance

The results recorded for modulus of elasticity, both for tensile and bending tests, tensile strength, flexural strength and impact strength are presented in Figure 9.

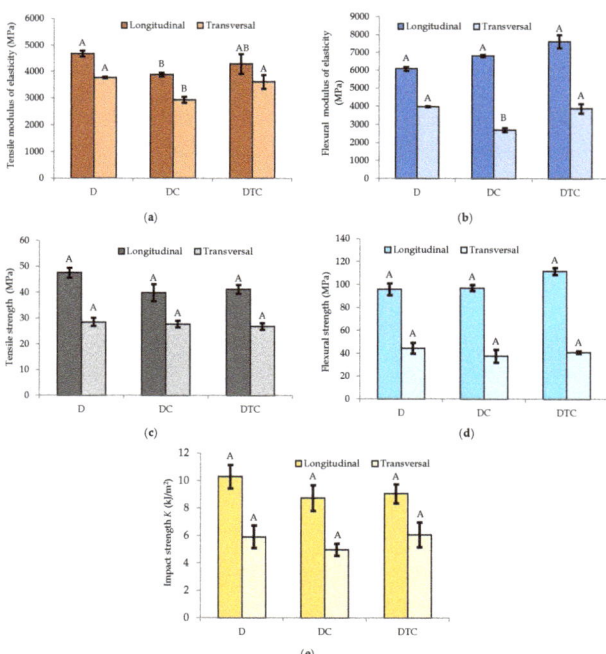

Figure 9. Comparison of the mechanical properties for the tested composite materials (variables with the same letter mean that the difference is not statistically significant): (**a**) tensile modulus of elasticity; (**b**) flexural modulus of elasticity; (**c**) tensile strength; (**d**) flexural strength; (**e**) impact strength.

For the longitudinal tested samples, the highest average values of tensile strength and tensile modulus of elasticity, respectively, were recorded for the three-layer structures D (47.5 MPa and 4678.6 Mpa, respectively), followed by five-layer structures with treated *Calotropis* DTC (41.2 MPa and 4289 Mpa, respectively) and untreated DC (39.8 MPa and 3877 Mpa, respectively). The same trend was noticed for impact strength (10.3 MPa for D composite, 9.1 MPa for DTC and 8.7 MPa for DC). Instead, the flexural modulus of elasticity and flexural strength values increased with the addition of *Calotropis* fibres, the highest ones being reached by DTC composites with treated *Calotropis* fibres (average values of 7623 MPa for flexural modulus of elasticity and 111.7 MPa for flexural strength). Average values recorded for DC/D structures were as follows: 6802 MPa/6072.5 MPa for the flexural modulus of elasticity and 97.1 MPa/95.7 for the flexural strength.

The mechanical strength values recorded for the three-layer laminate (D) tested on the longitudinal direction are higher (flexural strength was 1.5 times higher and the tensile strength was 1.2 times higher) than those recorded in another study [28] for panels manufactured with 50:50 as weight ratio of date-palm fibres to epoxy resin, using date-palm fibres from different parts of the plant. As regarding the flexural modulus of elasticity, with the better interfacial adhesion and dispersion of fibres and matrix, there are better flexural modulus results [29]. As found in the literature [28,30–32], alkali treatment applied to natural fibres improves the adhesion between the fibres and matrix, improving the mechanical performance of the resulted composite. The results obtained in this research confirm that the chemical treatment applied to the *Calotropis* fibres led to the increase of the mechanical characteristics of the composite material DTC compared with the ones of the composite DC. In Figure 9a, it is shown that the chemical treatment of the *Calotropis* fibres led to a tensile modulus of elasticity 10.66% and 22.95% higher than the one corresponding to the composite DC containing untreated *Calotropis* fibres in longitudinal and transversal directions of the fibres, respectively. Figure 9b,d show that the chemical treatment of the *Calotropis* fibres has increased the flexural modulus of elasticity by 12.1% and the flexural strength by 15% for the DTC composite compared to DC composites.

The flexural modulus of elasticity in the longitudinal direction of the fibres for DC and DTC composites is higher by 12.02% by 25.54%, respectively, than the flexural modulus of elasticity of composite D (Figure 9b). Contrarily, the flexural modulus of elasticity in the transversal direction of the fibres decreased by 32.07% or 2.52% by adding the two layers of untreated or treated *Calotropis* fibres, respectively, compared to the flexural modulus of elasticity obtained for composite D (Figure 9b).

The alkali treatment of *Calotropis* fibres proved to be beneficial also for the tensile modulus of elasticity and tensile strength (Figure 9a,c), but the increasing percentages turned out to be lower. Impact strength is included in the same category (Figure 9e). The impact strength of the composite material DTC improved by less than 5% compared to the composite DC when untreated *Calotropis* fibre sheets were replaced by the treated ones (Figure 9e).

Epoxy composites reinforced with 10 up to 40 wt% of the *Calotropis* fibres [44] recorded flexural strength values between 25 MPa and 29 MPa, tensile strength around 11 MPa and impact strength between 1 kJ/m^2 and 1.5 kJ/m^2, values approximately four times lower than those recorded for epoxy composites reinforced with date-palm fibres (D composite). Based on these results, the decrease of the mechanical strength of the laminates with the participation of *Calotropis* fibres (DC structures) can be explained by their low bearing capacity.

The flexural strength on longitudinal direction of the reinforcement fibres was 1.46% or 16.72% higher for the composite containing untreated or treated *Calotropis* fibres, respectively, than the one obtained for the composite material D, which did not contain *Calotropis* fibres, as shown in Figure 9d. Contrarily, the flexural strength evaluated in the transversal direction of the reinforcement fibres recorded a small decrease by adding either the untreated or treated *Calotropis* fibres, as shown in Figure 9d.

The impact strength in the longitudinal direction of the reinforcement fibres was 15.53% or 11.65% lower for the composite materials containing untreated or treated *Calotropis* fibres, respectively, than the impact strength determined for the composite material D, which did not contain *Calotropis* fibres, as shown in Figure 9e. This behavior was expected as long as the flexural modulus of elasticity had increased by adding fibres (Figure 9b) and the elastic failure energy stored by the impact was inversely proportional to the flexural modulus of elasticity. Just a small variation in the impact strength in the transversal direction of the reinforcement fibres was observed for the composites containing *Calotropis* fibres—composites DC and DTC—with respect to the impact strength determined for the composite material D, which did not contain *Calotropis* fibres.

The resulting values for tensile strength in the present research are similar to those of woven kenaf fibre-reinforced acrylonitrile-butadiene-styrene laminates [28]. As this study shows, the orientation of fibres in the structure influences the mechanical performance of the composite material. The results presented in Figure 9 show that having the orientation of the samples in the transversal direction when conducting the mechanical tests results in approximately half of the mechanical strength recorded for the longitudinal orientation of the samples for all tests.

As the SEM analysis performed by some researchers showed, the *Calotropis* fibres treated with NaOH for 5 min at room temperature tend to stick together [20], which can explain why the agglomeration of the fibres in some areas is to the detriment of the others (as seen in Figure 2d), thus affecting the strength of the composites in some areas and recording a larger field of data scattering. Fibres extracted from the stem of the *Calotropis* tree used as reinforcement in an epoxy composite [23] showed higher values of tensile strength when an alkali treatment was applied (46.21 N/mm^2 compared to 39.01 N/mm^2). An increase from 39.84 MPa to 41.17 MPa has been also recorded in the present research for structure DTC compared to structure DC, showing that the treatment applied to *Calotropis* fibres was beneficial to the tensile strength results when the samples were tested in the longitudinal direction. The flexural strength and impact strength had the same improvement trend for structure DTC compared to DC, a trend also noticed by [23] for alkali-treated fibres. The improvement of the mechanical properties of composite DTC compared to composite DC is attributed to a better impregnation of the treated *Calotropis* fibres compared to the untreated ones, an aspect revealed by the observations made through the microscopic analysis of these composites.

Natural fibres with higher mechanical strengths have higher cellulose content [30]. According to the FTIR analysis performed for treated and untreated *Calotropis* fibres, an apparent increase of cellulose content and free hydroxyl groups was recorded for the treated fibres, thus explaining the higher values recorded for tensile and bending strength and also for impact strength in the case of composites containing *Calotropis* treated fibres. From the values recorded for mechanical strengths, the influence of the direction of testing is evident, with lower values being recorded for the transversal direction in all cases and for all composites.

The existence of the natural waxy substance on the surface of date-palm fibre does not allow a strong bond with epoxy polymer matrix. Having the advantage of being lighter than other natural fibres, chemically modified date-palm long fibres can be used as a reinforcing component to improve the tensile and flexure strength in fibre-reinforced biocomposites, and these may be also used as substitutes for heavy-weight civil engineering materials [31]. Further research can be performed to improve the bond at the interface area between the date-palm fibres and epoxy polymer matrix.

3.6. Statistical Analysis

The addition of the treated and untreated *Calotropis* fibres into the structure of the composite materials has significantly ($p < 0.05$) affected both the tensile and flexural modulus of elasticity in the case of tests applied in the transversal direction of the fibres. Additionally, both impact strength and tensile strength were not statistically significantly affected by the applied method of testing (longitudinal or transversal) in the presence of *Calotropis* fibres in the composite structure. For the tensile modulus of elasticity in the longitudinal direction of fibres, the treatment of *Calotropis* fibres had a significant effect ($p = 0.03$). Additionally, the presence of *Calotropis* fibres in the structure DC showed a statistical significantly effect on the thickness swelling (TS) and water absorption (WA). The grouping of the statistical classes for each test's method are clearly presented separately in the comparative diagrams in Figures 4d and 9a–e. The variables with the same letter mean that the difference is not statistically significant.

4. Conclusions

This study showed that long date-palm fibres extracted from midribs are attractive natural resources to be used for structural composites based on laminar epoxy resin as binder.

The investigation of the mechanical properties of the *Phoenix dactylifera*/*Calotropis procera* fibre-reinforced epoxy hybrid composites showed that the presence of *Calotropis* fibres (treated and untreated) in five-layer composite structures (DTC and DC) resulted in lower values of tensile/impact strength compared to those of the three-layer composites made only from date-palm sheets (D). The same trend was noticed for the tensile modulus of elasticity. Instead, the flexural strength and flexural modulus of elasticity were improved with the contribution of *Calotropis* fibres. An increase in water absorption (WA) and thickness swelling (TS) values was recorded in the presence of *Calotropis* fibres in the five-layer composites.

The alkali treatment of the *Calotropis* fibres improved the tensile and impact strength of the composites compared to the ones made with untreated fibres, but not enough to reach the tensile and impact strength values recorded by the three-layer structures made only from date-palm fibres. The increase in strength in this case is explained by an apparent increase of cellulose content and free hydroxyl groups as result of the decrease in hemicellulose content caused by saponification and alkaline hydrolysis, while some chemical changes in the structure of lignin also seem possible.

The transversal direction of testing the samples resulted in the decreasing of the mechanical strength of the composites by about half of that recorded for the longitudinal direction of the fibres, in all mechanical tests (tensile test, bending test and Charpy impact test) involved in this research.

It must be mentioned that just the flexural modulus of elasticity and flexural strength are greater for composites DC and DTC, which contain *Calotropis* fibres, than for composite D, which does not contain *Calotropis* fibres. Moreover, it may be remarked that the chemical treatment of the *Calotropis* fibres led to the flexural modulus of elasticity being 12% and 43.5% higher than the one corresponding to the composite DC containing untreated *Calotropis* fibres in longitudinal and transversal directions, respectively. The increase of the flexural modulus of elasticity caused by adding treated or untreated *Calotropis* fibre sheets in composites is the cause of the decrease of the impact strength, which varies inversely proportionally to the flexural modulus of elasticity.

The potential applications of the composites developed might include interior components in the automotive industry or maritime industry, but further research is necessary.

Author Contributions: Conceptualization, H.Z.H., C.C. (Camelia Coșereanu) and M.H.M.; methodology, H.Z.H., C.C. (Camelia Coșereanu) and C.C. (Camelia Cerbu); software, M.C.T.; validation, H.Z.H., C.C. (Camelia Coșereanu) and C.C. (Camelia Cerbu); formal analysis, M.H.M. and H.Z.H.; investigation, M.C.T., S.V.G. and C.C. (Camelia Cerbu); resources, M.H.M. and C.C. (Camelia Coșereanu); data curation, M.C.T., C.C. (Camelia Cerbu), M.H.M. and H.Z.H.; writing—original draft preparation, C.C. (Camelia Coșereanu) and C.C. (Camelia Cerbu); writing—review and editing, M.C.T., C.C. (Camelia Cerbu) and H.Z.H.; visualization, M.H.M., S.V.G. and H.Z.H.; supervision, H.Z.H., C.C. (Camelia Coșereanu) and C.C. (Camelia Cerbu); project administration, C.C. (Camelia Coșereanu); funding acquisition, H.Z.H. and C.C. (Camelia Coșereanu). All authors have read and agreed to the published version of the manuscript.

Funding: This research received external funding from Armaghan e Tabiat e Makkoran Co. (National Identification No. 14007930167, Registration No. 1658, and Economic code 411638437544) by contract No. 99-7424475.

Acknowledgments: We hereby acknowledge the structural funds project PRO-DD (POS-CCE, O.2.2.1., ID 123, SMIS 2637, No. 11/2009) for providing the infrastructure used in this work (https://icdt.unitbv.ro/en/research-and-development-projects/the-r-d-institute-project.html, accessed on 7 November 2022).

Conflicts of Interest: The authors declare no conflict of interest.

References

1. El-Mously, H.; Darwish, E.A. Date Palm Byproducts: History of Utilization and technical Heritage. In *Composites Science and Technology*; Mohammad, J., Ed.; University Putra Malaysia: Serdang, Malaysia, 2020; pp. 3–71.
2. Bamaga, S.O. Physical and mechanical properties of mortars containing date palm fibers. *Mater. Res. Express* **2022**, *9*, 015102. [CrossRef]
3. Zaid, A.; de Wet, P.F. Origin, Geographical Distribution and Nutritional Values of Date Palm. In *Date Palm Cultivation, FAO Plant Production and Protection Paper 156, Rev. 1.*; Zaid, A., Ed.; Coordinated by E.J. Arias-Jiménez; Produced within the framework of the Date Production Support Programme in Namibia FAO-UTF/NAM/004/NAM; Food and Agricultural Organization of the United Nations: Rome, Italy, 2002; Available online: https://www.fao.org/3/Y4360E/y4360e06.htm#bm06.2 (accessed on 3 September 2022).
4. Hanieh, A.A.; Hasan, A.; Assi, M. Date palm trees supply chain and sustainable model. *J. Clean. Prod.* **2020**, *258*, 120951. [CrossRef]
5. Bamaga, S.O. A Review on the Utilization of Date Palm Fibers as Inclusion in Concrete and Mortar. *Fibers* **2022**, *10*, 35. [CrossRef]
6. Jonoobi, M.; Shafie, M.; Shirmohammadli, Y.; Ashori, A.; Zarea-Hosseinabadi, H.; Mekonnen, T. A Review on Date Palm: Properties, Characterization and Its Potential Applications. *J. Renew. Mater.* **2019**, *7*, 1055–1075. [CrossRef]
7. Elseify, L.A.; Midani, M.; Shihata, L.A.; El-Mously, H. Review on cellulosic fibers extracted from date palms (*Phoenix dactylifera* L.) and their applications. *Cellulose* **2019**, *26*, 2209–2232. [CrossRef]
8. Ghulman, H.A.; Metwally, M.N.; Alhazmi, M.W. Study on the benefits of using the date palm trees residuals in Saudi Arabia for development of the non-traditional wooden industry. *AIP Conf. Proc.* **2017**, *1814*, 020012. [CrossRef]
9. Elseify, L.A.; Midani, M.; Hassanin, A.H.; Hamouda, T.; Khiari, R. Long textile fibres from the midrib of date palm: Physiochemical, morphological, and mechanical properties. *Ind. Crops Prod.* **2020**, *151*, 112466. [CrossRef]
10. Lottermoser, B.G. Colonisation of the rehabilitated Mary Kathleen uranium mine site (Australia) by *Calotropis procera*: Toxicity risk to grazing animals. *J. Geochem. Explor.* **2011**, *111*, 39–46. [CrossRef]
11. Hassan, L.M.; Galal, T.M.; Farahat, E.A.; El-Midany, M.M. The biology of *Calotropis procera* (Aiton) W.T. *Trees* **2015**, *29*, 311–320. [CrossRef]
12. Al-Rowaily, S.L.; Abd-ElGawad, A.M.; Assaeed, A.M.; Elgamal, A.M.; El Gendy, A.E.-N.G.; Mohamed, T.A.; Dar, B.A.; Mohamed, T.K.; Elshamy, A.I. Essential Oil of *Calotropis Procera*: Comparative Chemical Profiles, Antimicrobial Activity, and Allelopathic Potential on Weeds. *Molecules* **2020**, *25*, 5203. [CrossRef]
13. Kaur, A.; Batish, D.R.; Kaur, S.; Chauhan, B.S. An Overview of the Characteristics and Potential of *Calotropis procera* from Botanical, Ecological, and Economic Perspectives. *Front. Plant Sci.* **2021**, *12*, 690806. [CrossRef] [PubMed]
14. Brandes, D. *Calotropis procera* on Fuerteventura. 2005. Available online: https://www.biblio.tu-bs.de/geobot/fuerte.html (accessed on 2 August 2022).
15. Sharma, R.; Thakur, G.S.; Sanodiya, B.S.; Savita, A.; Pandey, M.; Sharma, A.; Bisen, P.S. Therapeutic Potential of *Calotropis procera*: A giant milkweed. *IOSR J. Pharm. Biol. Sci.* **2012**, *4*, 42–57. [CrossRef]
16. Menge, E.O.; Greenfield, M.L.; McConchie, C.A.; Bellairs, S.M.; Lawes, M.J. Density-dependent reproduction and pollen limitation in an invasive milkweed, *Calotropis procera* (Ait) R.Br. (Apocynaceae). *Austral Ecol.* **2017**, *42*, 61–71. [CrossRef]
17. Al Sulaibi, M.A.M.; Thiemann, C.; Thiemann, T. Chemical Constituents and use of *Calotropis procera* and *Calotropis gigantea*—A Review (Part 1–The Plants as Material and Energy Resources). *Open Chem. J.* **2020**, *7*, 1–15. [CrossRef]
18. Batool, H.; Hussain, M.; Hameed, M.; Ahmad, R. A review on *Calotropis procera* its phytochemistry and traditional uses. *Big Data Agric.* **2020**, *2*, 56–58. [CrossRef]
19. Song, K.; Zhu, X.; Zhu, W.; Li, X. Preparation and characterization of cellulose nanocrystal extracted from *Calotropis procera* biomass. *Bioresour. Bioprocess.* **2019**, *6*, 45. [CrossRef]
20. Sakthivel, J.C.; Muchopadhyay, S.; Palanisamy, N.K. Some studies on Mudar Fibers. *J. Ind. Text.* **2005**, *35*, 63–76. [CrossRef]
21. Hilário, L.S.; dos Anjos, R.B.; de Moraes Juviniano, H.B.; da Silva, D.R. Evaluation of Thermally Treated *Calotropis procera* Fiber for the Removal of Crude Oil on the Water Surface. *Materials* **2019**, *12*, 3894. [CrossRef]
22. Dos Anjos, R.B.; Hilário, L.S.; de Moraes Juviniano, H.B.; da Silva, D.R. Crude Oil removal using *Calotropis procera*. *BioResources* **2020**, *15*, 5246–5263. [CrossRef]
23. Raghu, M.J.; Goud, G. Effect of surface treatment on mechanical properties of *Calotropis procera* natural fiber reinforced epoxy polymer composites. *AIP Conf. Proc.* **2020**, *2274*, 030031.

24. Mrazova, M. Advanced composite materials of the future in aerospace industry. *INCAS Bull.* **2013**, *5*, 139–150.
25. Bulut, M.; Alsaadi, M.; Erkliğ, A. A comparative study on the tensile and impact properties of Kevlar, carbon, and S-glass/epoxy composites reinforced with SiC particles. *Mater. Res. Express* **2018**, *5*, 025301. [CrossRef]
26. Sasikumar, K.; Manoj, N.R.; Mukundan, T.; Rahaman, M.; Khastgir, D. Mechanical Properties of Carbon-Containing Polymer Composites. In *Carbon-Containing Polymer Composites*; Rahaman, M., Khastgir, D., Aldalbahi, A., Eds.; Springer Nature Singapore Ltd.: Singapore, 2019; pp. 125–155.
27. Krushnamurty, K.; Srikanth, I.; Rangababu, B.; Majee, S.K.; Bauri, R.; Subrahmanyam, C. Effect of nanoclay on the toughness of epoxy and mechanical, impact properties of E-glass-epoxy composites. *Adv. Mater. Lett.* **2015**, *6*, 684. [CrossRef]
28. Saiman, M.P.B.; Wahab, M.S.B.; Wahit, M.U.B. The Effect of Compression Temperature and Time on 2D Woven Kenaf Fiber Reinforced Acrolynitrile-Butadiene-Styrene (ABS). *Appl. Mech. Mater.* **2013**, *315*, 630–634. [CrossRef]
29. Supian, A.B.M.; Jawaid, M.; Rashid, B.; Fouad, H.; Saba, N.; Dhakal, H.N.; Khiari, R. Mechanical and physical performance of date palm/bamboo fibre reinforced epoxy hybrid composites. *J. Mater. Res. Technol.* **2021**, *15*, 1330–1341. [CrossRef]
30. AL-Oqla, F.M.; Alothman, O.Y.; Jawaid, M.; Sapuan, S.M.; Es-Saheb, M.H. Processing and Properties of Date Palm Fibers and its Composites. In *Biomass and Bioenergy-Processing and Properties*, 1st ed.; Hakeem, K., Jawaid, M., Rashid, U., Eds.; Springer: Berlin, Germany, 2014; Chapter 1; pp. 1–25.
31. Ghori, W.; Saba, N.; Jawaid, M.; Asim, M. A review on date palm (*Phoenix dactylifera*) fibers and its polymer composites. The Wood and Biofiber International Conference. *IOP Conf. Ser. Mater. Sci. Eng.* **2018**, *368*, 012009. [CrossRef]
32. Abdellah, M.Y.; Seleem, A.-E.H.A.; Marzok, W.W.; Hashem, A.M.; Backer, A.H. Tensile and Imp act Properties of Hybrid Date Palm Fibre Composite Structures Embedded with Chopped Rubber. *Int. J. Eng. Res. Appl.* **2022**, *12*, 54–66.
33. EN 317; Particleboards and Fibreboards. Determination of Swelling in Thickness after Immersion in Water. European Committee for Standardization: Brussels, Belgium, 1996.
34. EN 310; Wood-Based Panels. Determination of Modulus of Elasticity in Bending and of Bending Strength. European Committee for Standardization: Brussels, Belgium, 1993.
35. EN ISO 527-4; Plastics—Determination of Tensile Properties—Part 4: Test Conditions for Isotropic and Orthotropic Fibre-ReinForced Plastic Composites. International Organization for Standardization: Geneva, Switzerland, 2021.
36. ISO 179-1; Plastics—Determination of Charpy Impact Properties, Part 1: Non-Instrumented Impact Test. International Organization for Standardization: Geneva, Switzerland, 2010.
37. Cerbu, C.; Cosereanu, C. Moisture effects on the mechanical behavior of fir wood flour/glass reinforced epoxy composite. *BioResources* **2016**, *11*, 8364–8385. [CrossRef]
38. Cerbu, C.; Ursache, S.; Botis, M.F.; Hadăr, A. Simulation of the hybrid carbon-aramid composite materials based on mechanical characterization by digital image correlation method. *Polymers* **2021**, *13*, 4184. [CrossRef]
39. Schwanninger, M.; Rodrigues, J.C.; Pereira, H.; Hinterstoisser, B. Effects of short-time vibratory ball milling on the shape of FT-IR spectra of wood and cellulose. *Vib. Spectrosc.* **2004**, *36*, 23–40. [CrossRef]
40. Narayanasamy, P.; Balasundar, P.; Senthil, S.; Sanjay, M.R.; Siengchin, S.; Khan, A.; Asiri, A.M. Characterization of a novel natural cellulosic fiber from *Calotropis gigantea* fruit bunch for ecofriendly polymer composites. *Int. J. Biol. Macromol.* **2020**, *150*, 793–801. [CrossRef] [PubMed]
41. Oun, A.A.; Rhim, J.-W. Characterization of nanocelluloses isolated from Ushar (*Calotropis procera*) seed fiber: Effect of isolation method. *Mater. Lett.* **2016**, *168*, 146–150. [CrossRef]
42. Faix, O. Classifcation of lignins from different botanical origins by FT-IR spectroscopy. *Holzforschung* **1991**, *45*, 21–27. [CrossRef]
43. Pandey, K.K.; Pitman, A.J. FTIR studies of the changes in wood chemistry following decay by brown-rot and white-rot fungi. *Int. Biodeterior. Biodegrad.* **2003**, *5*, 151–160. [CrossRef]
44. Yoganandam, K.; Ganeshan, P.; NagarajaGanesh, B.; Raja, B. Characterization studies on *Calotropis Procera* fibers and their performance as reinforcements in epoxy matrix. *J. Nat. Fibers* **2019**, *17*, 1706–1718. [CrossRef]

Article

Some Methods for the Degradation-Fragility Degree Determination and for the Consolidation of Treatments with Paraloid B72 of Wood Panels from Icon-Type Heritage Objects

Anamaria Avram [1,2], Constantin Ștefan Ionescu [2] and Aurel Lunguleasa [1,*]

[1] Wood Processing and Design of Wooden Product Department, Transilvania University of Brasov, 29 Street Eroilor, 500038 Brasov, Romania; anamaria.avram@unitbv.ro
[2] Laboratory of Restoration and Research "Restaurare Ionescu Constantin", Henri Coandă 12, 550234 Sibiu, Romania; ionescu.constantin.stefan@unitbv.ro
* Correspondence: lunga@unitbv.ro

Abstract: The main objective of this paper is to develop methods for assessing the deterioration of wooden panels of iconic heritage objects and the effectiveness of consolidation treatments, methods that are easy to apply to the field of wood restoration. During the research, four evaluation methods were identified, respectively: the density method, the excessive porosity method, the Brinell hardness method, and the Mark hardness method. Each method was exemplified on five wooden panels (icons), and when needed, degraded specimens were used and/or treated with Paraloid B72. One of the main conclusions of the research is that, although all methods are minimally invasive and do not require cutting of these heritage objects, the applicability of each is done depending on the type of degradation, often requiring a combined analysis between two or several methods. Additionally, the classification of the cultural good in one of the five degrees of embrittlement-degradation help to design a technological flow regarding the treatments of consolidation/restoration of the heritage object.

Keywords: fragility degree; restoration; icon; heritage objects; consolidation; wood; degradation evaluation

1. Introduction

Heritage objects are valuable objects for a certain community or a certain geographical space, from the cultural, artistic, historical, faith, etc., points of view. The concept of heritage is constantly evolving through its traditional, chronological, and geographical character. In parallel with this aspect, the selection criteria of a heritage object have been extended [1] adding to the historical and artistic value, other values such as the cultural one, the national identity one, and the memory interaction one. The new heritage concept was structured on two levels: material and immaterial heritage, respectively, tangible and intangible.

The degradation and fragility of wood from heritage objects are determined by the fact that wood is a biological material conducting the development of wood-decaying insects and fungi, these being the main factors of its degradation and fragility [2]. Damage can sometimes be so severe, especially for small heritage objects (icons) [3,4] that it jeopardizes their continued existence. Figure 1 shows the three important stages of a heritage object, namely the initial stage as a new object, the intermediate degradation stage (with several stages), and the final stage when the degradation is so strong that the object can no longer be restored.

Wood degradation depends on its structure, especially its density but also the content of secondary chemicals [4]. For example, dense wood of 1260 kg/m^3 such as *Guaiacum officinale* [5] is more difficult to be attacked by fungi and insects, and wood containing a considerable percentage of tannins or resins is also less susceptible to insect–fungus attack. Several types of wood degradation are known [4], the most important of which are: cracking, deformations, insect holes, galleries, mold, rot, etc. (Figures 2 and 3).

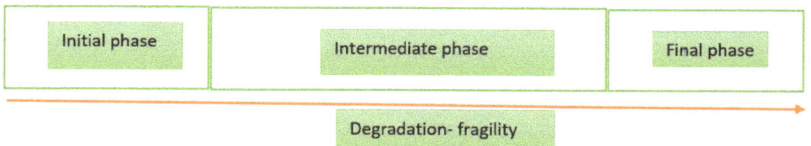

Figure 1. The stages of wood degradation-fragility from heritage objects.

Figure 2. Degradation of wood support: mold (a) and rot (b,c).

Figure 3. Wood degradation: (a) *Coniophora puteana*, (b) *Serpula lacrymans* and (c) *Anobidae* insects.

The main wood-decaying fungi that attack wood from heritage objects are *Coniophora puteana* (wet rot) and *Serpula lacrymans* (dry rot); they cannot grow at negative temperatures, but always at a wood moisture content of optimal 22–25%. The mycelium and hyphae of these wood-eating fungi can penetrate the masonry to reach the wood [4].

Macchioni et al. [6] highlighted the need to revise international standards for the evaluation of wooden structures from heritage objects. Macchioni et al. [7] studied the state of conservation of wood from artifacts. The research was subject to the guiding principles of the standard UNI 11161: 05 [8] on species determination, anatomical recognition of wood structure, mechanical properties, identification of biological attack, and assessment of the damage caused by fungi and insects. Fassina [9] described the legal regulation on specific European standardization work in the field of cultural heritage conservation as essential for a common, unified approach to the conservation and restoration of cultural heritage. Wood damage is a natural process that depends on a number of factors, be they biotic or abiotic. The temperature considered optimal for the installation and development of xylophagous insects has values between 18–25 °C and the relative humidity of 55–95% [10], both values are slightly variable for the different types of xylophagous insects that are frequently found in heritage assets with wooden support (*Anobium punctatum, Lyctus linearis, Xestobium rufovillosum*, etc.). The optimum moisture content of wood for insect development is 26–32%, and when it falls below 17%, insects can reduce their activity [11]. The larvae, in a life cycle of 1–2 years, produce galleries with diameters between 1.5 and 4 mm, decaying the wood along the fibers, the exit galleries become sinuous and with diameters of 0.5–2.2 mm.

Sakuno and Schniewind [12] showed the importance of the quality of the reinforcement materials used during the restoration, using old Douglas wood (*Pseutdosuga menziesii*). Three types of synthetic polymers were used in the experiments, namely Acrylic B72, Butylul B98 polyvinyl butyral, and Ayat polyvinyl acetate in 15% solution, each of the three with two types of solvents. For experimental research, old wood was taken in the form of poles, extracted from an old construction, where they stayed 70 years underground.

Cataldi et al. [13] investigated the use of other thermoplastic composites made of microcrystalline cellulose powder or even bamboo paper [14] in various percentages (up to 30%, mass basis) as a consolidator and Paraloid B72 as a matrix, in order to grow mechanical properties of the composite. During the experiment, microcrystalline cellulose powder (Sigma-Aldrich) with a specific gravity of 1.56 g/cm^3 was used as the reinforcing filler. Resin Paraloid B72 procured by Rohm and Haas (Germany) with a specific gravity of 1.15 g/cm^3 was also used as the polymer matrix. The works elaborated by Cataldi et al. [15–17], made an analysis of the use of microcrystalline cellulose as a filler-consolidator in composite materials that will be used to strengthen the wooden support of heritage objects. For the experimental analysis, two types of historical wood from the 18th century were used, walnut (*Juglans regia*) and white fir (*Abies alba*), which showed deep degradations. This old wood was reinforced with a composite material made from the commercial polymer Paraloid B72 which is often used to strengthen the wood, in combination with two different amounts of 5 and 30% (weight basis) microcrystalline cellulose. As a comparison, the same tests were performed on clean and fresh wood specimens from the same species. The presence of this filler in the composite has increased its resistance to static or impact tests. Mańkowski et al. [18] only used Paraloid B72 on old linden wood (*Tilia cordata*). The paper looked at the retention of the consolidating Paraloid B72 solution in butyl acetate. During the first cycle of impregnation, the retention of Paraloid B72 was double that of the second cycle of impregnation.

Charola et al. [19] state that the applied treatment will not protect the wood from further deterioration, but will slow down the deterioration process and give the heritage object a longer lifespan so that in the future a new treatment can be applied if it is necessary. Timar et al. [20] addressed the issue of consolidant retention, its depth of penetration, and uniform distribution on the surface and inside the wood. The reinforcing products were Paraloid B72, beeswax, and two other types of paraffin. The consolidant retention was low by 2–4% in the case of Paraloid B72 (dilute solutions 50–100 g/L), but much higher in wax and paraffin-based products, by 20–26%.

Deng et al. [21] proved how much electron tomography means in the analysis of the degree of degradation and fragility of heritage objects, by analyzing the missing areas, sometimes even on a sub-microscopic scale. The paper uses electron tomography and develops an algorithm to identify and reconstruct the missing areas and information in 3D space. Schniewind and Eastman [22] used the scanning method Scan Electron Microscopy (SEM) to observe the percentage of wood cells filling with the reinforcing material. Three different consolidants were used (Butvar B98, Acryloid B72, and Butvar B90) and applied on damaged wood by vacuum impregnation. The Douglas pillars from a 70-year-old house, the part buried in the ground, on the shore of a lake, with obvious bacteriological degradation, were used as wood specimens. It has been shown that the number of consolidants decreases from the surface to the core of the specimen. Pavlidis et al. [23] showed the importance of 3D digitization scanning of the heritage object. The study identified three main factors influencing 3D digitization: the complexity of shape and size, the level of detail, and the diversity of materials used. Rivers and Umney [24] analyze in their book the history of furniture restoration, as well as the classical and modern materials and techniques used in this case. Siau et al. [25] made a foray into the phenomena and processes that occur in wood and especially those related to moisture content transfer.

The most common treatment used to harden wood is acrylic polymers (often Paraloid B-72). The choice of solvent, its toxicity, explosiveness, and flammability must also be taken into account [18]. For example, the highest degree of wood saturation was obtained using Paraloid dissolved in methanol; however, due to the strong swelling caused by methanol, it could not be applied in conservation practice. High saturation causes a low penetration of the solvent into the depth of the wood [18]. In their research, they determined polymer content in damaged wood samples impregnated with Paraloid solution 20% B-72 in toluene and found that there is polymer at a depth greater than 7 mm in about 10% of wood vessels. A better supersaturation was obtained by dissolving the Paraloid in acetone [18], except that acetone produces dimensional instability of the wood, and its use in restoration should be judiciously observed. One of the main advantages of B-72 as a consolidator is that it is stronger and harder than others, without being extremely brittle. This consolidator is more flexible than many of the other typically used consolidators and tolerates more stress on jointing. Along with Paraloid B72, another synthetic product used to strengthen the wood support is polyethylene glycol. However, the disadvantages of reduced penetration in the cell membrane for concentrations higher than 10%, lack of antibacterial properties, increased acidity of the treated product, and reduced depth of penetration into the wood must be overcome [22].

The degree of impregnation will depend on the consolidating material, the solvent used, the concentration and the viscosity of the solution, the permeability of the wood to be consolidated, the technique used (brushing, injection, immersion, vacuum impregnation, etc.), and other treatment parameters such as duration and temperature [20]. Higher concentration solutions store more resin and will therefore give more strength. Another material used in restoration is Regalrez 1126—a saturated cyclic hydrocarbon similar to wax and paraffin. It was found that a concomitant mixture of the two substances (Paraloid B72 dissolved in toluene and ethyl acetate and Regalrez 1126) could cause precipitation, resulting in a suspension that would make the injection difficult, which could have a negative effect on the expected effect. Paduretu and Ghiorghita [3] stated that a restored object must remain original, without noticing the interventions that were operated on it. That is why the materials used for restoration must be compatible with the old wood from the heritage object. Walsh-Korbs and Avérous [26] argued that the reduction in density can also be explained as a decrease in the structural components of wood. Bucsa and Bucsa [4] showed that a humid space and a reduced air circulation are the main factors for the development of the attack of fungi and/or xylophagous, independent or combined. This attack causes rot or larval galleries, and these lead to chromatic changes or loss of mechanical strength of the wood. Other authors have analyzed the restoration process [27–29], others have noticed that the visual inspection of the surface is not enough [30], that a chemical cleaning of the surface is needed [31–33], a preliminary diagnosis is needed [34], even 3D digitization [35], the use of Paraloid B72 [36,37], or the use of new wood of the same species for comparison [38]. It has also been proposed to use heat-treated wood as an artificial degradation process [39,40] and to use modern FT-IR and X-ray tomography methods [41,42].

Many researchers in the field of wood restoration and conservation addressed in detail the analysis of the factors and mechanisms that produce the degradation, fragility, and even partial destruction of wood in heritage objects. They identify the problem and improve it, but do not establish a scale of these degradations, an index, or a percentage of the degradation. Following the critical analysis of the literature, three main objectives of the paper were identified. The first objective is to find some practical methods for assessing the degradation of the wooden support of the heritage object at the entrance to the restoration laboratory. The second objective is to evaluate the effectiveness of the wood reinforcement treatment, by the same methods specific to the first objective. The third objective is to develop a ranking of the level of wood degradation, simultaneously with the measures to consolidate the wood support for each level.

2. Materials and Methods

The experimental studies were carried out in the research and restoration laboratory ("Restaurare Ionescu", Sibiu, Romania), and the analyzed icons were 150–350 years old. Four new methods for wood degradation degrees have been developed:

- Determining the comparative density of healthy wood and degraded wood;
- Determination of excessive wood porosity caused by insect holes and galleries;
- Determination of Brinell hardness and comparison of the values obtained for healthy wood with degraded wood;
- Determination of hardness by means of the wood-pricking device, Mark 10, as a minimally destructive and alternative method.

Xylophagous insects, by the nature of the attack and the attacked wood species, can be classified as xylophagous insects that feed and grow on single wood and xylomycetophagous insects that grow in symbiosis with fungi. For many types of insects, which cause wood degradation, the action develops symbiotically (Figure 4) and is constantly preceded by the development of fungi [4].

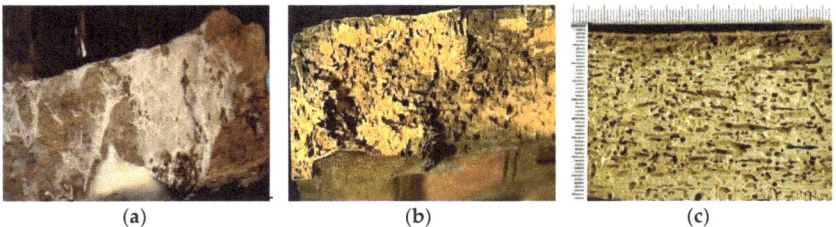

(a) (b) (c)

Figure 4. Degradation of the wood support; (**a**) fungal attack (*Serpula lacrymas*); (**b**) symbiotic attack; (**c**) xylophagous attack.

2.1. Determination of the Degree of Fragility by the Method of Comparative Densities

This method is intended to obtain an answer for the differences in mass (density) between the two panels (healthy and degraded), considering that the initial density of the degraded panel is the same as that of new, healthy wood specimens without defects and degradations of the same species. The density method determines the total density difference between the panels at the same percentage of wood moisture (degraded and the new reference). Moisture content measurement was performed with Gann HT65 Humidifier with M20 hammer (GANN Mess. Regeltechnik GmbH, Nürnberg, Germany), measuring range 4–60%, analytical balance, EWJ 600–2M Kern (Merck KGaA, Darmstadt, Germany), with an accuracy of 0.01 g. For this study, 5 icons on linden wood (*Tillia cordata* Mill.) degraded by intense and medium xylophagous attack (visible on the outside) were considered from the laboratory. Because we did not want to intervene in the moisture content of the wood panel, the following density transformation ratio was used at a certain moisture content (under fiber saturation point) to another one (Equation (1)):

$$\rho_{MC2} = \rho_{MC1} \frac{1 + MC_2}{1 + MC_1 + (MC_2 + MC_1) \cdot \rho_{MC1}} \left[\text{kg/m}^3 \right] \quad (1)$$

where: ρ_{MC2}—density at moisture content MC_2, in kg/m^3; ρ_{MC1}—density at moisture content MC_1, in kg/m^3; MC_2—Moisture content tip 2, in %; MC_1—Moisture content tip 1, in %.

The density of lime wood introduced into heritage objects 200 years ago is not precisely known and depends on the vegetation conditions of the tree and other biotic and abiotic factors. As there are no precise methods for evaluating it, it was considered the equivalence-approximation of density with that of the existing species (lime) to the required moisture content. Beyond these limitations, the assessment is within ±5% of current statistical analyzes. New 50 × 50 mm specimens, made of healthy wood, of the same essence, with the same thickness, conditioned at a moisture content of 6% or 8%, depending on the moisture content of the analyzed icon, were used for the sharing. The density of the degraded panel was determined, as a ratio between the mass and the degraded volume, obtaining the value of 290.69 kg/m³. For comparison, several rulers of timber of the same species were taken, brimstone linden, brought to the same moisture of 6% by conditioning, from which were cut 6 specimens with dimensions of 20 × 20 × 30 mm, at which the density was determined as the ratio between their mass and volume (EN 323: 1993). The determination of the degree of degradation-fragility by the density method is based on determining the density of the degraded wood support and the density of the original wood support, the non-degraded one, respectively, using the following relationship (Equation (2)):

$$G_{f\rho} = \frac{\rho_i - \rho_d}{\rho_i} \cdot 100 \ [\%] \quad (2)$$

where: $G_{f\rho}$—the degree of fragility-degradation by the density method, when the panels have the same moisture, in %; ρ_i—the density of the healthy wood specimen, at the same moisture, expressed in kg/m³; ρ_d—density of degraded wood specimens, at the same moisture content, expressed in kg/m³.

2.2. Determination of the Degree of Degradation by the Method of Excessive Porosity Caused by Xylophagous Attack

The excessive porosity of heritage objects is determined by the attack of wood-decaying insects inside the wood, causing holes and larval galleries, especially by the group of insects *Anobiidae*. The holes visible in the wood surface are multiplied inwards compared to the surface holes several times (Figure 5). Initially, an area of the outer surface was colored black and there was a color difference between the areas with holes and those without holes. The two surfaces were determined by color scanning. Then, the outer part was excavated on the same surface, revealing the holes and inner galleries, much more complex than those on the surface. The surface was dyed black, with darker and lighter areas appearing. The area of the two areas was determined by scanning. The ratio between the two surfaces was made, obtaining the coefficient of 3.89 for the 5 different surfaces taken into account. This value was used for all icons.

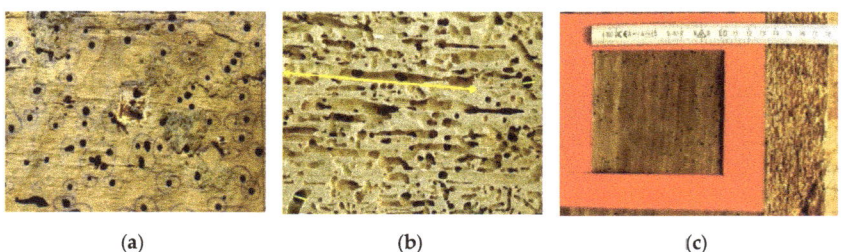

Figure 5. Insect holes and galleries: (**a**) at surface; (**b**) at interior; (**c**) comparative.

In the literature, this porosity is expressed as the number of holes per dm² [4], with current values of 40–200 holes/dm². Taking into account the volume/surface lost through these holes in the volume/surface of the clean wood, it is possible to quantify how much a consolidant solution is needed to restore the mechanical strength and hardness of the restored object. The principle of the method consists in observing a small perimeter, in shape and size, usually a square with a side of 100 mm (Figure 6) in order to analyze it. The flight holes are counted for at least 5 attacked areas. The average number of holes/dm² is multiplied by the area in dm² of a hole, depending on the average diameter of the holes. Diameters had to be measured at two perpendicular diameters for each hole and the average value was the reference.

Figure 6. Marking and counting of flight holes per dm².

The known number of flight holes is determined on this known surface, but also their average diameter is determined. In general, the intensity of the degradation is determined as the ratio between the area of the insect holes and the area taken into account (Equation (3)):

$$G_{fs} = \frac{k \cdot n \cdot \pi \cdot d^2}{4 \cdot A_{fh}} \cdot 100 \ [\%] \quad (3)$$

where: G_{fs}—degree of fragility-surface degradation as intensity of flight holes, in %; k—multiplication coefficient of the core holes, equal with 3.89; n—the number of flight holes on the entire analyzed surface; d—average diameter of flight holes, in mm; A_{fh}—the plane area of the surface that is taken into account, in mm².

In the conditions in which the number of holes per dm² was previously determined, the relation (4) was transformed into a simpler one, respectively:

$$G_{fs} = \frac{3.89 \cdot n_d \cdot \pi \cdot d^2}{400} \ [\%] \quad (4)$$

where: n_d—number of holes per dm².

The procedure was applied to 5 icons in the laboratory, each analyzing 5 distinct areas.

2.3. Determination of the Degree of Fragility by the Brinell Hardness Method

The degradation and fragility of wood are not uniform and can be on the surface, interior, or combined for the entire wooden support of the heritage object. Thus, one method of assessing the degradation of the wood surface is to compare the hardness of the degraded wood with that of the original wood introduced in the heritage object (equivalent to that of the wood species identified in the restored heritage object). If the Brinell hardness (Hardness Brinell) is noted with HB, the degree of degradation-fragility (expressed as a percentage loss of wood hardness) will be determined by the following relation (Equation (5)):

$$G_{fHB} = \frac{HB_i - HB_d}{HB_i} \cdot 100 \ [\%] \quad (5)$$

where: G_{fHB}—degree of fragility by Brinell hardness method, in%; HB_i—initial Brinell hardness, in N/mm²; HB_d—Brinell hardness of the panel damaged, in N/mm².

The purpose of Brinell hardness method is to determine the strength of the wood surface. Brinell hardness can be interpreted as the property of wood to resist the penetration of a 10 mm diameter penetrator (Figure 7), under the action of a constant force. The force tends to change its surface (EN 1534-2003) [43]. As the determination could not be made directly on icon surfaces because of the risk of their destruction, the research was organized in two directions. In the case of the first direction, more than 60 healthy specimens of lime and balsa wood were used, with or without Paraloid B72 consolidator, with a concentration of 10% in ethyl acetate and toluene 1:1, in order to determine the effectiveness of the treatment on new or degraded wood. In the second research direction, 30 pieces of 50 × 50 mm lime specimens with various degradations were used. These specimens were treated or untreated with 10% B72, being taken from old, abandoned panels, recovered, and kept in the laboratory archives. In this way, the degrees of fragility-degradation of the analyzed specimens were determined.

Figure 7. Brinell hardness tester: 1—device body; 2—threaded pin; 3—additional punch; 4—bushing; 5—metal ball.

The Brinell hardness calculation relationship was as follows (Equation (6)):

$$HB = \frac{2 \cdot F}{\pi \cdot D(D - \sqrt{D^2 - d^2})} \left[\frac{N}{mm^2}\right] \quad (6)$$

where: HB—Brinell hardness, in N/mm^2; F—pressure force, 100 N; D—diameter of the tip penetrator Ø10, in mm; d—the diameter of the imprint left on the wood, in mm.

For this research, a special stand with a magnifying glass of 5× magnification was used in the Brinell tests, and a copying indigo sheet was inserted between the penetrator and the test ball to highlight the diameters.

In order to compare, hardness determinations were performed for healthy new wood specimens for which consolidation treatment was applied with Paraloid B72, having a concentration of 10% in ethyl acetate, by immersion for one hour, for lime species and balsa (as a complement-consolidation wood).

The following formula was used to calculate the effectiveness of the hardening reinforcement treatment (Equation (7)):

$$E_{HB} = \frac{HB_{B72} - HB_i}{HB_i} \cdot 100 \; [\%] \quad (7)$$

where: E_{HB}—hardness efficiency after consolidation treatment, in %; HB_{B72}—hardness after consolidation with Paraloid B72; HB_i—initial Brinell hardness, before treatment.

2.4. Determination of Mark Hardness as a Minimally Invasive Method

The hardness test, with the Mark 10 dynamometer equipped with a pressing-extraction device (Mark-10 Corporation, Copiague, NY, USA), is based on the penetration into the wood of a thin cylindrical-conical tip with a diameter of 1.34 mm and a length of 6 mm (consisting of a conical area of 2 mm and a cylindrical area of 4 mm), with a pressing force indicated on the digital screen of the dynamometer, expressed in [N]. The principle of operation of the device is based on the stinging of the wood and the determination of the opposite force of the wood when the needle penetrates inside it. The Mark-10 digital camera (type M3-200) measures forces in N, in a range of 0–1000 N. The Mark 10 M3-200 series dynamometer and the S10 fixing-testing stand are manufactured by Mark-10 Corporation. For these tests, the 5 icons used in the first two methods (of excessive density and porosity) were used. A total of 5 areas with different degradations were chosen for each icon. As materials for comparative testing, the same specimens that were previously tested with the Brinell stand were used, and the determination of HM (Hardness Mark) was performed in the vicinity of the Brinell test. The total lateral area of the penetrator tip was calculated as the sum of the lateral areas of the two geometric bodies (cylinder and cone), obtaining the value of the total lateral area of the penetrator of 21.28 mm². Based on this area, Mark hardness was determined with the following relationship (Equation (8)):

$$HM = \frac{F}{A_{lt}} = \frac{F}{21.28} \quad \left[\frac{N}{mm^2}\right] \quad (8)$$

where: HM—Mark hardness, in N/mm²; F—Force read on dynamometer Mark 10, in N; A_{lt}—the total lateral surface of the tip penetrator; 21.28—lateral contact area between wood and the tip of the penetrator, in mm².

The degree of fragility by the Mark hardness method was determined by the following relationship (Equation (9)):

$$G_{fHM} = \frac{HM_i - HM_d}{HM_i} \cdot 100 \; [\%] \quad (9)$$

where: G_{fHM}—degree of embrittlement obtained with the Mark hardness method, in %; HM_d—Mark hardness of degraded wood, in N/mm²; HM_i—initial Mark hardness of new and healthy wood (reference), in N/mm².

Since the reporting area is a constant (21.28 mm²), the embrittlement-degradation formula can be simplified, taking into account only the compression force, respectively (Equation (10)):

$$G_{fFM} = \frac{FM_i - FM_d}{FM_i} \cdot 100 \; [\%] \quad (10)$$

where: G_{fFM}—degree of embrittlement obtained with the method of Mark force, in %; FM_d—Mark force of degraded wood, in N; FM_i—Mark force of initial new and healthy wood (reference), in N.

3. Results

3.1. Density Method Results

The degree of fragility by the density method (Gfρ) was determined for five examples of icons in the laboratory. The analyses of the results were performed separately for each icon, according to the European standard EN 17121 [44]. For example, the initial data of the icon were first recorded (example 1), at the entrance to the restoration laboratory, which was the following: mass of the panel degraded before restoration of 791 g and 818 g after the evaluation of the losses from the panel of 3.3%, the external dimensions 280 × 335 × 30 mm (V = 0.002814 m³), the moisture content of 6% and the wood species linden (*Tilia cordata* Mill.) was obtained [45]. An average density of 509.8 kg/m³ was obtained (Table 1). Using the formula (Equation (2)), applying the values of a new panel and for the degraded panel to 6% moisture content, total degradation of 42.97% was obtained. This degradation-fragility coefficient contained all wood losses, whether they are due to wood-decaying insects, wood-decaying fungi, fungal attach, or other physical damages.

Table 1. Degree of degradation-fragility obtained by the density method.

Icon	Initial Mass, g	Loss of Wood, %	Reconstituted Mass, g	Panel Dimensions, mm	Icon Density, kg/m^3	New Wood Density, kg/m^3	Loss of Density, %
1	791	3.3	818	280 × 335 × 30	290.69	509.8	42.97
2	1243	2.1	1270	415 × 345 × 28	316.7	509.8	37.87
3	1380	0	1380	415 × 345 × 28	344.3	509.8	32.46
4	9494	1.2	9610	970 × 730 × 34	399.1	514.4	22.41
5	5135	0	5135	780 × 630 × 24	435.4	514.4	15.35

The degree of fragility by the density method (Gfρ) was determined for five examples of icons, and they are recorded in Table 1, where all the values for the presented examples are highlighted.

Further analysis of the values was made and correlated with the visual aspect of the heritage objects, and classification was found from the point of view of the degrees of degradation-fragility in three categories: 15–25%—weak degradations, 25–35% medium degradation, and 35–45% as severe degradation.

3.2. Results of Xylophagous Insect Attack

The results obtained in the case of the attack of xylophagous insects were centralized in Table 2. The analysis was done for the same five icons used in the case of the density method, evaluating their back, without pictorial support. For example, for the number 1 icon, the analyzed surfaces had dimensions of 172 × 100 mm. An average number of 346 insect holes was obtained on the total surface and a specific one of 201 holes/dm^2. The diameter of the flight holes was between 1.4–2.6 mm, with an average value of 2.03 mm. Introducing in relation (5) a degree of fragility-degradation of 25.2% was obtained.

Table 2. Degree of fragility-degradation by the method of excessive porosity given by insect holes.

Icon	Planar Dimensions, mm	Total Number of Holes	Average Diameter, mm	Number of Holes/dm^2	Fragility, %
1	172 × 100	346	2.03	201	25.2
2	80 × 100	122	1.92	152	17.1
3	180 × 150	390	2.11	144	19.5
4	100 × 35	46	1.91	131	14.5
5	170 × 125	252	1.84	118	12.1

Based on the results obtained during the research of the five types of icons, but also on the mathematical modeling of the relation (4), the graph from Figure 8 was obtained. A slight increase in degradation was observed with an increase in the number of holes and an increase in the average diameter of the holes.

In practice, in real situations, these galleries can never be completely emptied of the sawdust inside them, so the degradation is not fully estimated, in fact, there are differences in mass and mechanical strength [4]. This explained the small amount of consolidant used during consolidation research, by about 10% less than the theoretical evaluations of consolidator Paraloid B72.

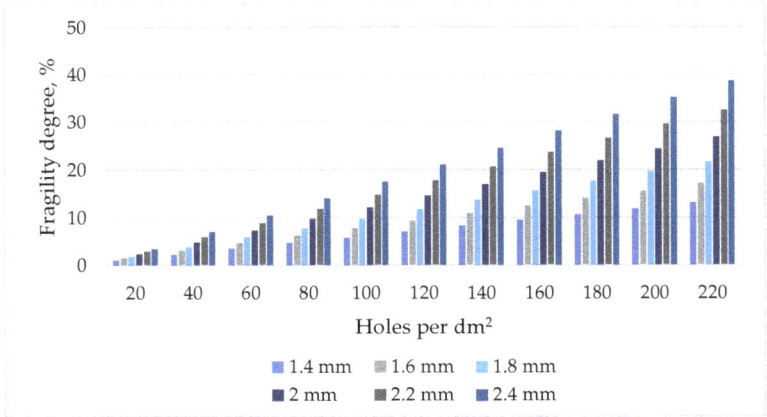

Figure 8. Fragility-degradation degree caused by xylophagous insects.

3.3. Results of Brinell Methods

In the first part of the research, namely that of establishing the effectiveness of the chemical consolidation treatment, the next Brinell hardness values were obtained: 17–22 N/mm^2 (with an average of 20.2 N/mm^2) on tangential surfaces in the case of untreated *Tilia cordata* and of 1.3–2.2 N/mm^2 (with an average of 1.8 N/mm^2) on the same radial-tangential surfaces in the case of *Ochroma pyramidale* (one of the usual species used in restoration works to complete the fragmentary losses, due to its very low density). The faces of the new wood were tangential, the direction of action of the forces being the radial one. The two species used in the research (balsa and lime) are frequently used in restoration. Analyzing the lime and balsa wood specimens, it can be seen that the new wood specimens that were treated with Paraloid B72 (the consolidation retention was about 3%, by dry mass), had an increased efficiency of Brinell hardness by 5.23% for lime and 7.22% for balsa. The growths are modest, which is why the treatment with the consolidant for new wood introduced in the heritage objects is not recommended. In the same way, the hardness of the wood degraded by xylophagous insects was determined, for the two types of specimens, respectively, degraded wood without treatment and with consolidation treatment with Paraloid B72 10%, for *Tillia cordata*. There were obtained the following values:

- 17–24 N/mm^2 (with an average of 21.5 N/mm^2) for new *Tillia cordata* wood;
- 15.1–18.7 N/mm^2 (with an average of 17.2 N/mm^2) for slightly degraded and untreated lime;
- 16.9–19.3 N/mm^2 (with an average of 18.3 N/mm^2) for lime treated with B72 10% in case of light xylophagous degradation;
- 10.8–15.2 N/mm^2 (with an average of 12.9 N/mm^2) for medium and untreated degraded lime;
- 12.5–17.5 N/mm^2 (with an average of 15.9 N/mm^2) for lime treated for an average degradation;
- 4.1–9.2 N/mm^2 (with an average of 6.4 N/mm^2) for strongly degraded and untreated lime;
- 7.7–13.1 N/mm^2 (with an average of 10.2 N/mm^2) for high xylophagous degradation and B72 treatment.

Analyzing the values in Figure 9, it is found that the slightly damaged wood had an improvement-efficiency of the hardness of 6.18% for medium degraded lime, then an increase in hardness by 23.33% for medium degraded lime wood, and in the case of highly degraded wood, the improvement was 60.46%. Even if this value (60.46%) seems very high, in fact, the Brinell hardness value of the very damaged wood increases from 6.40 N/mm^2 to 10.27 N/mm^2, but still does not reach the value of the average wood degraded before the treatment with B72. The Pearson coefficient of determination with values above 0.97 shows that the linear distribution of values best characterizes the modeling of values.

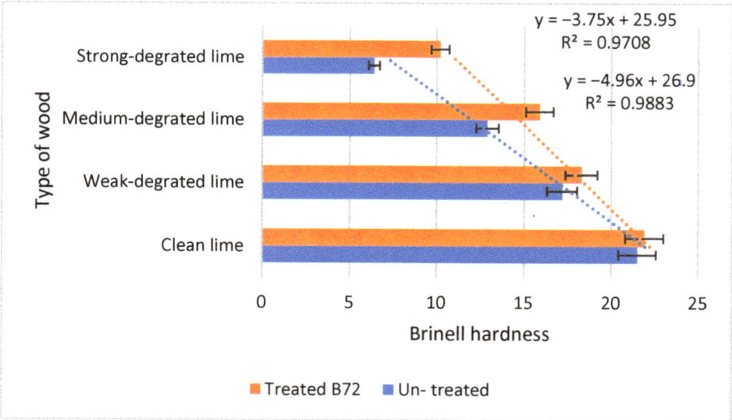

Figure 9. Brinell hardness of new and old degraded lime specimens, treated and untreated B72.

The degree of fragility-degradation in the case of Brinell hardness, determined with Equation (6), for slightly degraded lime wood was 20% for no-treated samples and 14.8% for lime treated with B72. In the case of medium-degraded specimens, the degree of fragility-degradation was 40% when the specimens were not treated and 26% when the specimens had B72 consolidation treatment. In highly degraded specimens, the degree of fragility-degradation was 70.2% when the specimens were not treated and 52.5% when treated with B72. In conclusion, the degradation-fragility values were identified by the Brinell method as 15–25% in case of light degradations, 30–45% in case of medium degradations, and 60–80% in case of strong degradations. Consolidated treatment reduced this degree of fragility by 26% in the case of light degradation, 35% in the case of medium degradations, and 25.2% in the case of strong degradations.

3.4. Mark Hardness Results

The results regarding the degree of fragility-degradation of the wood in the icons in the case of the Mark hardness were extended in two directions, respectively, for the degraded pieces of wood existing in the laboratory (at which the Brinell hardness was determined, for comparison) and for the five icons, which were analyzed in the first two methods. It was determined in both directions, Mark force as well as its corresponding strength, Mark hardness.

For the first part of the results, the obtained values are visible in Figure 10. This figure shows the Mark penetration force for both degraded and ungraded linden, in a probability plot statistical diagram. A degree of fragility-degradation was found, obtained with the relationship (11), of 74.1%, within the area of extended degradations, both the framing of the values between the limits on the graph, as well as two of the statistical parameters of the diagram, respectively, Anderson–Darling and p-value show the normality of the values in the 95% confidence interval. If the two values of the standard deviation and the 95% confidence interval are taken into account, the force variation intervals for the degraded lime and freshly clean lime of 153.5–242.7 N and 27.8–74.7 N, respectively, are found by calculation. These intervals are also found on the chart in its central area, for 50% values.

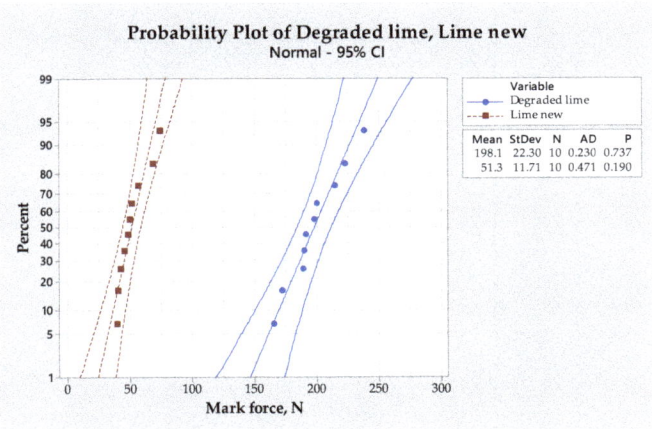

Figure 10. Pressure force obtained with the Mark10 dynamometer.

The results obtained in the second part of the research, respectively, based on the five analyzed icons, are visible in Table 3. In the example for icon 1, an area with medium and extended degradation was chosen on the back of the icons, obtaining a force of 73 N and 21 N, and a Mark hardness of 3.43 N/mm² and 0.98 N/mm². For comparison, healthy/new lime was used, without defects, on which measurements were made with the Mark 10 dynamometer, obtaining an average force of 205 N, respectively, a Mark hardness of 9.63 N/mm². Using the Formulas (10) and (11) a degree of fragility was determined of 64.38% for the area with medium degradation and 89.8% for the area with extended degradation.

Table 3. Degree of fragility with the Mark hardness method.

Icon	1		2		3	4	5	
Xylophagous Evaluation	Avg. *	Ext. **	Avg.	Ext.	Ext.	Ext.	Avg.	Low
Force, N	73	21	79	36	36	25	36	84	127
Mark hardness, N/mm²	3.43	0.98	3.71	1.69	1.69	1.17	1.69	3.94	5.96
Fragility degree, %	64.38	89.8	61.47	82.45	82.45	87.85	82.45	59.08	38.1

* Avg.—Average xylophagous damage; ** Ext.—Extend xylophagous damage; Low—low xylophagous damage.

Figure 11 shows a statistical graph made with the program Minitab 18, respectively, Empirical Cumulative Distribution Function (eCDF), which shows the curves in the form of "S" for clean lime, untreated degraded lime, and degraded lime—treated with B72. It is observed that the curve for degraded and treated lime has a more inclined curve, respectively, a larger variation (3.6–14.4 N/mm²) than the other two (0.4–6.4 N/mm², 15.9–19.5 N/mm²), this inclination gives the standard deviation of 2.7 N/mm², much higher than the other two groups of values of 1.5 N/mm² and 0.9 N/mm², respectively.

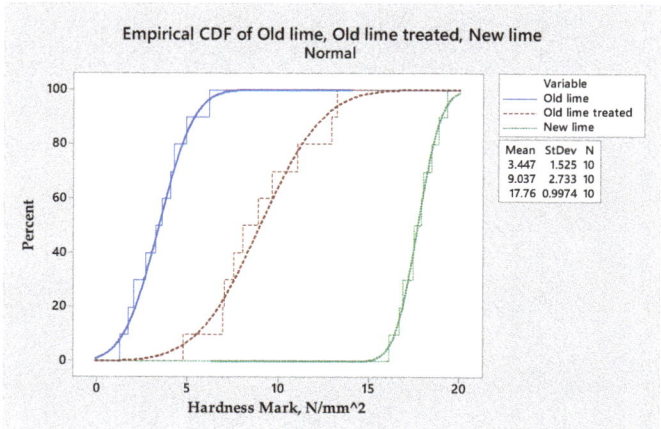

Figure 11. Mark hardness of 3 types of lime wood.

4. Discussion

Discussions about methods. In order to determine the destructive level of the methods that determine the hardness of the wood, the different areas of the imprint left by the penetrator were compared: for example, for the Janka hardness, it is found that the imprint left measures 100 mm^2, maximum 78 mm^2 for Brinell hardness, and at the Mark method the footprint is only 1.41 mm^2. Although all three methods can be considered destructive, it is still found that the footprint left by the penetrator adapted for the Mark dynamometer is very small and compared to Brinell is 98.19% smaller. The footprint of the Mark 10 penetrator is even smaller than that of an insect hole (Figure 12).

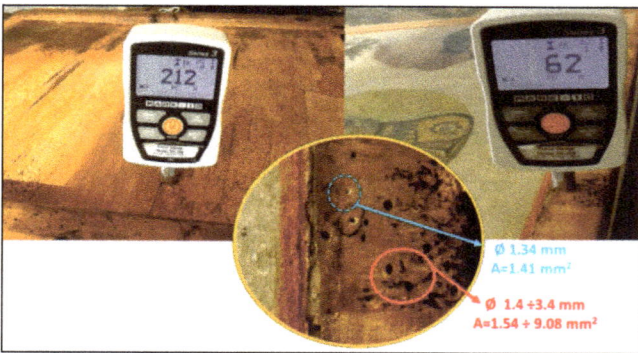

Figure 12. Comparison of the trace of the measuring tip with the flight hole caused by xylophagous insects: Mark 10 tip, Ø 1.34 mm A = 1.41 mm^2, flight hole, Ø 1.4–3.4 mm, A = 1.54–9.08 mm^2.

As Figure 10 shows, in most cases, the wood in the cultural goods is so degraded that, by pressing a force of even 20–80 N, the wood changes its flatness. For a force considered the minimum in the case of the Brinell hardness measurement of 100 N an imprint area eight times larger and a force ten times larger is used, causing deep damage, at least in the same ratio. As some of the recent studies stated [9,26,44] the strength of degraded wood must be evaluated. Even a 3D evaluation [21,23,35] should be performed, and Paraloid B72 is one of the best consolidators that can be used for wood strengthening in the restoration process [18,20] besides natural materials [22].

During the HB hardness determinations, for the degraded wood specimen, the values were measured at 100 N, and at the test of a force of 150 N, the wood split, not in the direction of the pressing force, but the sectioning occurred perpendicular to the direction of the pressing force; the minimal cohesion and strength between the anatomical elements of the wood gave way, producing the splitting of the specimen. If these measurements would take place for a cultural good, it would have resulted in its irreversible destruction [29].

Analyzing the average values between Brinell and Mark hardness, it is found that the ratios between the two methods of measuring hardness are below 1 N/mm^2, resulting in an overall average percentage difference of about ±3.34%. It can be seen that the least destructive method can be used, at least, to determine the hardness directly on the back of heritage assets, which have wooden support, especially when they are in different levels of degradation. Values of 17.2 N/mm^2 and 16.8 N/mm^2 were found in HB and HM for healthy lime, 4.6 and 4.7 N/mm^2 for highly degraded lime, 1.47 and 2.03 N/mm^2 for balsa wood, 9.66 and 10.64 N/mm^2 for old lime with B72 treatment and of 1.79 and 2.5 N/mm^2 for B72 treated balsa wood.

Using the formula for determining the Mark hardness [10,11], a degree of fragility was determined of 89.76% for the strongly degraded area, and 64.29% for the one with medium degradation; on the whole panel, the average degree of fragility was 77.02%, lower than a healthy wooden panel. For such a panel, whose hardness is reduced by 98.47% (on the most affected area, where HM = 1.97 N/mm^2) compared to a similar panel, made of healthy wood, there is the problem of the existence of the heritage object, whose support is extremely damaged. Such a value determines the time and level of the rescue intervention on the heritage object. When the Mark hardness was determined, a very degraded lime had a Mark hardness of 3.08 N/mm^2, compared to the healthy lime which had a Mark hardness of 10.39 N/mm^2, the ratio of these values being found in the case of Brinell hardness. The alternative to the Brinell hardness measurement, performed with the least destructive method MARK 10 starts from the premise that the degradation produced by the penetrator is very small.

Based on the values obtained with the Mark method, the lime degraded panel lost 81.18% of its hardness compared to a healthy panel; after the consolidation treatment, even if it produced an increase of 2.83 times compared to the strongly degraded panel, it can conclude that, in fact, the panel, after the treatment, improves in terms of hardness, only up to 46.59%, in relation with the reference panel.

It has been observed from Figure 13 that the weakening by the method of attack of xylophagous insects depends very much on the number of holes of xylophagous insects expressed per dm^2 [17], which is why the data from previous research have been centralized, obtaining a diagram that expresses the compression force related to flight holes. This diagram also has a linear regression equation, which best models the degradation phenomena, with a very good Pearson coefficient $R^2 = 0.989$, proving once again the dependence between the number of holes and the Mark hardness.

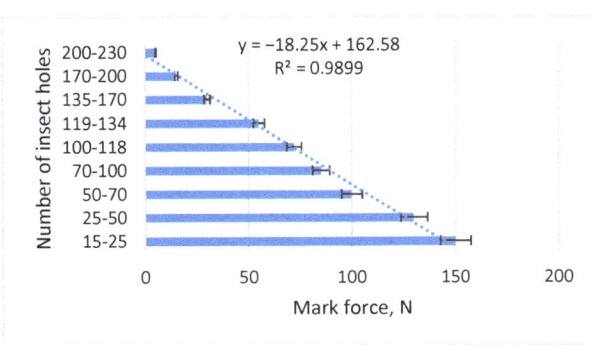

Figure 13. Influence of the number of holes on the Mark force stiffness.

When the intensity of the xylophagous attack exceeds 100 holes/dm^2, the wood becomes fragile [14,16,17] and the hardness is considerably reduced, the pressure of the penetrator falls below the average level, respectively, 70–80 N. As a general conclusion, it can be shown that if the number of holes flight increases, the hardness of Brinell and Mark decreases. Increasing the number of flight holes per dm^2 reduces the penetration resistance of the measuring tip. In this respect, a number of more than 60 flight holes/dm^2 reduces the puncture force to 110 N; at more than 100 holes/dm^2 the force decreases to 40 N, and at more than 220 holes/dm^2 the force is reduced below 10 N, the data refer to new, healthy and/or damaged linden wood.

The efficacy of Paraloid B72 treatment on medium degradation lime was analyzed from the point of view of the Brinell hardness method using ANOVA One-Way (Table 4), using null hypothesis, when all means are equal for a significance level alpha of 0.05. In this case, equal variances were assumed for the analysis. It is observed that the two values F-value and p-value correspond to significance assumed level alpha.

Table 4. Analysis of Variance (ANOVA) of Brinell Hardness for untreated and treated B72 medium-degraded lime.

Source	DF	Adj SS	Adj MS	F-Value	p-Value
Un-treated	9	0.8760	0.09733	*	*
Error	0	*	*		
Total	9	0.8760			

* These values are below 0.001.

On the same basis, by using the Minitab 18 statistical program, it found the interval plot of treated lime with Paraloid B72, comparative with un-treated lime, with medium degradation (Figure 14). There is a good correlation between the two groups of values when the confidence interval of the values is 95%, which shows once again the normality of the distribution of values.

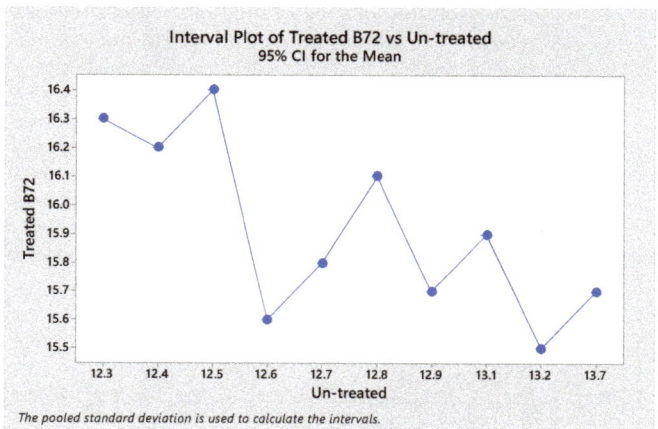

Figure 14. Interval plot of treated B72 vs. untreated medium-degraded lime values of Brinell hardness.

Establishing a hierarchy regarding the degree of degradation. According to the results obtained in this research, a differentiated ranking of the degradation level could be obtained. First of all, a visual instability was observed (visual method) with cracks, fissures, deformations, and gaps in the pictorial layer due to the successive swelling and contraction of the wooden panel in the icon. As other authors have noted before [3,7,30], simple visual inspection is not sufficient, requiring other methods of qualitative and quantitative quantification of degradation [41,42]. In this context, as was observed in the results, three degrees of fragility-degradation were obtained for all four methods. If it adds to these an incipient area in which the icon is not restored yet and a final one in which the heritage object can no longer be used, there were obtained the five levels of degradation-fragility of heritage objects (Table 5).

Table 5. Defining the degree and ranking level of degradation-fragility.

Degree	Level	Method	Fragility-Degradation Values	Icon Status
1.	Good	Density method, % Excessive porosity Number of holes/dm^2 Brinell hardness Hardness Mark	10–15 2–10 20–40 4–35 4–35	Acceptable. It is not being restored
2.	Weak	Density method, % Number of holes/dm^2 Excessive porosity Hardness Mark Brinell hardness	15.1–25 40.1–80 10.1–15 35–50 35.1–50	Worrying. Light restoration activities (surface consolidation treatments are needed)
3.	Average	Density method, % Number of holes/dm^2 Excessive porosity Hardness Mark Brinell hardness	25.1–35 80–160 15.1–20 36–50 50.1–65	Alarming. Medium restoration activities (surface and internal treatments are needed)
4.	Extended	Density method, % Number of holes/dm^2 Excessive porosity Brinell hardness Hardness Mark	35.1–45 160–200 20.1–25 65.1–80 51–60	Critical. Complex restoration activities, including replacement of wooden areas, are needed
5.	Exitus	Visual, all methods	Exceeding previous values	Unrecoverable icon. It is not being restored

A first observation obtained from Table 5 is that the two harnesses Brinell and Mark have very appropriate values for the first levels of degradation and quite appropriate for the following ones. A second observation is that there are different correlations of degradation levels between the four methods. Thus, there is a 5/1/2/2 ratio for the good level, 3/2/7/7 for the week level, 5/3/7/10 for the average level, and 7/4/13/10 for the extended level. These reports may amount to a certain method when there is no material possibility to carry it out effectively.

5. Conclusions

Each of the four simple and easy methods for assessing the degree of fragility-degradation has specific values and is characteristic of a certain type of degradation of the substrate. For example, the method of insect holes is specific to surface degradation, the density method is specific to the total evaluation of the wood support, the Brinell method for surface, and the Mark hardness method is specific to total complex degradation.

All four evaluation methods can be used both for the primary evaluation of the heritage object and also during the investigations and interventions of conservation and restoration of icons with wooden support.

The advantages of using the Mark 10 device are highlighted by the fact that the type of measurement is minimally invasive, and the hole left by the penetrating tip with a diameter of 1.34 mm is almost imperceptible (even smaller than a hole of wood-eating insects). Additionally, the operating time is short (a few seconds), the mobility of the device is higher with small dimensions, the device is portable, and the intervention has low costs.

The methods of assessing the fragility of wood panels are practical and concrete indicators for assessing their degradation and can indicate the nature and volume of consolidation materials that will be used in the time of restoration, but also the restoration techniques that will be put into practice.

In a general conclusion, it can be said that the use of these methods to determine the degree of fragility-degradation may be relevant in assessing the level of degradation or the effectiveness of the wood reinforcement treatment.

Author Contributions: Conceptualization, C.Ș.I. and A.A.; methodology, C.Ș.I.; software, A.L.; validation, A.L., C.Ș.I. formal analysis, A.L.; investigation, A.A.; resources, C.Ș.I.; data curation, A.A.; writing—original draft preparation, C.Ș.I.; writing—review and editing, A.L.; visualization, A.A.; supervision, A.L.; project administration, A.L.; funding acquisition, C.Ș.I. All authors have read and agreed to the published version of the manuscript.

Funding: This research received no external funding.

Data Availability Statement: Not applicable.

Acknowledgments: The authors would like to thank the management of the Transilvania University of Brasov and the Laboratory of Restoration and Research "Restaurare Ionescu Constantin" from Sibiu, for the administrative and technical support provided during this research.

Conflicts of Interest: The authors declare no conflict of interest.

References

1. Vecco, M. A definition of cultural heritage: From the tangible to the intangible. *J. Cult. Herit.* **2010**, *11*, 321–324. [CrossRef]
2. Ross, R.J. *Wood Handbook: Wood as an Engineering Material*; General Technical Report FPL–GTR–190; USDA Forest Service, Forest Products Laboratory: Madison, WI, USA, 2010; Volume 190, p. 510. [CrossRef]
3. Pădurețu, A.; Gheorghiță, V. Culture and connections preliminary research and preservation restoration interventions for wood icons. *Eur. Sci. J. ESJ* **2015**, *2*, 117–123.
4. Bucsa, L.; Bucsa, C. Degradari Biologice ale Structurilor de Lemn la Monumentele Istorice din Romania. Transsylvania Nostra, 2009, 2, pp. 22–30. Available online: http://www.transsylvanianostra.eu/download/05_livia_bucsa_degr_biologice_str_lemn.pdf (accessed on 4 January 2019).
5. The Wood Database. Available online: https://www.wood-database.com/wood-articles/wood-identification-guide/ (accessed on 23 January 2021).
6. Macchioni, N.; Bertolini, C.; Tannert, T. Review of Codes and Standards. In *In Situ Assessment of Structural Timber: State of the Art Report of the RILEM Technical Committee 215-AST*; Kasal, B., Tannert, T., Eds.; RILEM State of the Art Reports; Springer: Dordrecht, The Netherlands, 2011; pp. 115–121, ISBN 978-94-007-0560-9.
7. Macchioni, N. Diagnosis and Conservation of Wooden Cultural Heritage, 2015. Available online: https://www.ivalsa.cnr.it/en/research/diagnosis-and-conservation-of-wooden-cultural-heritage.htm (accessed on 13 June 2018).
8. UNI 11161; Cultural Heritage-Wooden Artefacts-Guideline for Conservation, Restoration and Maintenance. Ente Nazionale Italiano di Unificazione (UNI): Rome, Italy, 2005.
9. Fassina, V. CEN TC 346 Conservation of Cultural Heritage-Update of the Activity after a Height Year Period. In *Proceedings of the Engineering Geology for Society and Territory—Volume 8*; Lollino, G., Giordan, D., Marunteanu, C., Christaras, B., Yoshinori, I., Margottini, C., Eds.; Springer International Publishing: Cham, Switzerland, 2015; pp. 37–41.
10. Reinprecht, L. *Wood Deterioration, Protection and Maintenance*, 1st ed.; John Wileys & Sons: Chichester, UK, 2016; p. 376, ISBN 9781119106531.
11. Pournou, A. *Biodeterioration of Wooden Cultural Heritage*; Springer: Cham, Switzerland, 2020. [CrossRef]
12. Sakuno, T.; Schniewind, A.P. Adhesive Qualities of Consolidants for Deteriorated Wood. *J. Am. Inst. Conserv.* **1990**, *29*, 33. [CrossRef]
13. Cataldi, A.; Dorigato, A.; Deflorian, F.; Pegoretti, A. Effect of the Water Sorption on the Mechanical Response of Microcrystalline Cellulose-Based Composites for Art Protection and Restoration. *J. Appl. Polym. Sci.* **2014**, *131*. [CrossRef]
14. Chen, C.-P. The effects on bamboo paper from wood materials used in the conservation of Chinese wooden boxes. *J. Inst. Conserv.* **2017**, *40*, 212–225. [CrossRef]
15. Cataldi, A.; Deflorian, F.; Pegoretti, A. Microcrystalline cellulose filled composites for wooden artwork consolidation: Application and physic-mechanical characterization. *Mater. Des.* **2015**, *83*, 611–619. [CrossRef]
16. Cataldi, A.; Deflorian, F.; Pegoretti, A. Poly 2-ethyl-2-oxazoline/microcrystalline cellulose composites for cultural heritage conservation: Mechanical characterization in dry and wet state and application as lining adhesives of canvas. *Int. J. Adhes. Adhes.* **2015**, *62*, 92–100. [CrossRef]
17. Cataldi, A.; Dorigato, A.; Deflorian, F.; Pegoretti, A. Innovative microcrystalline cellulose composites as lining adhesives for canvas. *Polym. Eng. Sci.* **2015**, *55*, 1349–1354. [CrossRef]

18. Mankowski, P.; Kozakiewicz, P.; Krzosek, S. Retention of Polymer in Lime Wood Impregnated with Paraloid B-72 Solution in Butyl Acetate. *Ann. Wars. Univ. Life Sci.* **2015**, *92*, 263–267.
19. Charola, A.E.; Tucci, A.; Koestler, R.J. On the Reversibility of Treatments with Acrylic/Silicone Resin Mixtures. *J. Am. Inst. Conserv.* **1986**, *25*, 83–92. [CrossRef]
20. Timar, M.C.; Sandu, I.C.A.; Beldean, E.; Sandu, I. FTIR Investigation of Paraloid B72 as Consolidant for Old Wooden Artefacts Principle and Methods. *Mater. Plast.* **2014**, *51*, 382–387.
21. Deng, Y.; Chen, Y.; Zhang, Y.; Wang, S.; Zhang, F.; Sun, F. ICON: 3D reconstruction with 'missing-information' restoration in biological electron tomography. *J. Struct. Biol.* **2016**, *195*, 100–112. [CrossRef] [PubMed]
22. Schniewind, A.P.; Eastman, P.Y. Consolidant Distribution in Deteriorated Wood Treated with Soluble Resins. *J. Am. Inst. Conserv.* **1994**, *33*, 247–255. [CrossRef]
23. Pavlidis, G.; Tsiafakis, D.; Koutsoudis, A.; Arnaoutoglou, F.; Tsioukas, V.; Chamzas, C. Preservation of Architectural Heritage through 3D Digitization. *Int. J. Arch. Comput.* **2007**, *5*, 221–237. [CrossRef]
24. Rivers, S.; Umney, N. *Conservation of Furniture*; Routledge: London, UK, 2003; ISBN 978-0-08-052464-1.
25. Siau, J.F. *Transport Processes in Wood*; Springer: Berlin/Heidelberg, Germany, 1984.
26. Walsh-Korb, Z.; Avérous, L. Recent developments in the conservation of materials properties of historical wood. *Prog. Mater. Sci.* **2018**, *102*, 167–221. [CrossRef]
27. Zhang, T.; Gao, T.; Wu, Z.; Sun, T. Reinforced Strength Evaluation of Binding Material for the Restoration of Chinese Ancient Lacquer Furniture. *BioResources* **2019**, *14*, 7182–7192. [CrossRef]
28. Fierascu, R.C.; Doni, M.; Fierascu, I. Selected Aspects Regarding the Restoration/Conservation of Traditional Wood and Masonry Building Materials: A Short Overview of the Last Decade Findings. *Appl. Sci.* **2020**, *10*, 1164. [CrossRef]
29. Lahanier, C.; Preusser, F.; Van Zelst, L. Study and conservation of museum objects: Use of classical analytical techniques. *Nucl. Instrum. Methods Phys. Res. Sect. B Beam Interact. Mater. Atoms* **1986**, *14*, 1–9. [CrossRef]
30. Madhoushi, M. Species and Mechanical Strengths of Wood Members in a Historical Timber Building in Gorgan (North of Iran). *BioResources* **2016**, *11*, 5180. [CrossRef]
31. Mohamed Hamed, S.A.; Ali, M.F.; Nabil Elhadidi, N.M. Assessment of Commonly Used Cleaning Methods on The Anatomical Structure of Archaeological Wood. *Int. J. Conserv. Sci.* **2013**, *4*, 153–160.
32. Teacă, C.-A.; Roşu, D.; Mustaţă, F.; Rusu, T.; Roşu, L.; Roşca, I.; Varganici, C.-D. Natural bio-based products for wood coating and protection against degradation: A Review. *BioResources* **2019**, *14*, 4873–4901. [CrossRef]
33. Zhou, K.; Li, A.; Xie, L.; Wang, C.-C.; Wang, P.; Wang, X. Mechanism and effect of alkoxysilanes on the restoration of decayed wood used in historic buildings. *J. Cult. Herit.* **2020**, *43*, 64–72. [CrossRef]
34. Vitali, F.; Caldi, C.; Benucci, M.; Marzaioli, F.; Moioli, P.; Seccaroni, C.; De Ruggieri, B.; Romagnoli, M. The vernacular sculpture of Saint Anthony the Abbot of Museo Colle del Duomo in Viterbo (Italy). Diagnostic and Wood dating. *J. Cult. Herit.* **2021**, *48*, 299–304. [CrossRef]
35. Neamţu, C.; Bratu, I.; Măruţoiu, C.; Măruţoiu, V.; Nemeş, O.; Comes, R.; Bodi, B.Z.; Popescu, D. Component Materials, 3D Digital Restoration, and Documentation of the Imperial Gates from the Wooden Church of Voivodeni, Sălaj County, Romania. *Appl. Sci.* **2021**, *11*, 3422. [CrossRef]
36. Crisci, G.M.; La Russa, M.F.; Malagodi, M.; Ruffolo, S.A. Consolidating properties of Regalrez 1126 and Paraloid B72 applied to wood. *J. Cult. Herit.* **2010**, *11*, 304–308. [CrossRef]
37. Salem, M.Z.M.; Mansour, M.M.A.; Mohamed, W.S.; Ali, H.M.; Hatamleh, A.A. Evaluation of the antifungal activity of treated Acacia saligna wood with Paraloid B-72/TiO2 nanocomposites against the growth of *Alternaria tenuissima*, Trichoderma harzianum, and Fusarium culmorum. *BioResources* **2017**, *12*, 7615–7627. [CrossRef]
38. Dogu, D.; Yilgör, N.; Mantanis, G.; Tuncer, F.D. Structural Evaluation of a Timber Construction Element Originating from the Great Metéoron Monastery in Greece. *BioResources* **2017**, *12*, 2433–2451. [CrossRef]
39. Olarescu, C.M.; Campean, M.; Cosereanu, C. Thermal Conductivity of Solid Wood Panels Made from Heat-Treated Spruce and Lime Wood Strips. *Pro Ligno* **2015**, *11*, 377–382.
40. Ulker, O.; Hiziroglu, S. Thermo Mechanical Processing of Cappadocian Maple (Acer C.). *Pro Ligno* **2018**, *14*, 13–20.
41. Popescu, C.-M.; Popescu, M.-C.; Vasile, C. Characterization of fungal degraded lime wood by FT-IR and 2D IR correlation spectroscopy. *Microchem. J.* **2010**, *95*, 377–387. [CrossRef]
42. Paris, J.L.; Kamke, F.A.; Xiao, X. X-ray computed tomography of wood-adhesive bondlines: Attenuation and phase-contrast effects. *Wood Sci. Technol.* **2015**, *49*, 1185–1208. [CrossRef]
43. EN 1534:2003; Wood and Parquet Flooring. Determination of Resistance to Indentation (Brinell)—Test Method. European Committee for Standardization: Brussels, Belgium, 2003.
44. EN 17121:2019; Conservation of Cultural Heritage—Historic Timber Structures—Guidelines for the On-Site Assessment. European Committee for Standardization: Brussels, Belgium, 2009.
45. EN 323:1993; Wood-Based Panels—Determination of Density. European Committee for Standardization: Brussels, Belgium, 1993.

Article

Three Adhesive Recipes Based on Magnesium Lignosulfonate, Used to Manufacture Particleboards with Low Formaldehyde Emissions and Good Mechanical Properties

Gabriela Balea (Paul), Aurel Lunguleasa *, Octavia Zeleniuc and Camelia Coşereanu

Wood Processing and Design of Wooden Product Department, Transilvania University of Brasov, 29 Street Eroilor, 500038 Brasov, Romania; gabriela.balea@unitbv.ro (G.B.); zoctavia@unitbv.ro (O.Z.); cboieriu@unitbv.ro (C.C.)
* Correspondence: lunga@unitbv.ro

Abstract: Adhesives represent an important part in the wood-based composite production, and taking into account their impact on the environment and human health, it is a challenge to find suitable natural adhesives. Starting from the current concerns of finding bio-adhesives, this paper aims to use magnesium lignosulfonate in three adhesive recipes for particleboard manufacturing. First, the adhesive recipes were established, using oxygenated water to oxidize magnesium lignosulfonate (Recipe 1) and adding 3% polymeric diphenylmethane diisocyanate (pMDI) crosslinker (Recipe 2) and a mixture of 2% polymeric diphenylmethane diisocyanate with 15% glucose (Recipe 3). The particleboard manufacturing technology included operations for sorting particles and adhesive recipes, pressing the mats, and testing the mechanical strengths and formaldehyde emissions. The standardized testing methodology for formaldehyde emissions used in the research was the method of gas analysis. Tests to determine the resistance to static bending and internal cohesion for all types of boards and recipes were also conducted. The average values of static bending strengths of 0.1 N/mm^2, 0.38 N/mm^2, and 0.41 N/mm^2 were obtained for the particleboard manufacturing with the three adhesive recipes and were compared with the minimal value of 0.35 N/mm^2 required by the European standard in the field. Measuring the formaldehyde emissions, it was found that the three manufacturing recipes fell into emission classes E1 and E0. Recipes 2 and 3 were associated with good mechanical performances of particleboards, situated in the required limits of the European standards. As a main conclusion of the paper, it can be stated that the particleboards made with magnesium-lignosulphonate-based adhesive, with or without crosslinkers, can provide low formaldehyde emissions and also good mechanical strengths when crosslinkers such as pMDI and glucose are added. In this way magnesium lignosulfonate is really proving to be a good bio-adhesive.

Keywords: particleboard; formaldehyde emission; magnesium lignosulphonate; adhesive recipe; MOR; MOE; IB

Citation: Balea, G.; Lunguleasa, A.; Zeleniuc, O.; Coşereanu, C. Three Adhesive Recipes Based on Magnesium Lignosulfonate, Used to Manufacture Particleboards with Low Formaldehyde Emissions and Good Mechanical Properties. *Forests* **2022**, *13*, 737. https://doi.org/10.3390/f13050737

Academic Editors: Antonios Papadopoulos and Christian Brischke

Received: 17 March 2022
Accepted: 6 May 2022
Published: 9 May 2022

Publisher's Note: MDPI stays neutral with regard to jurisdictional claims in published maps and institutional affiliations.

Copyright: © 2022 by the authors. Licensee MDPI, Basel, Switzerland. This article is an open access article distributed under the terms and conditions of the Creative Commons Attribution (CC BY) license (https://creativecommons.org/licenses/by/4.0/).

1. Introduction

Formaldehyde emission is a current problem for wood-based composite materials, including particleboards. In addition to irritation of the nose and eyes and respiratory cancer, long exposure to formaldehyde emission has been shown to have a carcinogenic effect on human blood, causing irreversible mutations [1,2]. Therefore, since 1970, the limit of emissions of formaldehyde had to be continuously reduced, falling within tight tolerances [3]. It should be recognized that wood, due to its chemical composition, has a certain formaldehyde content that is acceptable to human breathing, which could be taken as a reference in the classification of formaldehyde emissions from wood-based boards. Common European wood species, such as beech, spruce, oak, pine, etc., have values of formaldehyde emission of 0.114–0.431 mg/m^{2}h for a moisture content of over 50% (decreasing moisture content also decreases formaldehyde emissions) [3]. The most used methods for testing the formaldehyde emissions and contents of wood composites are the following [3]: large and normal chamber method according to EN 717-1: 2004 [4] and ASTM E 1333: 2014 [5], the gas analysis method according to standard EN 717-2: 1995 [6] and ISO 12460-3: 2015, the perforator method according to EN 120: 1992 [7], and the desiccator method according to JIS A 1460: 2001 [8]. Less used is the vial method according to EN 717-3: 1996 [9]. The permissible emission limits and formaldehyde contents are given in Table 1 [3]. In European countries, only the E1 class is allowed for wood-based panels, and, in addition, other countries already require the E0 class, which is half of E1.

Table 1. Permissible limits of formaldehyde emission for particleboards.

Methods	Standard	Permissible Limits	Emission Class
Chamber method	EN 717-1: 2004	≤ 0.1 ppm or ≤ 0.12 mg/m^3	E1
		>0.1 ppm	E2
Large chamber method	ASTM E 1333:2014	0.14 ppm	E1
		0.18 ppm	F ***
		0.09 ppm	F ****
Perforator method	EN 120: 1992	≤ 8 mg/100 g	E1
		8–30 mg/100 g	E2
Gas analysis method	EN 717-2: 1995	3.5 mg/m^{2}h	E1
		1.4 mg/m^{2}h	E0
Desiccator method	JIS A 1460:2001	1.5 mg/L	F **
		0.5 mg/L	F ***
		0.3 mg/L	F ****

It is well-known that formaldehyde-based adhesives, such as urea-formaldehyde (UF) and phenol-formaldehyde (PF) are not environmentally friendly because of formaldehyde release. In this context, some solutions must be found to modify or replace them. Because the main material that increases the amount of formaldehyde in a particleboard is the adhesive (except pMDI—polymeric diphenylmethane diisocyanate), some concerns about reducing formaldehyde emissions/content and obtaining bio-adhesives or green adhesives have had and continue to have three main directions:

- Reducing emissions by using formaldehyde catchers in the adhesive recipe of urea-formaldehyde [UF] or phenol-formaldehyde [PF] adhesives;
- The use of bio-adhesives with very low formaldehyde emissions/contents to replace the classical synthetic adhesives;
- The use of crosslinking agents, both for classic adhesives and for bio-adhesives.

The first method of reducing emissions by using formaldehyde catchers has been used for a long time and has gained the attention of many researchers [10,11]. The emission of formaldehyde from urea-formaldehyde adhesives depends on the molar ratio between urea and formaldehyde and also on the environmental conditions during the manufacturing process (temperature, air humidity, etc.) [12]. Usually, these catchers are obtained from natural materials such as bio-oils, tannin, soy, or lignin [3,13] and are used in recipes with UF and UMF resins for MDF (medium-density fiberboard), with formaldehyde emissions of 0.04–0.08 ppm or 2.5–3.0 mg/100 g. Various adhesives were obtained by gradually replacing the phenol in phenol-formaldehyde adhesives with lignosulfonates. The best mechanical strengths were obtained by adding 20% lignosulfonate to the PF resin [14]. The use of the same recipes in particleboard manufacturing led to emission class E0 (equivalent to class F ***), with values of 0.02–0.04 ppm or 3–3.5 mg/100 g [3]. Another formaldehyde catcher was soy protein or rapeseed, which were used in the case of UF resins for the manufacture of particleboard with a density of 680 kg/m^3 and a formaldehyde emission belonging to class E1 [15–18]. Other catchers of formaldehyde were wheat gluten as well as the leftovers from the production of starch, pea protein, or cottonseed protein [19–21].

The tannins used in the production of adhesives for wood-based composites reduce the emission of formaldehyde, but in a pure state they do not provide good strengths when used in combination with classical UF or PF adhesives [22]. Particleboard made with tannin and formaldehyde adhesives had very low formaldehyde emissions and induced good mechanical strength for indoor use panels [23–26]. Another study [27] investigated the role of nano-clay (cloisite Na$^+$) in reducing formaldehyde emissions from particleboards. The specimens without the addition of nano-clay emitted about three times more formaldehyde compared to the boards that contained only 1% nano-clay. Other acid catalysts, such as oxalic or citric acid, have reduced the formaldehyde content from 4.1 mg/100 g to 3.5 mg/100 g. Moreover, the addition of pMDI reduced this formaldehyde content by half. The addition of 1.5% silicon nano-dioxide (SiO$_2$) improved the strength of UF resin-bonded boards and substantially improved the emission of formaldehyde [28].

Usually, bio-adhesives use waste or other natural materials, such as plant extract, wood biomass, vegetable and animal proteins, etc. However, they have a low reactivity, which is why more studies are needed to increase their reactivity, but the raw materials used could increase the price of the final board. Moreover, the mechanical strengths of the boards with bio-adhesives are lower than those obtained with the classical synthetic adhesives UF or PF. One of these bio-adhesives is lignin in the form of ammonium lignosulfonate [29], alone or in combination with pMDI resin. Lignin from lignosulphonate has the capacity of adhesion, as the native lignin has in time of tree growth. Some researchers in the field of bio-adhesives have obtained an adhesive based on soy protein with modest hydrophobic and mechanical performance. Soy protein has been mixed with lignin in other studies [30], thereby improving water resistance. The use of other types of proteins, such as wheat gluten, which is a by-product of starch production, pea protein, or cottonseed protein, has also been considered. The most used adhesive in the lignocellulosic boards is the urea-formaldehyde adhesive UF; it has the highest mechanical strength (well above the minimum required limit), which is why in the recipe of this adhesive solution filler is included to reduce the costs of adhesion. This adhesive also has the highest formaldehyde emissions, requiring formaldehyde catchers, which greatly increase the cost of adhesion. This is why bio-adhesives are the cheap solution to replace UF adhesive. Even if the mechanical performance of the boards is weaker, it can fall within the minimum required limit of European standards [10].

Crosslinking agents promoted the formation of intermolecular bonds between polymer chains (glyoxal, pMDI, sugars, furfuryl and furfuryl alcohol, citric acid, maleic anhydride, etc.). One of these agents is glyoxal [29]; it was noticed that when using a mixture of 0.5% glyoxal and 50% lignin as an adhesive in the production of particleboard and plywood, the panels recorded good mechanical strengths and were also environmentally friendly. Another crosslinking agent used in the particleboard manufacturing is pMDI [31], used in proportions of 4, 6, and 8%, with favorable effects on water resistance and mechanical strength. Moreover, the tannins can be mixed in the adhesives used for the production of plywood, chipboard, OSB, MDF, glulam, and LVL. Tannin extract offers a significant potential for resorcinol replacement, thus reducing the costs of gluing in the production of fiberboard-type wood composites [32]. Some studies have shown that the addition of nano-clays contributes to an improvement in the mechanical strength of boards with lignin-based adhesives [33,34].

Some researchers have shown that addition of different share rates of phenolated Kraft lignin [34] or lignosulfonates in PF and UF resins could reduce the formaldehyde emission. Industrial lignin (Kraft lignin) or in the form of lignosulfonate (such ammonium or magnesium lignosulfonates) is a residual by-product that is obtained after the production of cellulose (pulp and paper) following the chemical separation of cellulose from lignin. Therefore, lignosulfonate is a residual and natural product, as it is obtained from wood, which is why it is a good bio-adhesive. Moreover, being a chemical constituent of wood, it has a low formaldehyde emission. In its pure state, lignin has low adhesion, and its reactivity needs to be improved. In this regard, hydrolyzed lignin treated with oxygen plasma was obtained [35] with good results on changing the chemical structure of lignin and increasing reactivity [36], increasing reactivity by hydroxy methylation [37,38] and oxidation of this [36,39]. The oxidation of lignin was a practical way to weaken its structure, being more easily polymerizable [40]. The simplest oxidation procedure is the addition of oxygen peroxide, especially due to its ecological characteristics [41,42]. Oxidation was performed both in an acidic medium using $FeSO_4$ as reagents and in a basic medium with NaOH. Research on the use of magnesium lignosulfonate as a bio-adhesive has been conducted for veneered fiberboards [43–46]. Actual research [47–52] has shown that the introduction of a small percentage of ammonium lignosulfonate in the boards leads to good mechanical properties and a formaldehyde content of only 1.1 mg/100 g of dry plate, super quality E0. Other research has examined the use of recycled wood [53], the use of magnesium lignosulfonate for wood composites [54], the availability of wood and vegetable resources for the next 20 years in the Czech Republic [55], and the main bio-adhesives used in the last decade [56].

From the analysis of the research works in the field, a first conclusion can be drawn, namely, that the bio-adhesives based on lignosulfonates occupied an important place in international research on reducing the emission/content of formaldehyde in particleboard, but their performance has not been fully investigated. There are also a lot of studies oriented to increase the reactivity of lignin, as its reactivity is low in a pure state, the most widely used method being that of oxidation. Last but not least, there are very few studies on the use of magnesium lignosulfonate in the manufacture of bio-adhesives for particleboard. The general objective of the present research work is to manufacture particleboards with low formaldehyde emissions, using three adhesive recipes based on magnesium lignosulfonate that is oxidized with hydrogen peroxide. These boards will have to meet the mechanical strength conditions (MOR, MOE, and IB) imposed by the European standard EN 312: 2003 [57] for indoor uses in a dry environment (type P2). Moreover, the boards will have to fall into the E1 class of formaldehyde (or lower), corresponding to the European standard EN 717-2: 1995 [6] with a limit of 3.5 mg/m^{2}h obtained by the gas analysis method. In order s to fall into emission classes E0 (corresponding to emission class F ****) using the gas analysis method, the measured value for the particleboard's formaldehyde emissions must be below 1.4 mg/m^{2}h.

2. Materials and Methods

Magnesium lignosulphonate. There are five types of water-soluble lignosulfonates: magnesium lignosulfonate, calcium lignosulfonate, sodium lignosulfonate, potassium lignosulfonate, and ammonium lignosulfonate. They are differentiated by the method of cellulose extraction (sulfite, Kraft, basic, acid, etc.) and contain small amounts of magnesium, calcium, sodium, potassium, or ammonia. The most common ones are ammonium and magnesium sulphonate, with no differences in the formaldehyde emissions and mechanical properties of wooden composites [11,13,14]. Magnesium lignosulphonate (Lignex MG) was purchased from Sappi Biotech GmbH (Düsseldorf, Germany) in an unmodified state and was used in the preparation of the bio-adhesive for particleboard manufacturing. The physical appearance of magnesium lignosulfonate was in the form of a yellowish-brown powder. It was obtained by the process of purification, evaporation, chemical treatment, and drying of the black liquid waste resulting from the cellulose (pulp and paper) manufacturing process. The characteristics of magnesium lignosulfonate (Lignex MG), as stated in the data sheet issued by the manufacturer, are presented in Table 2.

Table 2. Characteristics of magnesium lignosulphonate.

No.	Characteristics	Values
1	Dry mater content	93 ± 2%
2	Magnesium content, minimum	6 ± 1%
3	pH	5.5 ± 1
4	Bulk density	400 kg/m^3
5	Ignition temperature	530 °C
6	Water insolubility, maximum	1%
7	Moisture content, maximum	7%

Lignin oxidation. In order to obtain the recipes, the lignin powder was oxidized with 30% hydrogen peroxide (H_2O_2). The final H_2O_2 content was 5.7% by weight of the resin. NaOH was used to raise the pH to 9. The recipe for preparation was as follows: 460 g of magnesium lignosulfonate; 35 g of 30% hydrogen peroxide (7.6% from magnesium lignosulfonate); 246 mL of distilled water (53.5% by mass from magnesium lignosulfonate), and 66 mL of sodium hydroxide at a concentration of 50% (14.3% by mass of magnesium lignosulfonate). The dry matter content of the bio-adhesive thus obtained was 57%, and the flow time through the viscometry cup with a diameter of 6 mm was 3 min and 11 s.

Wood particles. The particleboards were made of beech wood particles (30%) (*Fagus sylvatica*) and softwood, usually spruce (*Picea abies*) and pine (*Pinus sylvestris*) (70%). The moisture content of the dried particles was of 6.8%. Particles were selected from each sample for the entire quantity, using the quarter method, and scanned at 1200 dpi on an HP ScanJet 7650 scanner along with a dimensional reference (in mm). For a dimensional evaluation of the scanned images, they were imported into AutoCAD (2018) and adjusted to a 1: 1 scale. Measurements of length and width were then performed in the AutoCAD program. The size of the fine particles was determined using a Nikon YS100 microscope, which was made in Japan, with an accuracy of 0.01 mm. Samples of 25 g were extracted from particle bags provided from the Kastamonu Romania SA company (Reghin, Romania).

Granulometry analysis. Coarse and fine wood particles were analyzed as participation rates using the Retsch-type vibrating screen, which was made in Germany, setting the vibration of the sieves for 10 min. For coarse particles, the apparatus was equipped with five sieves placed from top to the bottom, starting with the sieve with the largest meshes of 4.00 mm × 4.00 mm, then 3.15 mm × 3.15 mm, 2.00 mm × 2.00 mm, 1.25 mm × 1.25 mm, 1.00 mm × 1.00 mm, and ending with the collector cylinder for the rest. The mass of each sample was 25 g. The particles were screened at a frequency of 60 oscillations/min, then they were collected from each sieve and weighed to the nearest 0.01 g. In order to analyze the fine chips from the point of size distribution, they were screened with the same electrical equipment using 1.00 mm × 1.00 mm, 0.80 mm × 0.80 mm, 0.53 mm × 0.53 mm, 0.40 mm × 0.40 mm, and 0.16 mm × 0.16 mm mesh screens. The share distribution of the particles was calculated for each type of particle (coarse and fine) and for each sieve separately, with the next relation (Equation (1)):

$$Cd = m_f : m \times 100 \, [\%] \quad (1)$$

where Cd is the particle distribution in %; m_f is the mass of the fraction collected in the sieve in g; and m is the mass of the whole sample expressed in g.

For each average value of the particle size, five experimental tests were made. Particles smaller than those obtained on sieves 0.4 mm × 0.4 mm, 0.16 mm × 0.16 mm, and the rest were considered dust and were removed from those for making particleboards.

Particleboard manufacturing. Based on the experience of the manufacturer Kastamonu of particles and boards and also on the research results of other groups of researchers [35,40], in order to obtain well-compacted boards, after weighing the remaining chips in each sieve and removing dust, the following proportion was established within the boards: 35% coarse (large) particles (those sifted and left in the 2 mm × 2 mm sieve) and 65% fine chips (those sifted and left in the 0.5 mm × 0.5 mm sieve). A mixture of 15% adhesive solution and 85% in-factory particles (percent by mass, particles absolutely dry) was made using a paddle drum by spraying the adhesive on the particles. The particleboard mat was placed on a 3 mm thick steel sheet covered with heat-resistant paper. For the formation of the particle mat, there were used beech wood frames with internal dimensions of 420 mm × 420 mm × 50 mm and steel stopers with a thickness of 16 mm to limit the thickness of the boards to this value. The target density established for all particleboard was 650 kg/m^3 (comparable to that of agglomerated boards with urea-formaldehyde adhesive manufactured at the Kastamonu company), with the boards having the dimensions of 450 mm × 450 mm at a temperature of 180 °C, a pressure of 2.5 MPa, and a pressing time of 16 mm (i.e., 1 min for each mm of thickness). After evacuation from the hot press, the boards were conditioned at a temperature of 20 °C and a relative humidity of 65% for 7 days in order to homogenize the stresses and to stabilize the moisture content to 10%. A minimum of 10 chipboards were manufactured for each type of recipe.

The three adhesive recipes. For the manufacture of particleboard, there was a basic recipe including oxidized magnesium lignosulfonate (15% of the weight of the dry wood particles) that was marked as Recipe 1 and two others, which differed by the addition of two crosslinking agents, respectively, pMDI resin (for Recipe 2, in a proportion of 3% by mass of the dry particles) and a combination of pMDI resin with glucose (for Recipe 3, in a total proportion of 5% of the amount of particles or separately as 2.9% pMDI and 2.2% glucose related to the dry particle mass). Some preliminary research as well as the bibliographic study [11] showed that non-oxidized lignosulfonate has very poor adhesive properties, which is why all adhesive recipes contained hydrogen peroxide. The percentage of dry adhesive was also considered to be 15% of the amount of dry particle mass. Sodium hydroxide was introduced in order to obtain a basic pH of 8–9 to add a high level of adhesiveness. The three manufacturing recipes are presented in Table 3.

Table 3. Composition of experimental panels.

Specifications	Recipe 1	Recipe 2	Recipe 3
Particles, 7% Mc	1838 g	1838 g	1838 g
Magnesium lignosulphonate	460 g	275 g	275 g
Hydrogen peroxide, 30%	35 g	21 g	22 g
Distilled water	246 g	147 g	145 g
Sodium hydroxide, 50%	66 mL	39 mL	38 mL
pMDI	-	55 g	54 g
Glucose	-	-	41 g
Dry matter content	60.1%	59.9%	55.7%

Gas analysis method for formaldehyde emissions. The formaldehyde emissions, determined by the gas analysis method, used specimens with dimensions of 400 mm × 50 mm × 16 mm. For testing, the specimens were sealed on the edge with self-adhesive foil, leaving only the faces of the specimens free. The specimens were placed in a chamber with a constant temperature of 60 ± 0.5 °C and an air humidity of less than 3%, in the direction of the air current. The formic aldehyde released from the samples and the air with formaldehyde was passed through washing dishes with distilled water that absorbs this formaldehyde. The determination of formaldehyde emission is based on the Hantzsch reaction, according to which formaldehyde reacts with acetylacetone and ammonium acetate, resulting in diacetyl-dihydrouridine (DDL), with a maximum absorption spectrum of 412 nm. Therefore, in each of the 4 h of testing, 10 mL of aqueous solution was taken and placed in a vial to which 10 mL of acetylacetone solution and 10 mL of ammonium acetate solution were added. The absorption of this mixed solution was determined at 412 nm using a Jenway spectrophotometer (Staffordshire, UK). In parallel, the absorption of the control sample with distilled water was determined. Based on these data, the emission of formaldehyde was determined with the following relation (Equation (2)):

$$FE_i = ((A_s - A_b) \times f \times V) : F \; [mg/m^2 h] \qquad (2)$$

where FE_i is the formaldehyde emission in mg/m^{2}h; i = 1, 2, 3, 4 are indicative of the time in h at which the determination was made; A_s is the absorption from the washing dishes with formic aldehyde; A_b is the absorption etalon for distilled water; f is the slope of the calibration curve in mg/mL; V is the volume of the volumetric flask in ml; and F is the emitting unsealed surface in m^2.

An average value was determined as the arithmetic mean of the values obtained in a time of 4 h and represented the final value of formaldehyde. The comparison of the formaldehyde emission was made with the ISO 12460-3: 2015 standard [58].

Mechanical properties. After conditioning, the boards were cut into specimens on a universal circular saw. The minimum number of specimens for the mechanical tests was chosen according to the European standard EN 310: 1993 [59] for MOR and MOE and according to EN 319: 1993 [60] for the internal bond. The mechanical tests were performed on a universal Zwick/Roell Z010 machine (Ulm, Germany), with the bending test quantifying the strength and elasticity of the boards and the internal bond test showing the adhesion strength. The bending test consisted of applying a static load in the middle of the specimens supported on two supports, and the internal cohesion test was performed by applying a tensile force on the specimens glued with epoxy adhesive on two metal supports. The calculation relationships used by the universal test machine software were as follows (Equation (3)):

$$MOR = (3 \times F_{bmax} \times l1) : (2 \times b \times t^2) \; [N/mm^2]$$

$$MOE = (l1^3 (F_2 - F_1)) : (4 \times b \times t^2 (f_2 - f_1)) \; [N/mm^2]$$

$$IB = F_{cmax} : (l_2 \times l_3) \; [N/mm^2] \qquad (3)$$

where MOR is the modulus of resistance in N/mm^2; F_{bmax} is the maximum of bending force in N; l1 is the distance between the fulcrums in mm; b is the width of the sample in mm; t is the thickness of the sample in mm; MOE is the modulus of elasticity in N/mm^2; F_2 is the force of 40% from the maximum force in N; F1 is the force of 10% from the maximum force in N; f_2 is the deformation corresponding to the F_2 force in mm; f1 is the deformation corresponding to the F_1 force in mm; F_{cmax} is the maximum of the compression (breaking) force in N; l_2 is one of the plane dimensions of the IB sample in mm; and l_3 is the second plane dimension of the IB sample in mm.

The values of the obtained mechanical properties were compared with those of the standard EN 312: 2004 [57] for the interior-type P2 boards, including that for furniture uses. The increase in the real values compared to the reference ones was determined with the help of a relation as follows (Equation (4)):

$$Ip = (V_{max} - V_{min}) : V_{ref} \times 100 \; [\%] \qquad (4)$$

where Ip is the increasing of properties in %; V_{max} is the maximum value; V_{min} is the minimum value; and V_{ref} is the reference value (can be V_{max} or V_{min}).

The density profile on thickness. The density profile was determined using an X-ray machine, DPX 300 (IMAL group, San Damaso, Italy). Square-shaped specimens with sides of 50 mm were weighed with an EU-C-LCD precision scale (Gibertini Elettronica, Novate Milanese, Italy), and their dimensions were measured with the same profilometer. Because the visibility and clarity of the density profile left something to be desired, the values were imported into an Excel sheet and the respective graph was recreated.

Statistical analysis. The obtained values were statistically processed by determining the survey median and the standard deviation for a confidence interval of 95% and an alpha error of 0.05. The standard deviation was also applied to the graphs obtained in Microsoft Excel (2018). A one-way ANOVA was used to observe the dependence of one median on other one when the null hypothesis was considered. The statistical analysis program Minitab 18 (2018) was used to obtain the empirical cumulative distribution function (CDF) graphs under the same 95% confidence interval.

3. Results

3.1. The Particle Dimensions

Overall, the particle sizes ranged from 2.4 mm to 34.1 mm in length, from 0.5 mm to 10.6 mm in width, and from 0.1 mm to 4.1 mm in thickness. Correspondence between the size of the sorting sites and the dimensions of the particles were as follows:

- For the 4.00 mm × 4.00 mm sieve, the length range was 7.6–25.8 mm, the width was 4.1–10.6 mm, and the thickness was 0.2–4.1 mm;

- For the 3.13 mm × 3.15 mm sieve, the length range was 6.1–34.1 mm, the width was 4.1–5.7 mm, and the thickness was 0.2–4.1 mm;
- For the 2.00 mm × 2.00 mm sieve, the length range was 4.2–20.1 mm, the width was 1.1–5.2 mm, and the thickness was 0.2–1.8 mm;
- For the 1.25 mm × 1.25 mm sieve, the length range was 3.7–19.5 mm, the width was 0.9–3.4 mm, and the thickness was 0.2–1.6 mm;
- For the 1.00 mm × 1.00 mm sieve, the length range was 2.4–18 mm, the width 0.5–1.7 mm, and the thickness 0.1–0.9 mm.

Figure 1 clearly shows the high dimensional difference between the length and the other two dimensions of the particles, proving that they have a needle shape, a characteristic necessary for the production of quality particleboards. It was observed (Figure 1) that once the dimensions of the sieves decreased, the maximum dimensions of the particles also decreased, so there is a relationship of direct proportionality between them. The high Pearson R^2 coefficient values, higher than 0.9 in all three cases, show that the linear regression equation for estimating the variation is very good. It was also observed that the smallest variation was in the case of particle thickness, and the largest variation was in the case of particle width. The same analysis can be conducted in the case of the minimum or average dimensions of the used particles, with the results being the same.

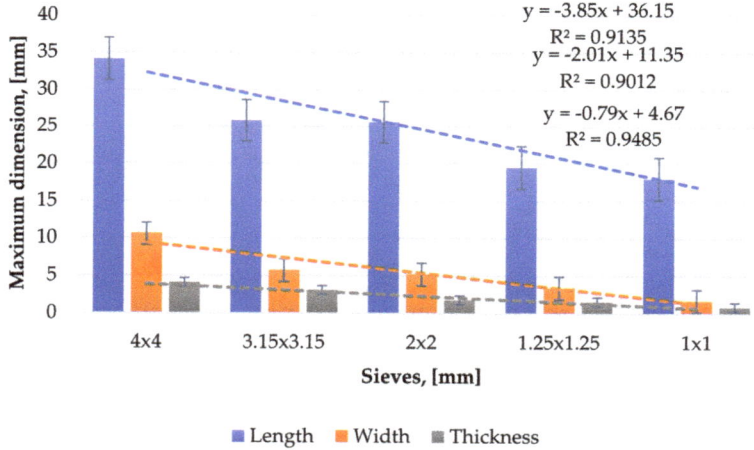

Figure 1. Correlation between maximum of coarse chip dimensions and the sieve dimensions.

3.2. Granulometry of Wooden Particles

The particle size analysis was performed differently on large and fine particles, because the percentage of their mass participation in the board was different, in order to obtain a superior compaction. From the point of view of the coefficient of determination, R^2 (0.58 in the case of large particles and 0.84 in the case of fine particles) (Figure 2a,b), a certain dimensional non-uniformity and a poor correlation of the participation rate was observed when obtaining these maximum values by a 2nd degree polynomial equation in the case of large particles and an exponential equation in the case of fine particles. From the same point of view, the fine particles have a granulometry superior to the large ones.

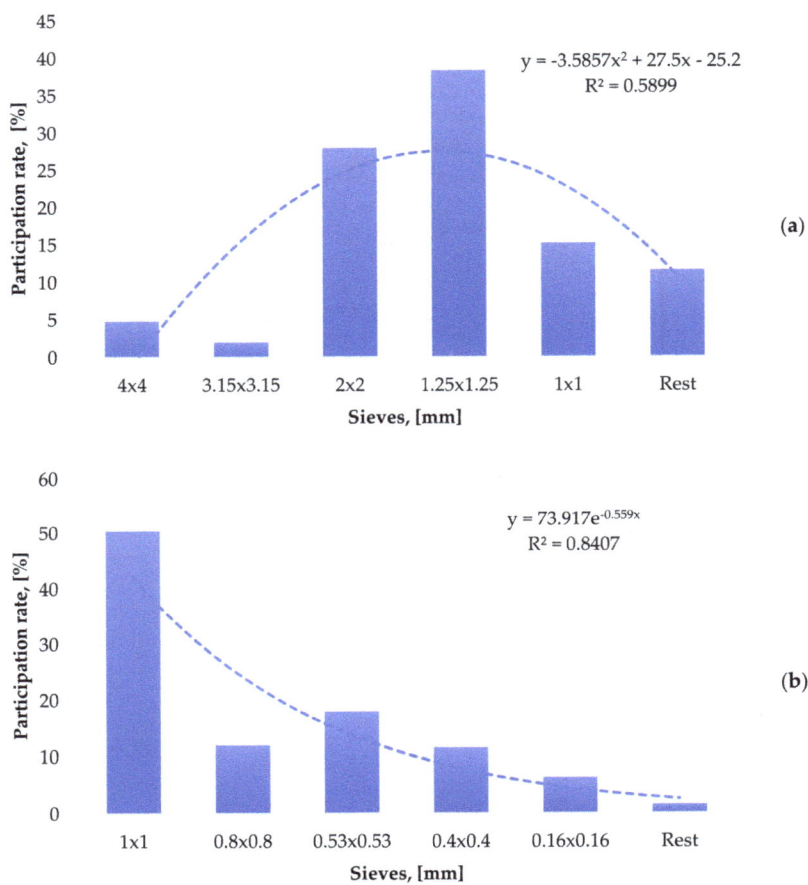

Figure 2. Granulometry of particles: (**a**) for coarse particles; (**b**) for fine particles.

A central value of 1.21 mm was found in the case of coarse particles and 0.64 mm in the case of fine ones. A displacement in the values of coarse particles was also observed (Figure 2a) to the right beyond the 1.25 mm × 1.25 mm sieve, with a rest value over 10%, in addition to a displacement in the values of fine particles (Figure 2b) to the left beyond the 0.53 mm × 0.53 mm sieve. From this tendency of particle positioning, it can be determined that the median value of the total particle dimensions was around the 1 mm × 1 mm sieve.

3.3. Density Profile on Thickness

The average densities of each of the three types of particleboards, obtained based on the EN 323: 1993 standard [61], were around 650 ± 20 kg/m^3, similar to the density of particleboard manufactured with UF resin at a medium moisture content of 10%, determined based on EN 322: 1993 [62]. In general, all three density profile diagrams (corresponding to the three manufacturing particleboards based on three different recipes) followed the same rules:

- they had two symmetrical maximum peaks that were arranged a few millimeters from the faces as well as a minimum located in the middle area of the board;

- the lowest density values were arranged in the exterior areas of the boards due to the thickness relaxation of this area after evacuation from the press;
- a slight deviation of the densities towards one of the particleboard faces was determined by the fact that during the formation of the particle mat the small particles tended to migrate to the bottom of the mat, and the fine particles have a higher degree of densification than the coarse particles.

The vertical density profiles are presented in Figure 3. It could be observed that the in cases of Recipes 1 and 2, the diagrams are quite similar, with low surface densities of about 400 kg/m^3 (Recipe 1) and 370 kg/m^3 (Recipe 2), which increased to 750 kg/m^3 and 780 kg/m^3 at 1.1 mm and 2.9 mm from the surface. In the core layer, the densities were reduced to values of about 500 kg/m^3 and 595 kg/m^3, respectively. The uniform and nearly flat shape of the vertical profile of boards (785 kg/m^3 on the faces and 610 kg/m^3 in the core) were obtained for Recipe 3.

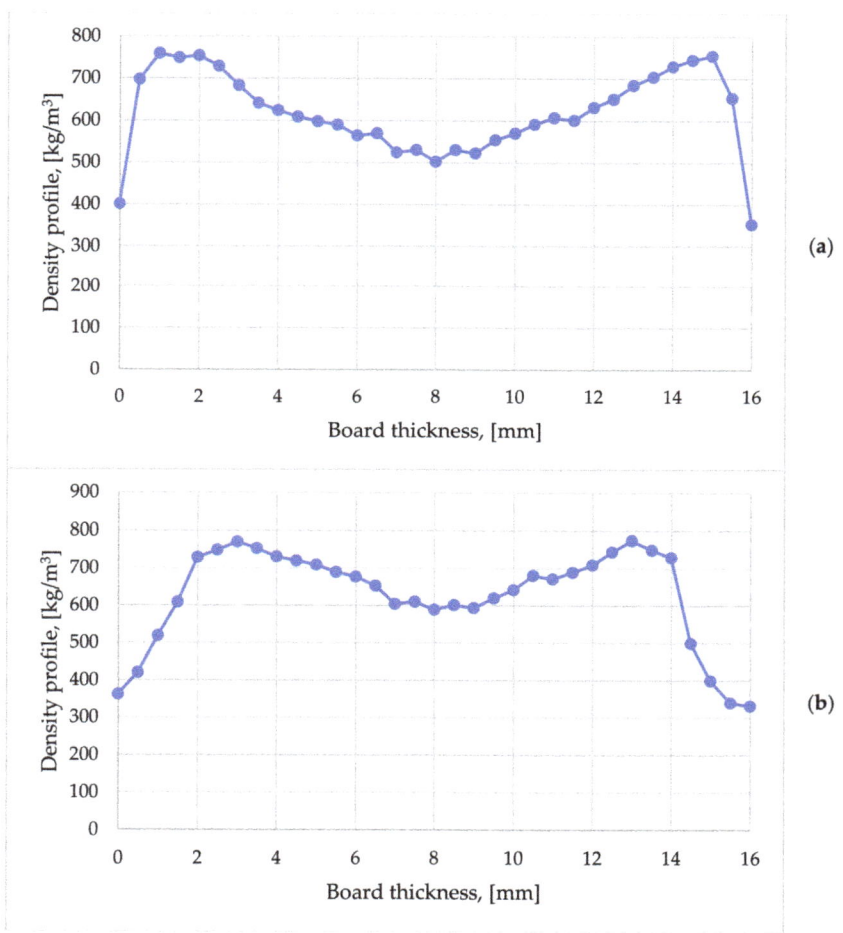

Figure 3. *Cont.*

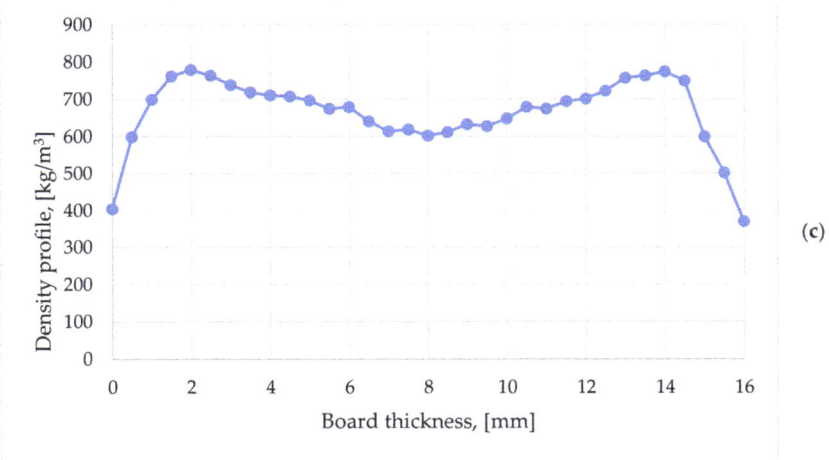

Figure 3. Particleboard density profiles: (**a**) for Recipe 1; (**b**) for Recipe 2; (**c**) for Recipe 3.

3.4. Internal Bond (IB)

The adhesiveness quality of each adhesive recipe is usually given by the value of the resistance to an internal bond test (which indicates the strength of the core area and is influenced by the core density, particle geometry, and adhesive quantity and distribution). From this point of view, Recipe 2 and Recipe 3 were clearly better than Recipe 1 (Figure 4). This means that the introduction of the two crosslinkers lead to an increase in the cohesive property of the particles. Thus, the introduction in the recipe of the pMDI crosslinker had a medium effect, namely, 3.8 times higher than in the case of the boards without this crosslinker (Recipe 1). The addition of glucose in the adhesive recipe (Recipe 3) led to a slight increase in IB by 7.8% compared to Recipe 2 but 4.1 times higher than the board without crosslinkers (Recipe 1).

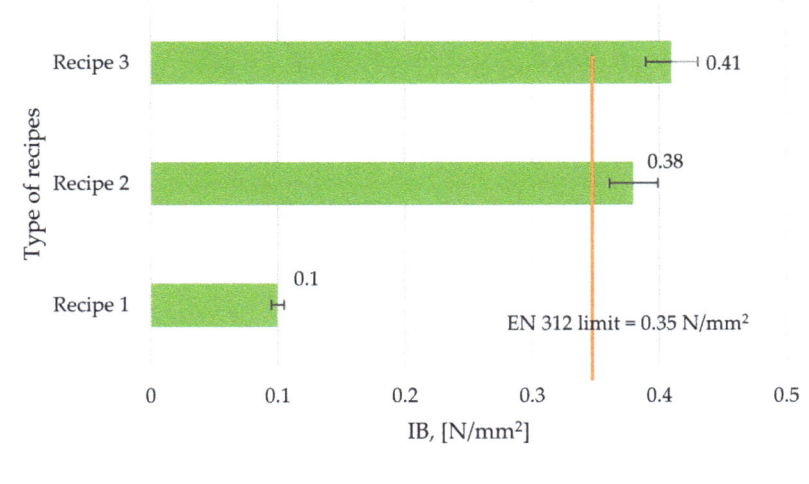

Figure 4. Internal bond of all three experimental particleboards.

If the comparison with the minimum value of 0.35 N/mm² of the European standard EN 312 [57] is made, it was observed that the boards obtained with Recipe 1, without crosslinkers, did not correspond to the requirements of the standard, with the difference being very high: 2.91 times. Therefore, the boards obtained using only oxidized magnesium lignosulfonate are not suitable in terms of internal bond. They are recommended to be used in situations where the cohesion of the boards may be neglected. The other two types of boards had an IB above the minimum allowable value, exceeding it by 8.5% when 3% pMDI crosslinker was used or by 17.1% when a combination of glucose with pMDI (about 5%) crosslinkers (Recipe 3) was used. It is obvious that pMDI and glucose as crosslinkers had a positive influence on the IB strength of the experimental particleboards. Other studies showed improvements in IB by using a combination of bio-based materials (glyoxalase lignin, ammonium lignosulfonates, tannin, and soya) or urea-based resin with pMDI [3,14,16].

3.5. Modulus of Elasticity (MOE)

The moduli of elasticity that were obtained with each of the three recipes had very good values. The best elasticity values (Figure 5) belonged to that boards obtained with Recipe 2 (with the addition of 3% pMDI), which were 26% higher than the boards without crosslinkers (Recipe 1) and 25.9% higher than the boards with the addition of pMDI and glucose crosslinkers (Recipe 3). It turns out that the addition of separate pMDI crosslinker and a combination of pMDI and glucose added a positive value to the particleboard elasticity.

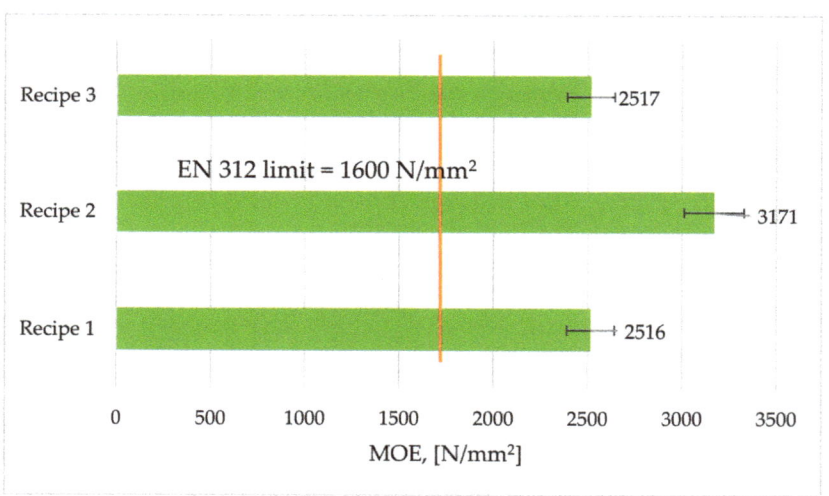

Figure 5. Moduli of elasticity for all three types of experimental particleboards.

Compared to the limit value of 1600 N/mm² specified by the European standard EN 312, it was found that all MOE values exceeded the limit. The boards obtained with Recipe 1 exceeded it by 57%, the boards obtained with Recipe 2 exceeded it by 98.1%, and the boards obtained with Recipe 3 exceeded it by 57.3%. The most elastic panels were those obtained with the pMDI crosslinker (Recipe 2).

3.6. Modulus of Resistance (MOR)

The modulus of resistance for static bending resulting from laboratory tests differed from one adhesive recipe to another, with Recipe 3 on top (having both crosslinkers), providing the best board strength (Figure 6). Compared to the other two recipes, the boards obtained with Recipe 3 had an MOR 37.5% higher than the boards without crosslinkers

(Recipe 1) and only 0.2% higher than the boards with Recipe 2 (with the pMDI crosslinker). Thus, it was shown that the introduction of the two crosslinkers (separately or in combination) significantly increased the resistance to static bending of the boards.

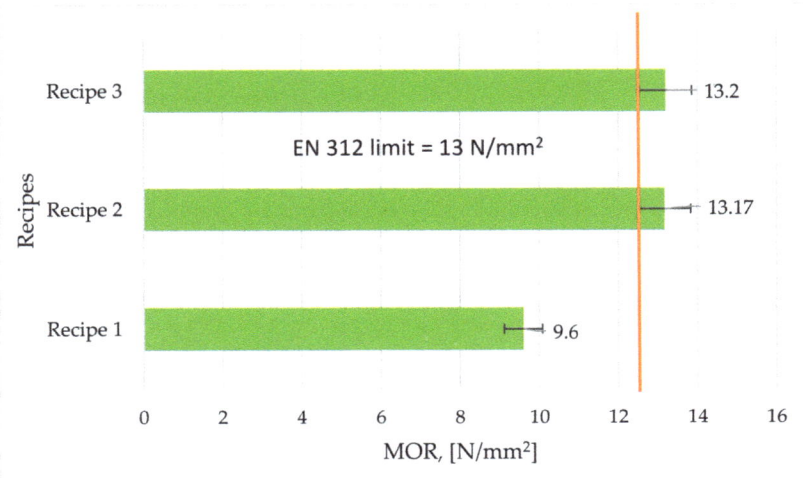

Figure 6. Modulus of Resistance (MOR) for all experimental particleboards.

Comparing the real MOR values with the limit value of 13 N/mm^2, imposed by the EN 312 standard, it was found that the boards bonded with Recipe 1 (without crosslinkers) are not recommended, having a value 27.1% below the allowable limit. Boards made with Recipe 2 (with the pMDI crosslinker) had a value with above the standard limit, and boards made with Recipe 3 (with the pMDI and glucose crosslinkers) had values 1.5% above the standard limit. Therefore, from the point of view of the MOR, the boards obtained with Recipes 2 and 3, for which crosslinkers were used, are recommended for P2-type panels (jointed panels used for indoor conditions). The addition of pMDI obviously improved MOE and MOR values (MOR: 11.1 N/mm^2, MOE: 2056 N/mm^2) as was also observed by other authors [54,55].

3.7. Emission of Formaldehyde

Regarding the formaldehyde emissions, the values obtained for the particleboards with the three adhesive recipes were compared with the average reference value of 3.5 mg/m^{2}h imposed by EN 717-2: 1995 for emission class E1 and with the value of 1.4 mg/m^{2}h for emission class E0 (Figure 7). All results were in the E1 and E0 emission classes, but the particleboards obtained with Recipe 3 had the lowest emissions, which meant that the addition of crosslinkers significantly reduced the formaldehyde emissions. In the case of Recipe 1, the value obtained for formaldehyde emission was low, 6.4 times lower than the standardized E1 limit and 2.5 times lower than the standardized E0 limit, which places these boards in the category of particleboards with ultra-low formaldehyde emission, close to that of natural wood [20]. By combining this value with the mechanical strengths (MOR and IB) of this type of particleboard, this board is recommended to be used in the field where no high-strength particleboards are required (i.e., paneling, decorative furniture, etc.).

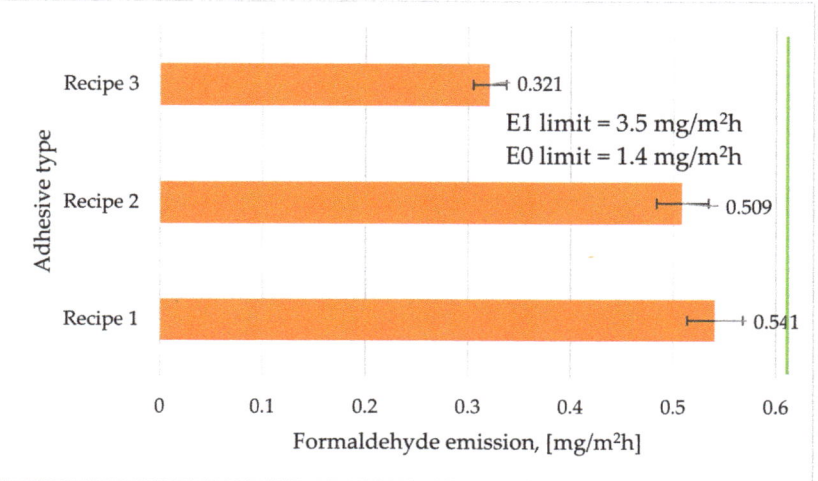

Figure 7. Formaldehyde emission for the three types of experimental particleboards.

It can also be observed in Figure 7 that the addition of only 3% pMDI crosslinker in Recipe 2 decreased the formaldehyde emission by 5.9% compared to boards with Recipe 1. This showed that the introduction of pMDI crosslinker is beneficial, with this crosslinking being considered to have zero formaldehyde emission and major improvements in particleboard strength [29]. The average free formaldehyde emission for magnesium lignosulfonate adhesive modified by oxidation with H_2O_2 and the addition of pMDI and glucose (Recipe 3) was below the maximum standardized limit (beyond 10 times lower, more precisely 10.9 times) for emission class E1 (3.5 mg/m²h limitative value) and 4.36 times lower compared to class E0 (1.4 mg/m²h limitative value), which is required by the European standard EN 717-2: 1995 [7]. This demonstrated that the introduction of the PMDI-glucose crosslinker combination strongly decreases formaldehyde emissions, remaining, as in the case of Recipe 1, below the E0 class limit of 1.4 mg/m²h and getting closer to that of solid wood [20].

4. Discussion

The geometry of wood particles has a positive influence and could give good resistance to the particleboards if they have a needle shape where the length is much larger than the width and thickness [45]. Large differences were observed between coarse and fine chips, as observed using granulometry (Figure 2a,b) and as the variation equations from the figures specified. Moreover, in Figure 1 it was observed that the width and thickness of the particles are very close to each other when the dimensions of the sieves decrease, and at a certain moment (when the sieves are smaller than 1 mm × 1 mm) it is no longer possible differentiate between width and thickness. Therefore, in the research, a clear separation was made between the two categories of particles (coarse and fine), as other authors have specified before [44,45]. Referring to the granulometry analysis, large differences were observed between coarse and fine particles in terms of participation rate. The participation rate analysis for particles with diverse geometry is justified by the fact that the particles are mostly inhomogeneous in terms of size and shape. As a result, two particles going through the same sieve with the same mesh size may differ in shape [17].

The particleboard vertical profile of density obtained with Recipe 2 differs from the one with Recipe 1 by the distance of the peak density, which is situated at 2.9 mm from the surface compared to 1 mm in Recipe 1, which means that the calibration operation requires the removal of thicker layers to have a good density of faces. It is known that not all this area should be removed during calibration and only the relaxation area should be removed after pressing, about 0.2–0.4 mm. Some authors [63,64] observed that in order to obtain high strengths the peaks of the density profile must be around 900 kg/m^3, a value that can also be observed from the density profile of the boards in Figure 3. It was also found that the short pressing time brings the two peaks very close to the board surface, and the high pressing time sends the two peaks to the central area of the board. From this point of view, the boards obtained in the research had a high pressing time of 16 min, with the two peaks of density being slightly away from the surface of the board. More distance was obtained with Recipe 2. The vertical density profiles are presented in Figure 3. It could be observed that in case of Recipes 1 and 2, the shapes are quite similar, with low surface densities of about 400 kg/m^3 (Recipe 1) and 370 kg/m^3 (Recipe 2), which increased to 750 kg/m^3 and 780 kg/m^3 at 1.1 mm and 2.9 mm from the surface. In the core layer, the densities were reduced to values of about 500 kg/m^3 and 595 kg/m^3, respectively. The uniform and nearly flat shape of vertical profile of boards (785 kg/m^3 on the faces and 610 kg/m^3 in the core) was obtained for Recipe 3. This demonstrates higher mechanical properties that were obtained in the case of Recipe 3 compared to Recipes 1 and 2. Similar results in the density profile were obtained by other authors [64]. The addition of glucose to Recipe 3 determined the decrease in the differences between the densities for the core and faces and also the distance between the peak of maximum density, which leads to more uniform density along the thickness of a board. The addition of pMDI and glucose as crosslinking agents improved the internal cohesion of particles [13].

Even if the average board densities of the experimental panels were around 650 kg/m^3, the variation in the values differed from one board to another. This is given by the different standard deviations of the three values of 4.196, 2.060, and 2.065 (Figure 8), as other authors have stated before [15]. Other authors [15,18] have established that the optimal density of particleboard could be 680 kg/m^3.

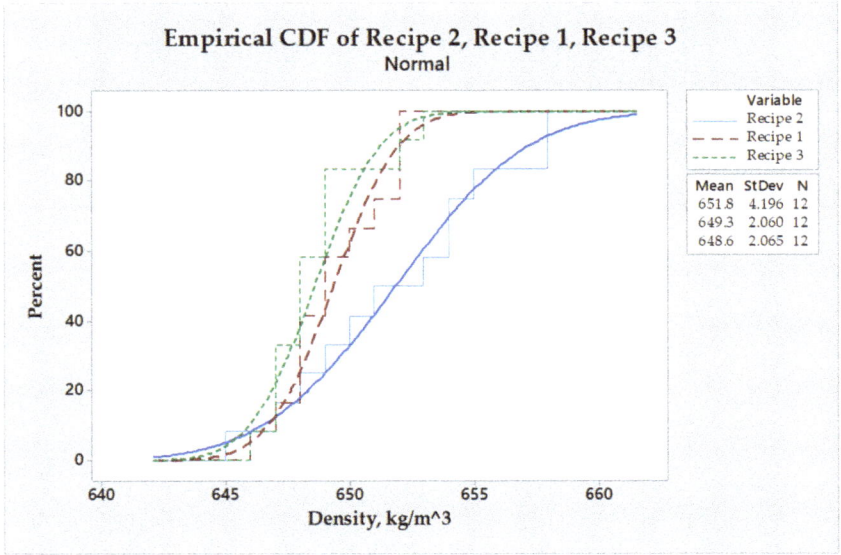

Figure 8. Empirical CDF for density of experimental particleboard.

The empirical cumulative distribution function (CDR) in Figure 8 shows the statistical distribution of density values. Moreover, if the confidence interval of 95% is taken into account, the intervals of density variation were found to be 645–653 kg/m^3 for Recipe 1, 643–660 kg/m^3 for Recipe 2, and 644–652 kg/m^3 for Recipe 3. The intervals calculated statistically and based on the median and the standard deviation overlapped perfectly with those visible in Figure 8 in the middle area.

Table 4 shows the one-way ANOVA statistical analysis for the density of the boards obtained with Recipes 1 and 2. It was found that, out of the 12 values on the basis of which the means of the densities of the two different types of boards were determined as a manufacturing recipe, only 1 value does not match. Moreover, the low values of the F-value and p-value highlighted the normality of the distribution for 95% confidence and the correctness of the statistical analysis.

Table 4. Analysis of Variance (ANOVA) for particleboard densities.

Source	DF	Adj. SS	Adj. MS	F-Value	p-Value
Recipe 1	10	97.83	16.31	0.85	0.0582
Error	1	95.83	19.17		
Total	11	193.67			

The quality of adhesion was assessed by an internal cohesion test that indicated the strength of the core region. This is influenced by the core density, particle geometry, and adhesive quantity and distribution. A lower value of IB (of 0.18 N/mm^2 and below 0.16 N/mm^2) was also obtained by Da Silva et al. [44] in the case of solely using calcium and magnesium lignosulfonates as adhesives for particleboard manufacturing without crosslinkers. It is obvious that pMDI and glucose as crosslinkers had a positive influence on the IB strength of the experimental particleboards. Other studies showed improvements in IB by using a combination of bio-based materials (glyoxalase lignin, ammonium lignosulphonates, tannin, and soya) or urea-based resin with pMDI [26,48].

Static bending strength is one of the most important mechanical properties of particleboards, along with the internal bond [29–31]. Figure 9 shows the graphical empirical CDF for all three manufacturing board types, which highlights the small variation in values. This is evidenced by the small standard deviations of 0.09, 0.25, and 0.18 N/mm^2 for Recipes 1, 2, and 3, respectively. Taking into account the 95% confidence interval, the variation ranges of the MOR of 10.41–10.80 N/mm^2 for Recipe 1, of 15.43–16.44 N/mm^2 for Recipe 2, and of 14.93–15.66 N/mm^2 for Recipe 3 were obtained, which are the same as those observable on the CDF chart. The addition of pMDI obviously improved MOE and MOR values (MOR: 11.1 N/mm^2, MOE: 2056 N/mm^2) as was also observed by Hemmila et al. [65] (MOR: 5 N/mm^2, MOE: 2000 N/mm^2). Similar results for MOR and lower values for MOE compared to those obtained with Recipe 1 were found by Bekhta et al. [66] by replacing a urea-formaldehyde adhesive with 10–20% magnesium lignosulfonate and by Hemmila et al. [51,65] for boards with particleboards produced with urea-melamine-formaldehyde. An increase in the content of lignosulfonate solution in the adhesive recipe can lead to an increase in the moisture content, a greater fragility of the boards, and a higher content of the steam–gas mixture in the process of pressing, determining in this way a decrease in the MOE and MOR values [65–67].

Many studies [3,13] have shown a decrease in formaldehyde emissions when using lignosulfonates in the case of composite boards, such as MDF or plywood, using percentages greater than 20% in order to obtain the E1 emission class [14]. In the research, the use of 15% lignosulfonate compared to the amount of chips in the particleboard manufacturing led to obtain the emission class E0 and even below to that of wood formaldehyde emission. Moreover, all research using lignosulfonates, whether based on magnesium or ammonium [25,29,33], has had a beneficial effect on reducing formaldehyde emissions from composite boards. It was shown in this way that the small differences between the two lignosulfonates, of percentages of less than 6% of magnesium or ammonium, do not have an influence on the formaldehyde emissions. The low formaldehyde emissions of experimental boards with pMDI and glucose participation may be explained by the short time of adhesive film curing the adhesive on the particle surfaces, which can block the release of formaldehyde from the surface of wood particles. This theory is supported by the microscopic images, where a good adhesion between wood particles can be noticed.

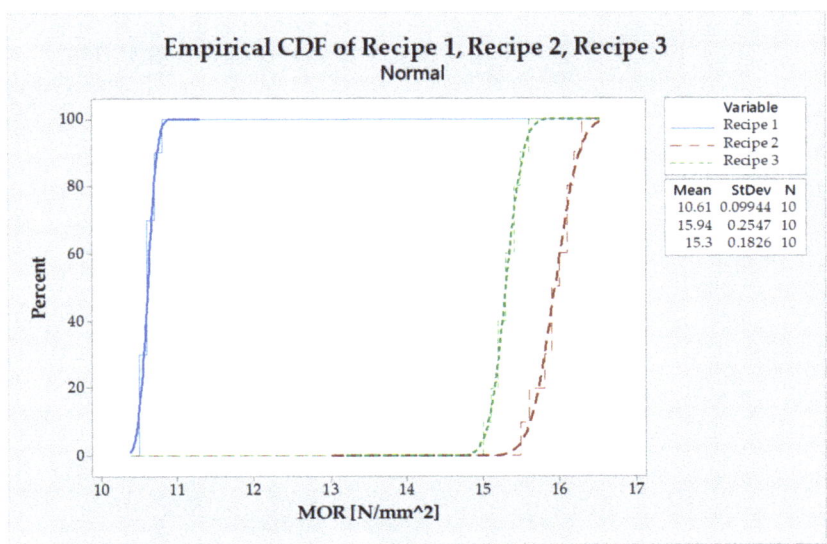

Figure 9. Empirical CDF of modulus of resistance (MOR) for the three types of chipboards.

Several studies [44,50,67] concluded that substances with hydroxyl groups positively influence the internal bond of the lignocellulosic composites and their low formaldehyde emission. Thus, D-glucose, which has hydroxyl groups in its composition, can decrease the emission of formaldehyde. Recent research [54] has shown that magnesium lignosulfonate (15% by weight of wood particles) can be used in fiberboard manufacturing with good mechanical and physical properties. Their low formaldehyde emission (1.1 mg/100 g oven dry panel, super E0 class), recommend them for indoor uses.

Figure 10 presents a sketch of the main research activities, putting the stress on raw materials (lignosulphonate and chips) and the pressing of chip mats.

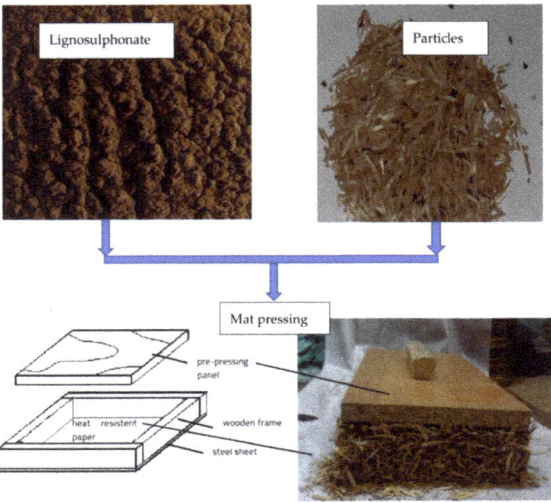

Figure 10. Scheme of research activities.

The images of microscopic analysis (Figure 11) of the particleboard structure show the interconnections between particles. Larger gaps between the bonded wood particles and the adhesive agglomeration in some areas in the structures without crosslinkers were observed (Recipe 1) (Figure 11a). The structures of particleboards with Recipes 2 and 3 (Figure 11b,c) were characterized by small and rare gaps and a more uniform distribution of adhesive, which led to higher values of MOR, MOE, and IB compared to panels with Recipe 1.

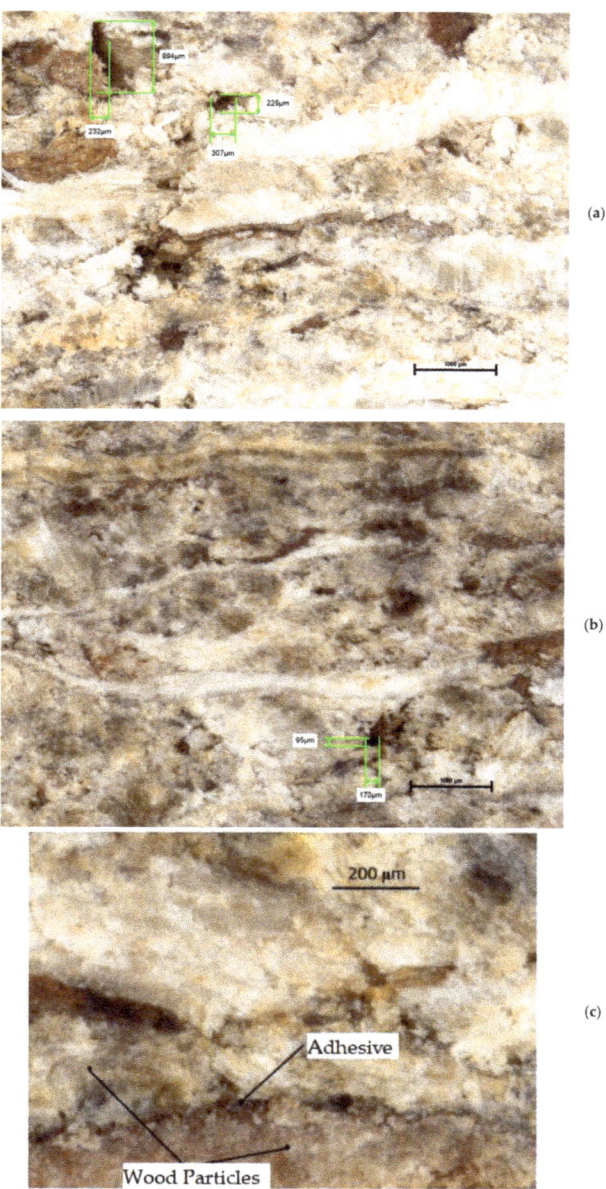

Figure 11. Images with 30× magnification taken on core of panel: (**a**) without crosslinker (Recipe 1); (**b**) with pDMI (Recipe 2); (**c**) with pMDI and glucose (Recipe 3).

5. Conclusions

The use of all three adhesive recipes that were based on the magnesium lignosulfonate for particleboard manufacturing had significantly impacts in reducing formaldehyde emissions by classifying them in the E0 and super E0 classes. This demonstrated that magnesium-lignosulfonate-based bio-adhesives can successfully replace the classic urea-formaldehyde or phenol-formaldehyde adhesives (with high formaldehyde emission).

The granulometric analysis of the wood particles demonstrated the importance of the different participation rates of the coarse and fine particles in the structure of the panels. The participation rates of the two chip fractions led to a good internal cohesion of the chipboard and resistance to static bending.

The particleboards glued with magnesium lignosulfonate in combination with pMDI and glucose (Recipe 2 and Recipe 3) had very good strengths. The recorded values of internal bond, MOR, and MOE exceeded the limit values of the EN 312 standard for P1-type panels. In this way, it was shown that in addition to magnesium lignosulfonate, particleboard manufacturing recipes must also contain some crosslinkers, which will increase the strengths of the boards.

The density profiles of the experimental panels showed a good adhesion of particles by the higher IB obtained with Recipes 2 and 3, which complied with the general principles and proved that the parameters of the pressing process are decisive in this direction.

The hydroxyl groups of D-glucose (Recipe 3) might lower the emission of formaldehyde to the level of some natural wood species.

Author Contributions: Conceptualization, A.L. and G.B.; methodology, O.Z.; software, A.L.; validation, C.C., O.Z. and A.L.; formal analysis, C.C.; investigation, G.B.; resources, G.B.; data curation, G.B.; writing—original draft preparation, A.L.; writing—review and editing, C.C.; visualization, O.Z.; supervision, C.C.; project administration, A.L.; funding acquisition, A.L. All authors have read and agreed to the published version of the manuscript.

Funding: This research received no external funding.

Data Availability Statement: Not applicable.

Acknowledgments: The authors would like to thank the management of the Transilvania University of Brasov for the administrative and technical support provided during this research.

Conflicts of Interest: The authors declare no conflict of interest.

References

1. Nelson, N.; Levine, R.J.; Albert, R.E.; Blair, A.E.; Griesemer, R.A.; Landrigan, P.J.; Stayner, L.T.; Swenberg, J.A. Contribution of Formaldehyde to Respiratory Cancer. *Environ. Health Perspect* **1986**, *70*, 23–35. [CrossRef] [PubMed]
2. Casteel, S.W.; Vernon, R.J.; Bailey, E.M., Jr. Formaldehyde: Toxicology and Hazards. *Vet. Hum. Toxicol.* **1987**, *29*, 31–33. Available online: https://europepmc.org/article/MED/3824872 (accessed on 3 February 2022).
3. Athanassiadou, E.; Tsiantzi, S.; Markessini, C. Producing Panels with Formaldehyde Emission at Wood Levels. COST E49 paper. 2009. Available online: https://chimarhellas.com/wp-content/uploads/2021/06/athanassiadou-tsiantzi-markessini-paper-2.pdf (accessed on 26 January 2022).
4. *EN 717-1:2004*; Wood-Based Panels. Determination of Formaldehyde Release. Formaldehyde Emission by the Chamber Method. European Committee for Standardization: Brussels, Belgium, 2004.
5. *ASTM E1333-14*; Standard Test Method for Determining Formaldehyde Concentrations in Air and Emission Rates from Wood Products Using a Large Chamber. ASTM International: West Conshohocken, PA, USA, 2010.
6. *EN 717-2: 1994/AC:2002*; Wood-Based Panels—Determination of Formaldehyde Release—Part 2: Formaldehyde Release by the Gas Analysis Method. European Committee for Standardization: Brussels, Belgium, 2002.
7. *EN ISO 12460-5:2015*; Wood-Based Panels—Determination of Formaldehyde Release—Part 5: Extraction Method (Called the Perforator Method). European Committee for Standardization: Brussels, Belgium, 2015.
8. *JIS A 1460:2001*; Building Boards Determination of Formaldehyde Emission—Desiccator Method. Japanese Industrial Standard: Tokyo, Japan, 2001.
9. *EN 717-3:1997*; Wood-Based Panels—Determination of Formaldehyde Release—Part 3: Formaldehyde Release by the Flask Method. European Committee for Standardization: Brussels, Belgium, 1997.
10. Dunkey, M. Urea Formaldehyde (UF) Adhesive Resins for Wood. *Int. J. Adhes. Adhes.* **1998**, *18*, 95–107. [CrossRef]

11. Akhtar, A.; Lutfullah, G.; Zahoorulah, U. Lignosulfonate—Phenolformaldehyde Adhesive: A Potential Binder for Wood Panel Industry. *J. Chem. Soc. Pak.* **2011**, *33*, 535–538.
12. Gavrilović-Grmuša, I.; Dunky, M.; Milijković, J.; Djiporović-Momčilović, M. Influence of the Degree of Condensation of Urea-Formaldehyde Adhesives on the Tangential Penetration into Beech and Fir and on the Shear Strength of the Adhesive Joints. *Eur. J. Wood Prod.* **2012**, *70*, 655–665. [CrossRef]
13. Pizzi, A. Bioadhesives for Wood and Fibres: A critical Review. *Rev. Adhes. Adhes.* **2013**, *1*, 88–113. [CrossRef]
14. Alonso, M.V.; Oliet, M.; Rodriguez, F.; Garcia, J.; Gilarranz, M.A.; Rodriguez, J.J. Modification of Ammonium Lignosulfonate by Phenolation for Use in Phenolic Resins. *Biores. Tech.* **2005**, *96*, 1013–1018. [CrossRef]
15. Santoni, I.; Pizzo, B. Evaluation of Alternative Vegetable Proteins as Wood Adhesives. *Ind. Crops Prod.* **2013**, *45*, 148–154. [CrossRef]
16. Fapeng, W.; Jifu, W.; Chunpeng, W.; Fuxiang, C.; Xiaohuan, L.; Jiuyin, P. Fabrication of Soybean Protein-Acrylate Composite Mini-Emulsion toward Wood Adhesive. *Eur. J. Wood Wood Prod.* **2017**, *76*, 305–313. [CrossRef]
17. Dukarska, D.; Derkowski, A. Rape Straw-Wood Particleboards Resinated with UF Resin and Supplemented with Nano-SiO$_2$. *Ann. Wars. Univ. Life Sci.—SGGW. For. Wood Technol.* **2014**, *85*, 49–52.
18. Pradyawong, S.; Qi, G.; Li, N.; Sun, X.S.; Wang, D. Adhesion Properties of Soy Protein Adhesives Enhanced by Biomass Lignin. *Int. J. Adhes. Adhes.* **2017**, *75*, 66–73. [CrossRef]
19. Eom, Y.-G.; Kim, H.-J.; Kim, J.-S.; Kim, S.-M.; Kim, J.-A. Reduction of Formaldehyde Emission from Particleboards by Bio-Scavengers. *Makchae Konghak* **2006**, *34*, 29–41.
20. Salem, M.Z.M.; Böhm, M. Understanding of Formaldehyde Emissions from Solid Wood: An Overview. *BioResources* **2013**, *8*, 4775–4790. [CrossRef]
21. Ghaffar, S.H.; Fan, M. Lignin in Straw and Its Applications as an Adhesive. *Inter. J. Adhes. Adhes.* **2014**, *48*, 92–101. [CrossRef]
22. Neimsuwan, T.; Hengniran, P.; Siramon, P.; Punsuvon, V. Effect of Tannin Addition as a Bio-Scavenger on Formaldehyde Content in Particleboard. *J. Trop. For. Res.* **2017**, *1*, 45–56.
23. Xi, X.; Pizzi, A.; Delmotte, L. Isocyanate-Free Polyurethane Coatings and Adhesives from Mono- and Di-Saccharides. *Polymers* **2018**, *10*, 402. [CrossRef]
24. Pichelin, F.; Nakatani, M.; Pizzi, A.; Wieland, S.; Despres, A.; Rigolet, S. Structural Beams from Thick Wood Panels Bonded Industrially with Formaldehyde-Free Tannin Adhesives. *For. Prod. J.* **2006**, *56*, 31–36.
25. Bertaud, F.; Tapin-Lingua, S.; Pizzi, A.; Navarette, P.; Petit-Conil, M. Development of Green Adhesives for Fiberboard Manufacturing, Using Tannins and Lignin from Pulp Mill Residues. *Cell. Chem. Technol.* **2012**, *46*, 449–455.
26. Mansouri, H.R.; Navarrete, P.; Pizzi, A.; Tapin-Lingua, S.; Benjelloun-Mlayah, B.; Pasch, H.; Rigolet, S. Synthetic-Resin-Free Wood Panel Adhesives from Mixed Low Molecular Mass Lignin and Tannin. *Eur. J. Wood Wood Prod.* **2011**, *69*, 221–229. [CrossRef]
27. Ismita, N.; Ranjan, M.; Semwal, P.; Prakash, A. Reduction in Formaldehyde Emission Liberation from Urea Formaldehyde Due to Closite Na$^+$ Addition. *Inter. J. Chem. Stud.* **2018**, *6*, 3502–3504.
28. Roumeli, E.; Papadopoulou, E.; Pavlidou, E.; Vourlias, G.; Bikiaris, D.; Paraskevopoulos, K.; Chrissafis, K. Synthesis, Characterization and Thermal Analysis of Urea-Formaldehyde/Nano SiO$_2$ Resins. *Thermochim. Acta* **2012**, *527*, 33–39. [CrossRef]
29. Hemmila, V.; Adamopoulos, S.; Hosseinpourpia, R.; Ahmed, S.I. Ammonium Lignosulfonate Adhesives for Particleboards with pMDI and Furfuryl Alcohol as Crosslinkers. *Polymers* **2019**, *11*, 1633. [CrossRef] [PubMed]
30. Núñez-Decap, M.; Ballerini-Arroyo, A.; Alarćon-Enos, J. Sustainable Particleboards with Low Formaldehyde Emissions Based on Yeast Protein Extract Adhesives *Rhodotorula rubra*. *Eur. J. Wood Wood Prod.* **2018**, *76*, 1279–1286. [CrossRef]
31. Younessi-Kordkheili, H.; Pizzi, A. Improving the Physical and Mechanical Properties of Particleboards Made from Urea-Glyoxal Resin by Addition of pMDI. *Eur. J. Wood Wood Prod.* **2018**, *76*, 871–876. [CrossRef]
32. Boran, S.; Usta, M.; Ondaral, S.; Gümüşkaya, E. The Efficiency of Tannin as a Formaldehyde Scavenger Chemical in Medium Density Fiberboard. *Compos. Part B* **2012**, *43*, 2487–2491. [CrossRef]
33. Antov, P.; Mantanis, G.I.; Savov, V. Development of Wood Composites from Recycled Fibres Bonded with Magnesium Lignosulfonate. *Forests* **2020**, *11*, 613. [CrossRef]
34. Younesi-Kordkheili, H.; Pizzi, A.; Niyatzade, G. Reduction of Formaldehyde Emission from Particleboard by Phenolated Kraft Lignin. *J. Adhes.* **2015**, *92*, 485–497. [CrossRef]
35. Zhou, X.; Zheng, F.; Lv, C.; Tang, L.; Wei, K.; Liu, X. Properties of Formaldehyde-Free Environmentally Friendly Lignocellulosic Composites Made from Poplar Fibres and Oxygen-Plasma-Treated Enzymatic Hydrolysis Lignin. *Compos. Part B Eng.* **2013**, *53*, 369–375. [CrossRef]
36. Hu, L.; Pan, H.; Zhou, Y.; Zhang, M. Methods to Improve Lignin's Reactivity as a Phenol Substitute and as a Replacement for Other Phenolic Compounds: A Brief Review. *BioResource* **2011**, *6*, 3515–3525. [CrossRef]
37. Malutan, T.; Nicu, R.; Popa, V.I. Contribution to the Study of Hydroxymethylation Reaction of Alkali Lignin. *BioResource* **2008**, *3*, 13–20.
38. Aro, T.; Fatehi, P. Production and Application of Lignosulfonates and Sulfonated Lignin. *Chem. Sus. Chem.* **2017**, *10*, 1861–1877. [CrossRef]
39. Klapiszewski, L.; Jamrozik, A.; Strzemiecka, B.; Matykiewicz, D.; Voelkel, A.; Jesionowski, T. Activation of Magnesium Lignosulfonate and Kraft Lignin: Influence on the Properties of Phenolic Resin-Based Composites for Potential Applications in Abrasive Materials. *Int. J. Mol. Sci.* **2017**, *18*, 1224. [CrossRef] [PubMed]

40. Fernandes, M.R.C.; Huang, X.; Abbenhuis, H.C.I.; Hensen, E.J.M. Lignin Oxidation with an Organic Peroxide and Subsequent Aromatic Ring Opening. *Int. J. Biol. Macromol.* **2019**, *123*, 1044–1051. [CrossRef] [PubMed]
41. Junghans, U.; Bernhardt, J.J.; Wollnik, R.; Triebert, D.; Unkelbach, G.; Pufky-Heinrich, D. Valorization of Lignin via Oxidative Depolymerization with Hydrogen Peroxide: Towards Carboxyl-Rich Oligomeric Lignin Fragments. *Molecules* **2020**, *25*, 2717. [CrossRef] [PubMed]
42. Geng, X.; Li, K. Investigation of Wood Adhesives from Kraft Lignin and Polyethylenimine. *J. Adhes. Sci. Technol.* **2006**, *20*, 847–858. [CrossRef]
43. Khan, M.A.; Ashraf, S.M. Development and Characterization of Groundnut Shell Lignin Modified Phenol Formaldehyde Wood Adhesive. *Indian J. Chem. Technol.* **2006**, *13*, 347–352.
44. Da Silva, M.A.; Dos Santos, P.V.; Silva, G.C.; Costa Lelis, R.C.; Do Nascimento, A.M.; Brito, E.O. Using Lignosulfonate and Phenol-Formaldehyde Adhesive in Particleboard Manufacturing. *Sci. For.* **2017**, *45*, 423–433.
45. Cetin, N.; Özmen, N. Use of Organosolv Lignin in Phenol-Formaldehyde Resins for Particleboard Production II. Par-Ticleboard Production and Properties. *Int. J. Adhes. Adhes.* **2002**, *22*, 481–486. [CrossRef]
46. Cetin, S.; Özmen, N. Studies on Lignin-Based Adhesives for Particleboard Panels. *Turk. J. Agric. For.* **2003**, *27*, 183–189.
47. Duan, H.; Qiu, T.; Guo, L.; Ye, J.; Li, X. The Microcapsule-Type Formaldehyde Scavenger: The Preparation and the Application in Urea-Formaldehyde Adhesives. *J. Hazard. Mater.* **2015**, *293*, 46–53. [CrossRef]
48. Mansouri, N.-E.; Pizzi, A.; Salvado, J. Lignin-Based Polycondensation Resins for Wood Adhesives. *J. Appl. Polym. Sci.* **2007**, *103*, 1690–1699. [CrossRef]
49. Ghorbani, M.; Liebner, F.; van Herwijnen, H.W.G.; Pfungen, L.; Krahofer, M.; Budjav, E.; Konnerth, J. Lignin Phenol Formaldehyde Resoles: The Impact of Lignin Type on Adhesive Properties. *BioResource* **2016**, *11*, 6727–6741. [CrossRef]
50. Ibrahim, M.N.M.; Zakaria, N.; Sipaut, C.S.; Sulaiman, O. Chemical and Thermal Properties of Lignins from Oil Palm Biomass as a Substitute for Phenol in a Phenol Formaldehyde Resin Production. *Carbohydr. Polym.* **2011**, *86*, 112–119. [CrossRef]
51. Kim, S.; Kim, H.-J.; Kim, H.-S.; Lee, H.H. Effect of Bio-Scavengers on the Curing Behavior and Bonding Properties of Melamineformal-Dehyde Resins. *Macromol. Mater. Eng.* **2006**, *291*, 1027–1034. [CrossRef]
52. Mancera, C.; Ferrando, F.; Salvado, J.; El Mansouri, N.E. Kraft Lignin Behaviour during Reaction in Alkaline Medium. *Biomass Bioenergy* **2011**, *35*, 2072–2079. [CrossRef]
53. Ramesh, M.; Rajeshkumar, L.; Sasikala, G.; Balaji, D.; Saravanakumar, A.; Bhuvaneswari, V.; Bhoopathi, R.A. Critical Review on Wood-Based Polymer Composites: Processing, Properties, and Prospects. *Polymers* **2022**, *14*, 589. [CrossRef]
54. Antov, P.; Savov, V.; Krišťák, L'.; Réh, R.; Mantanis, G.I. Eco-Friendly, High-Density Fiberboards Bonded with Urea-Formaldehyde and Ammonium Lignosulfonate. *Polymers* **2021**, *13*, 220. [CrossRef]
55. Procházka, P.; Honig, V.; Bouček, J.; Hájková, K.; Trakal, L.; Soukupová, J.; Roubík, H. Availability and Applicability of Wood and Crop Residues for the Production of Wood Composites. *Forests* **2021**, *12*, 641. [CrossRef]
56. Ramesh, M.; Rajesh Kumar, L. Green Adhesives: Preparation, Properties and Applications. In *Bioadhesives*, 1st ed.; Inamuddin, R.B., Mohd, I.A., Abdullah, M.A., Eds.; Elsevier: Toronto, ON, Canada, 2020. [CrossRef]
57. *EN 312*; Particleboards. Specifications. European Committee for Standardization: Brussels, Belgium, 2003.
58. *ISO 12460-3:2015*; Wood-Based Panels—Determination of Formaldehyde Release—Part 3: Gas Analysis Method. International Organization for Standardization: Geneva, Switzerland, 2015.
59. *EN 310*; Wood-Based Panels. Determination of Modulus of Elasticity in Bending and of Bending Strength. European Committee for Standardization: Brussels, Belgium, 1993.
60. *EN 319*; Particleboards and Fibreboards. Determination of Tensile Strength Perpendicular to the Plane of the Board. European Committee for Standardization: Brussels, Belgium, 1993.
61. *EN 323:1993*; Wood-Based Panels—Determination of Density. European Committee for Standardization: Brussels, Belgium, 1993.
62. *EN 322:1993*; Wood-Based Panels—Determination of Moisture Content. European Committee for Standardization: Brussels, Belgium, 1993.
63. Suo, S.; Bowyer, J. Simulation Modeling of Particleboard Density Profile. *Wood Fiber Sci.* **1994**, *26*, 397–411.
64. Gamage, N.; Setunge, S. Modelling of Vertical Density Profile of Particleboard, Manufactured from Hardwood Sawmill Residue. *Wood Mater. Sci. Eng.* **2014**, *9*, 157–167. [CrossRef]
65. Bekhta, P.; Noshchenko, G.; Réh, R.; Kristak, L.; Sedliacik, J.; Antov, P.; Mirski, R.; Savov, V. Properties of Eco-Friendly Particleboards Bonded with Lignosulfonate-Urea-Formaldehyde Adhesives and pMDI as a Crosslinker. *Materials* **2021**, *14*, 4875. [CrossRef]
66. Hu, J.-P.; Guo, M.-H. Influence of Ammonium Lignosulfonate on the Mechanical and Dimensional Properties of Wood Fiber Biocomposites Reinforced with Polylactic Acid. *Ind. Crop. Prod.* **2015**, *78*, 48–57. [CrossRef]
67. Costa, S.; Costa, C.; Madureira, J.; Valdiglesias, V.; Teixeira-Gomes, A.; de Pinho, P.G.; Laffon, B.; Teixeira, J.P. Occupational Exposure to Formaldehyde and Early Biomarkers of Cancer Risk,1 Immunotoxicity and Susceptibility. *Environ. Res.* **2019**, *179*, 108740. [CrossRef] [PubMed]

Article

A Sustainable Approach to Build Insulated External Timber Frame Walls for Passive Houses Using Natural and Waste Materials

Sergiu-Valeriu Georgescu [1], Daniela Șova [2], Mihaela Campean [1] and Camelia Coșereanu [1,*]

[1] Faculty of Furniture Design and Wood Engineering, Transilvania University of Brasov, B-dul Eroilor nr. 29, 500036 Brasov, Romania; sergiu.georgescu@unitbv.ro (S.-V.G.); campean@unitbv.ro (M.C.)
[2] Faculty of Mechanical Engineering, Transilvania University of Brasov, B-dul Eroilor nr. 29, 500036 Brasov, Romania; sova.d@unitbv.ro
* Correspondence: cboieriu@unitbv.ro

Abstract: This paper presents structures of timber-framed walls designed for passive houses, using natural and waste resources as insulation materials, such as wool, wood fibers, ground paper, reeds (*Phragmites communis*), and Acrylonitrile Butadiene Styrene (ABS) wastes. The insulation systems of stud walls composed of wool–ABS composite boards and five types of fillers (wool, ABS, wood fibers, ground paper, and reeds) were investigated to reach U-value requirements for passive houses. The wall structures were designed at a thickness of 175 mm, including gypsum board for internal wall lining and oriented strand board (OSB) for the exterior one. The testing protocol of thermal insulation properties of wall structures simulated conditions for indoor and outdoor temperatures during the winter and summer seasons using HFM-Lambda laboratory equipment. In situ measurements of U-values were determined for the experimental wall structures during winter time, when the temperature differences between outside and inside exceeded 10 °C. The results recorded for the U-values between 0.20 W/m^{2}K and 0.35 W/m^{2}K indicate that the proposed structures are energy-efficient walls for passive houses placed in the temperate-continental areas. The vapour flow rate calculation does not indicate the presence of condensation in the 175 mm thick wall structures, which proves that the selected thermal insulation materials are not prone to degradation due to condensation. The research is aligned to the international trend in civil engineering, oriented to the design and construction of low-energy buildings on the one hand and the use of environmentally friendly or recycled materials on the other.

Keywords: timber frame walls; thermal insulation; passive house; natural materials; waste materials

1. Introduction

According to the European Commission, 40% of energy consumption and 36% of greenhouse gas emissions in the European Union belong to buildings, mainly from their construction, usage, renovation, and demolition. By renovating existing buildings, the total energy consumption in the EU could be reduced by 5–6% and CO_2 emissions by about 5% [1]. About 35% of EU buildings are now over 50 years old and their energy efficiency is low [2]. In order to increase the energy efficiency of buildings, it is necessary to reduce energy consumption and CO_2 emissions, aiming for the goal of carbon-neutrality by 2050, an essential priority in tackling the climate change and environmental degradation [3]. The materials of energy-efficient houses must reach certain values of the heat transfer coefficient depending on the climatic zone in which they are used, according to the criteria of Passive House Institute (PHI) for low energy buildings [4]. Passive houses are a concept developed in Darmstadt, Germany for sustainable architecture, and their design involves adapting to the climate. A sustainable home involves the characteristics of both green and passive houses, as well as environmentally friendly technologies. Passive houses are characterized by a very good thermal insulation of the exterior walls, expressed by the heat transfer coefficient or by thermal transmittance. For the continental temperate zone specific to Romania, the recommended thermal transmittance (U) for exterior walls is between 0.30 W/m^{2}K and 0.50 W/m^{2}K according to the criteria of passive houses issued by certification by PHI [4].

Current trends in civil engineering are the design and construction of low-energy buildings on the one hand and the use of environmentally friendly or recycled materials on the other. Natural or recycled materials are more and more investigated for their use as insulation materials. Reeds, bagasse, kenaf, cattail fibers, corn's cob, cotton stalk fibers, date palm, and sunflower are just few examples of natural resources used as raw materials for thermal-insulating structures [5–12]. The natural materials have the advantage of being renewable and ecological resources with low densities, good thermal properties, and low costs. Instead, they have drawbacks, such as hygroscopic properties and low anti-fungal and fire resistance, which can be attenuated by their chemical treatments [13]. Recently, the public perception of the presence of plastic in the environment has changed, and there is a growing international concern for improving the management of this material at the end of the life cycle of the products that use it. One of the recommended solutions is the use of bio-plastics [14], another is recycling [15]. ABS and polyethylene are known as thermoplastics with a long period of biodegradation. Therefore, these materials are on the list of those for which recycling options are sought [16].

The trend of bringing environmentally friendly materials back into building construction is becoming more pronounced in the international community. Greenery systems are considered passive tools for energy savings in buildings and for acoustic insulation [17–19]. Reed is considered a renewable resource with high potential as thermal insulator [20,21], even if in situ measurements of moisture content in various climate conditions correlated with thermal conductivity coefficient demonstrated the sensitivity to moisture of this material, to the detriment of thermal insulating performance [22]. Reed is also a carbon-neutral raw material and a CO_2 sink and it has been used for centuries for various uses, including vernacular architecture [23]. Spruce bark, available in large quantities and not used with a high value added, was investigated as insulating filling material in external timber frame wall structure, resulting in potential thermal insulator [24] and mixed pine tree bark and cannabis residues glued with two different adhesives formed boards with good insulation properties [25]. Tree bark is suitable for thermal insulation material due to its low density, very good thermal insulation properties, a high proportion of cork cells, high chemical extractives content serving as a protection against microorganisms, and low flammability [13,26]. Wood fibers resultant from low-quality wood are also a part of the insulation materials market, being used to produce wood fiber insulation boards through various methods, such as incubating with laccase–mediator system and then hardened with steam–air mixture [27], or applying the wet process for the low-density bark fibre boards without using additives or adhesive [28]. Wool is another natural material that is insufficiently introduced into the economic circuit. Wool is a performant thermal insulation material, due to its unique property of accumulating the moisture fully in the central part and simultaneously keeps the hydrophobic surface shelves dry, thus offering dry and warm surfaces with high thermal resistance [29]. Being also a biodegradable material, wool is an alternative solution to green buildings construction [30,31]. The high ability of wool to absorb moisture up to 35% prevents condensation and recommends this material to be used as an insulated material both for building walls and roofs [32]. One of the most important advantages of natural materials, such as fibers originating from vegetable sources, or animals fibers including wool, is that they involve low carbon emissions and energy for manufacture and transport. They can also buffer the humidity, mitigate moisture, and improve indoor air quality, allowing the building to "breathe" [33]. The environmental and health impact of the natural materials is also low, as long as they allow binding into thermal insulation boards without chemical adhesives, as in the case of using agro-wastes [34–36] or other unconventional resources [37]. Paper waste and recycled paper are also alternative solutions for a sustainable market of thermal insulation boards [38].

In addition to the determination by laboratory measurement of thermal resistance and thermal conductivity coefficients, in situ experimental methods for determining the heat transfer and for measuring thermal transmittance are also applied in some research works. Thermocouples and sensors are placed inside and outside the walls, and heat flow meters are used on the inner surface, or inside the wall structures for in situ measurements [39–42]. Another research work compared the in situ measurement method and a method based on simulation software applied on timber frame wall with several insulation materials and investigated characteristics such as humidity and condensation [43]. Additionally, the vapour transfer through various types of materials should not be neglected and is the object of investigation in several research works in order to assess the condensation risk inside the structure [44] and mould growth [45,46]. Vapour transfer through various materials and at various surfaces using classical numerical method is also addressed [47]. Coupled vapour and heat transfer under real climate conditions [48] combined with classical numerical methods have shown that the heat flow through the wall was influenced by the vapour flow crossing the insulation materials with high hygroscopic characteristics (wood fibre and cellulose).

The investigation of insulation materials for building construction comprised a large amount of natural resources and recycled wastes in order to achieve both resource and energy sustainability in this sector [49]. All investigated materials for the building sector may have a role in the construction of panels, walls, and roofs, depending on the individual properties of each material. The present study investigates the thermal insulation properties and the presence of condensation in the 175 mm thick external timber frame walls designed for passive houses, for which the recommended thermal transmittance (U) of the exterior walls ranges between 0.30 W/m^{2}K and 0.50 W/m^{2}K. Wood fibers and renewable natural materials insufficiently introduced into the economic circuit, such as wool and reeds, but also recyclable materials such as plastic or paper are used as insulation materials inside the wall structures. The study's approach is based both on laboratory and in situ measurements of the thermal behaviour of the experimental walls and the comparison of their thermal performances.

2. Materials and Methods

2.1. Insulation Materials

Five materials were used in this study as insulation materials in the structure of external timber frame walls, composed (beginning from the inner face to the outer face) of 12 mm gypsum board (GB), 0.2 mm thick aluminium foil used as a vapor barrier, insulation materials, and oriented strand board (OSB) of 12.5 mm. The insulation materials were as follows: Acrylonitrile Butadiene Styrene (ABS) waste, wool, wood fibers, reed, and recycled paper. ABS waste was in the form of small particles, which resulted from the edge banding machine, after edging the sides and ends of wood-based materials used in furniture production, such as medium density fiberboard (MDF) or particleboard. A mixture of wt. 40% wool (as reinforcement) and wt. 60% ABS (as matrix) was used to form thermal insulation panels with sizes of 564 mm × 564 mm × 20 mm. The thermal insulation panels were obtained by hot pressing at a temperature of 160 °C and without pressure for 20 min. First, the two components were mixed together, and then the mixture has been arranged inside a wooden frame and covered with a heat-resistant foil on both sides, to avoid sticking the composition to the press plates (Figure 1). The panels were afterwards conditioned, for 48 h, at constant temperature of 20 °C and relative humidity of air of 65% and afterwards sized to the final dimensions.

(a) (b) (c)

Figure 1. ABS-wool panel formation: (**a**) ABS (60%) and wool (40%) prepared to be mixed; (**b**) Mat formation in wooden frame; (**c**) Panel after sizing to the final dimensions.

Wood fibers (FL) were provided by MDF manufacturer S.C. Kastamonu S.A. from Reghin, Romania and the common reed stems (R) were collected from the Danube Delta. Sheep wool (W) was purchased directly from a local sheep farmer. Strips of approx. 15 cm of wool were subjected to manual combing. Recycled paper (P) was soaked in water, then collected into small pieces and squeezed, then dried and ground with a hammer mill. All these materials, including ABS waste were used as fillers, one by one, inside the structure of the experimental wall. Moisture content was around 6.1% for wood fibers, 6% for the reed, and 3.2% for the ground paper. Bulk density (ρ) and thermal conductivity coefficient (λ) were determined for all these materials, including ABS–wool panel, and presented in Table 1. The thermal conductivity coefficient values were measured under the laboratory conditions by using HFM 436 Lambda equipment (NETZSCH-Gerätebau GmbH, Selb, Germany).

Table 1. Density and thermal conductivity coefficient of the insulation materials.

Type of Material	Density, in kg/m^3	Thermal Conductivity Coefficient, λ, in W/mK
Wool (W)	23.3	0.0424
Paper (P)	56.1	0.0338
Wood fibers (WF)	66.0	0.0391
ABS	77.6	0.0410
Reed (R)	79.4	0.0524
ABS-Wool composite (ABS-W)	300	0.0420

2.2. Timber Frame Wall Structure

Five experimental wall structures with length of 600 mm, width of 600 mm, and thickness of 175 mm were designed and built for thermal insulation properties investigation. The designed structures of the walls are presented in Figure 2.

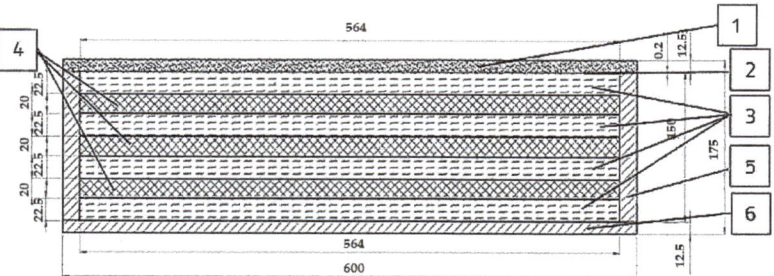

Figure 2. The structure of the timber frame wall: 1—gypsum board; 2—aluminum foil; 3—fillers; 4—ABS-wool composites; 5—particleboard; 6—OSB.

As seen in Figure 2, the insulation materials used inside the wall are structured into seven layers. Thus, three ABS–W composites were placed equidistantly between the two faces of the wall. Empty spaces were loosely filled in successively with the five materials selected for the experiment. The design of the experimental walls is presented in Table 2. Two replicates were built for each type of structure, one for in situ measurements and the other for measurements under laboratory conditions.

Table 2. Experimental design of the wall structures.

Wall Code	GB (pcs.)	Fillers (no. of Layers)					ABS-W (pcs.)	OSB (pcs.)
		ABS	Wool	Paper	Wood Fiber	Reed		
W1	1	4					3	1
W2	1		4				3	1
W3	1			4			3	1
W4	1				4		3	1
W5	1					4	3	1

2.3. Thermal Transmittance

The five types of experimental wall structures were subjected to thermal transmittance (U) measurements in real conditions (in situ) and the other five were tested under laboratory conditions for thermal resistance (R) measurements. The laboratory tests were conducted on HFM436 Lambda equipment (NETZSCH-Gerätebau GmbH, Selb, Germany), according to ISO 8301:1991 [50] and DIN EN 12667:2001 [51]. This testing method is based on the determination of the quantity of heat that is passed from a hot plate to a cold plate through the sandwich composite structure of the experimental wall (in our case), based on Fourier's Law. Before the samples were tested, the equipment was calibrated and the temperature configuration was set up, as presented in Table 3.

Table 3. Temperature configuration setup.

Test Number	T_1, in °C for Bottom Plate	T_2, in °C for Top Plate	$\Delta T = T_2 - T_1$ in °C	Mean Temperature $(T_1 + T_2)/2$
1	−10	20	30	5
2	−5	20	25	7.5
3	0	20	20	10
4	5	20	15	12.5
5	10	20	10	15
6	15	20	5	17.5

The temperature configuration intended to simulate the difference of temperature between indoor and outdoor environment, assumed to be equal to the temperature difference ΔT (Table 3). Thus, the top plate was set to a temperature (T_2) of 20 °C, to represent indoors, whilst the temperature of the bottom plate (T_1) was set to vary between −10 °C and 15 °C, simulating the outdoor environment. Six measurements of thermal conductivity coefficient (λ) and thermal resistance (R) were conducted in the experiment for each wall structure, for the six temperature differences (ΔT) varying from 5 °C up to 30 °C. The thermal resistance (R) values resulted from the laboratory measurements of the wall structure were used to calculate the theoretical thermal transmittance or overall heat transfer coefficient (U) using Equation (1), according to ISO 6946:2017 [52].

$$U = \frac{1}{R}, \left(W/m^2 K \right) \tag{1}$$

For the thermal transmittance measured in real conditions (in situ), temperature and heat flow sensors were used, using the Thermozig BLE system from OPTIVELOX, Province of Prato, Italy. Thermozig BLE is professional equipment which measures thermal transmittance (U-value) through the use of a Bluetooth Low Energy sensors network. This system can be configured to a network up to eight measurement nodes and has applications in temperature, thermal flux, or humidity measurements. It uses the automatic recording mode and downloading the data in PC, being supplied with a software package that easily allows generating technical reports with graphics and numerical values. Each test was performed for a minimum period of seven days. According to the user manual recommendations, for accurate results a temperature difference of at least 10 °C between the indoor and outdoor environment was needed, so the tests were conducted during the winter. The positioning of the structures and the sensors are highlighted in Figure 3.

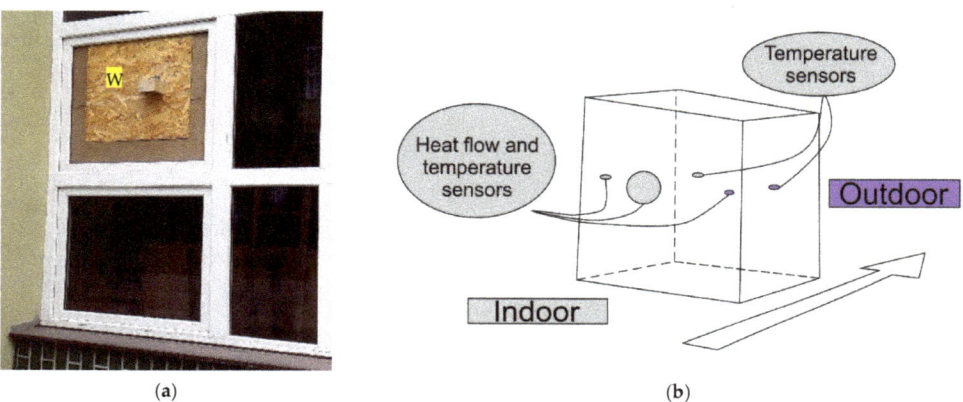

Figure 3. In situ experimental test: (**a**) Positioning of the experimental wall (W); (**b**) Placement of the sensors on the structure.

2.4. Vapour Transfer

The method of calculation the vapour flow rate per unit area (vapour flux) was used to verify if condensation occurs in the tested wall structures. If the variation lines of the vapour partial pressures (p_v) intersect the variation curves of the saturation pressures (p_s), it means that there is an area in the wall where water vapour can condense. This phenomenon can lead to degradation of fillers and increased heat exchange, which causes a decrease in thermal insulation properties.

The determination of the heat flux through a flat wall will allow the determination of linear temperature variations of the vapour partial pressures for each layer thickness of the wall. The variations of the partial pressures of water vapour on the thickness of each layer are compared with the variations of the saturation pressures on the thickness of each layer. If they do not intersect, it means that the temperature in any plane of the wall is higher than the dew point.

Heat flux, unit vapour flux, variations of temperatures, and partial pressures were calculated for each tested wall structure based on a climatic scenario, and will be exemplified by graphical representations.

If the wall delimits the indoor environment (which is air) from the outdoor environment (air), which have different temperatures, a heat transfer occurs from the higher temperature environment to the lower temperature environment. In this case, an overall heat transfer occurs, and the heat flux is expressed by the Equation (2):

$$q = \frac{t_i - t_e}{R_{tot}} = \frac{t_i - t_e}{R_{si} + R_p + R_{se}} \left[\frac{W}{m^2}\right] \quad (2)$$

where t_i is the average temperature of the indoor environment, in °C; t_e is the average temperature of the external environment, in °C; R_{tot} is the total thermal resistance, in m²K/W; R_{si} is the thermal resistance of the inner surface, in m²K/W; R_{se} is the thermal resistance of the outer surface, in m²K/W; and Rp is the thermal resistance of the wall, in m²K/W.

The vapour flux (q_v) is the amount of water vapour in the air transferred per unit time through the unit area of a wall. It can be expressed in terms of the vapour partial pressure on the boundary surfaces of the wall and the resistance to vapour transfer on the wall thickness. The vapour transfer is achieved through the wall in the direction of decreasing the partial vapor pressures and is calculated using Equation (3).

$$q_v = \frac{Pv_1 - Pv_2}{Rv} \left[\frac{\text{kg}}{\text{m}^2\text{s}}\right] \qquad (3)$$

where Pv_1 is the vapour partial pressure on the inner surface of the wall, in Pa; Pv_2 is the vapour partial pressure on the outer surface of the wall, in Pa; and Rv is the resistance to vapour transfer, in m/s.

2.5. Statistical Analysis

The obtained values were statistically processed, by determining the standard deviation, for a confidence interval of 95% and an alpha error of 0.05. The standard deviation was also applied to the graphs obtained in Microsoft Excel. One-way Analysis of Variance (ANOVA) was performed with Microsoft® Excel package in order to analyse the significance of the factors affecting the thermal performance of the investigated structures in terms of thermal resistance (R) and thermal transmittance (U-value).

3. Results and Discussions

3.1. Thermal Transmittance

The results of the thermal transmittance (U_i) generated by in situ measuring system with temperature and heat flow sensors are presented in Figure 4. The measurements were recorded for seven days, when differences between indoor and outdoor temperatures exceeded 10 °C. In the first hours of testing performance, there was a balancing of the measuring sensors, resulting in a greater variation of the U values. These hours were needed for the calibration of the system. After the system was stabilized, and the graph lines tended to a constant value (after approximately 1 day) the recorded U_i values were used to calculate the mean U_i value.

The U_i mean values of all experimental wall structures and their standard deviations are presented in Figure 5 in ascending order. As observed, the lowest U-values were obtained by the structures with wood fiber and wool fillers. They were followed by the structures with ground paper, ABS, and reed fillers.

The calculated thermal transmittance value (U_c) with Equation (1) was based on the measurement of thermal resistance (R) under the laboratory conditions by means of Netzsch Lambda heat flow meter. The values were recorded for all six temperature configuration setups and ΔT ranging between 5 °C (corresponding to warm season simulation) and 30 °C (corresponding to cold season simulation). The results are presented in Figure 6. As the graph shows, the thermal insulation performance of the fillers is better in warm seasons. Similar results were obtained for wood fibers filler [53].

The structures are not homogeneous, so they did not have a predictable behaviour (increase or decrease) of thermal resistance considering the variation of the experimental setup conditions. The negative temperatures impact on the structures caused the lowering of the thermal resistance. As the graph in Figure 6 shows, the behaviours of the wall structures are similar, regardless of filling material. Instead, the thermal resistance is influenced by the thermal insulation performance of the fillers. For the laboratory measurements, wool filler in the wall structure determined the higher thermal resistance of the investigated timber frame wall and implicitly a better U_c value.

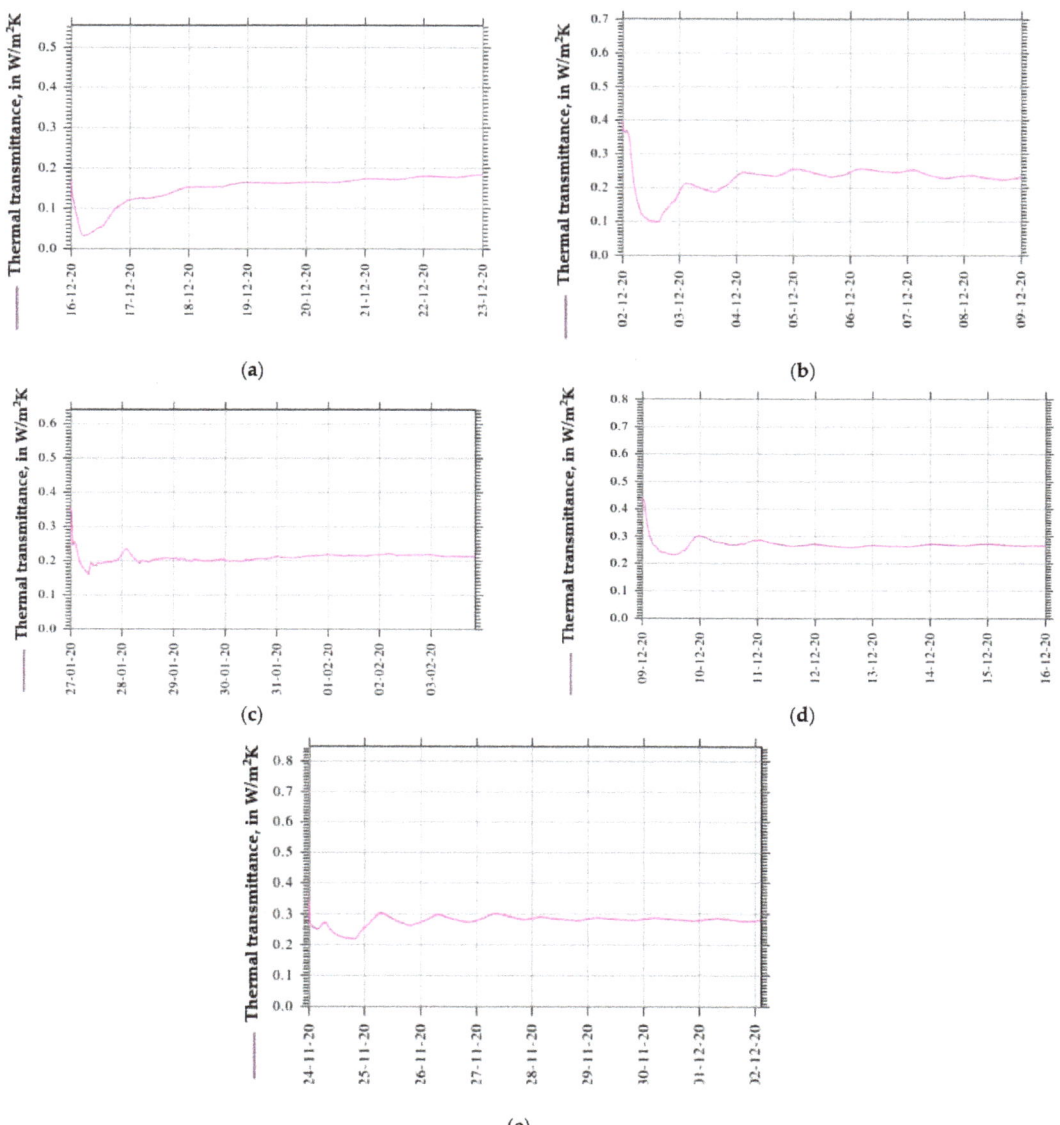

Figure 4. In situ measurements diagrams of thermal transmittance of the structures with the following fillers: (**a**) wood fibers; (**b**) ground paper; (**c**) wool; (**d**) ABS waste; (**e**) reed.

In order to compare the calculated U values (U_c) with measured in situ values (U_i), the thermal resistance (R) recorded for $\Delta T = 10$ °C was used for U_c calculation. This thermal resistance was selected due to the close values of the temperature differences with those from the measurements taken in real conditions (in situ). The comparison between U_c and U_i for all experimental wall structures is shown in Figure 7. As seen from the diagram, the calculated U values are higher than those measured in real conditions for seven days. These differences ranged between 7.64% (structure with ABS waste filler) and 56.3% (structure with wood fibers filler). This result contradicts the conclusions of other research work [54], where the calculated U value was lower than the measured in situ one (with one exception). The differences between calculated and in situ values of thermal transmittance in the mentioned research ranged between 15% and 43%. The theoretical calculation in this case was made using the thermal resistance of each component of the wall structure.

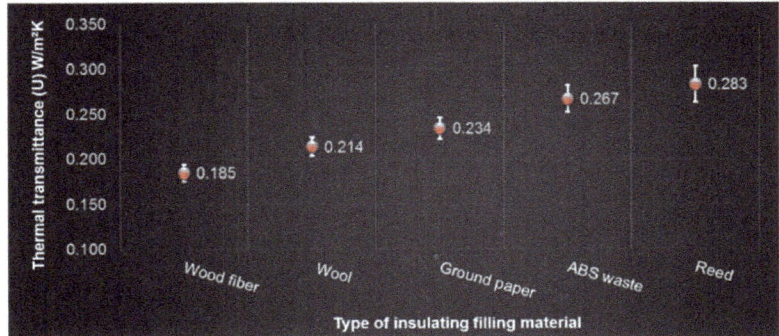

Figure 5. In situ measured thermal transmittance (*U*) values for the experimental wall structures.

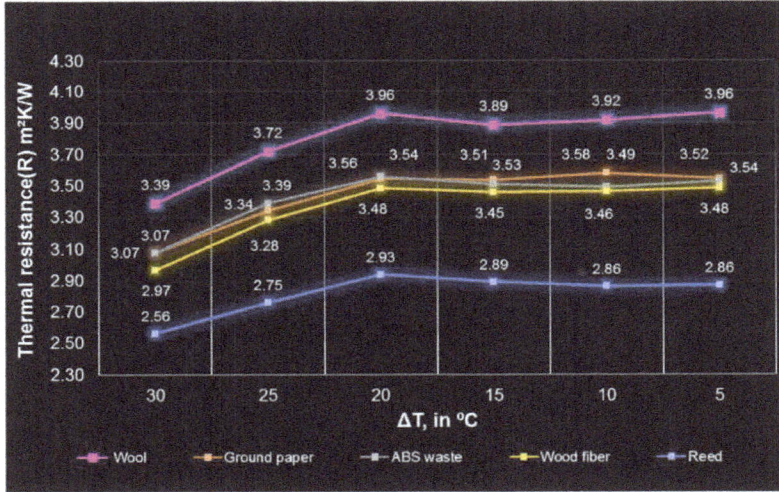

Figure 6. Comparison between thermal resistance values of investigated wall structures in the conditions of cold and warm season simulation, for different fillers.

There are several factors that can determine the differences between U_c and U_i values. There are more variables that cannot be controlled in the real climatic conditions. One of these variables is the temperature. The differences between inside and outside temperatures are not constant in real conditions and they could vary from one hour to another. Instead, the measuring protocol with HFM LAMBDA 436/6 installation is based on constant differences of temperatures. Another factor that can influence in situ measurements is the accuracy of the heat flow sensors. The measurement accuracy of the thermal flow sensor of the Ble thermosig device is ±5% (T = 20 °C), whilst the HFM Lambda 436/6 installation accuracy of the entire measurement is ±1% to 3% according to the user manual. An important variable not considered in the present experimental research was the relative humidity of air. In this direction, research on 336 mm thick wall structures, with mineral wool, basalt wool, and hemp fillers as insulation materials [55] proved that, in the closed vapour diffusion layers (mineral and basalt wool), the thermal conductivity coefficient was kept almost constant, while in the hemp layer (considered to be open diffusion), the thermal conductivity coefficient increased by 30% due to moisture transport. As found in the literature, natural materials such as fibers originated from vegetable sources or animal fibers, including wool, can buffer the humidity, mitigate moisture, and allow the building to "breathe" [33]. This advantage may be constrained when the wall structure is placed between the plates of the heat flow meter installation and the humidity is practically trapped inside the structure, possibly influencing the heat transfer. This phenomenon was observed inside the walls with spruce wood faces and 140 mm thick wood fiber and hemp core. As the temperature increased, the wood in the structure released free water and the humidity of the air also increased [56]. In Figure 7, it can be seen that for the structure with ABS (considered to be closed diffusion) filler, the difference between U_i and U_c is smaller than for the other materials, which can be considered open diffusion. As mineral and basalt wool [55], ABS can be less sensitive to humid air, perhaps explaining the low difference between U_i and U_c of only 7.64%.

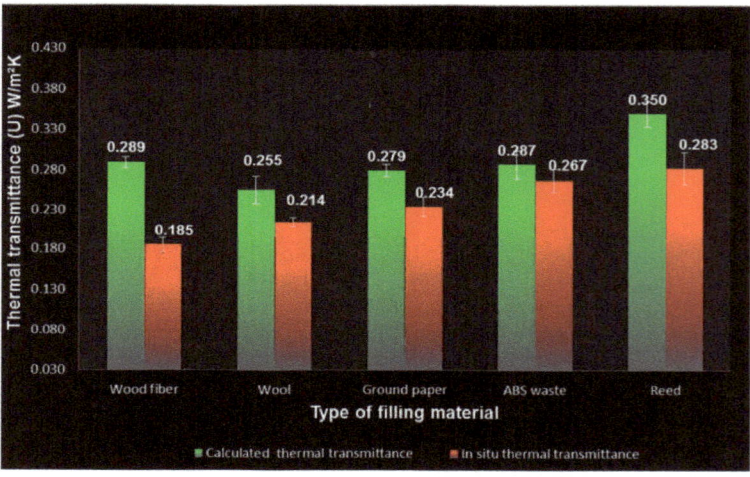

Figure 7. Mean values of thermal transmittance for the entire test cycle.

The external timber frame walls with thickness of 175 mm investigated in this paper have better U-values (0.185–0.283 W/m^{2}K) than those measured in situ for 510 mm thick walls (from which 350 mm were represented by reed panels) [20] and for which the recorded U-values ranged between 0.207 W/m^{2}K and 0.383 W/m^{2}K. Better U-values were obtained for wall structures consisting of wheat straw bales [21] with a thickness of 900 mm, which recorded in situ measured U-value of 0.125 W/m^{2}K. Comparing the ratios between the thickness of the walls and corresponding U-values with those presented in [20,21], it can be stated that the wall structures proposed in the present research have good thermal performances. Laboratory readings of U-values for light-frame timber wall structures with various insulation materials and their combinations (extruded and expanded polystyrene, stone wool, mineral wool, and glass wool) and a total thickness of 160 mm [43], in the conditions of indoor temperature of 20 °C and outdoor temperature around 8 °C ranged between 0.394 W/m^{2}K and 0.419 W/m^{2}K. These results can be compared with those calculated at $\Delta T = 10$ °C in the present research, for which U-values varied between 0.289–0.350 W/m^{2}K, but for a thicker structure of wall of 175 mm.

3.2. Statistical Analysis

Thermal transmittance (U) was statistically significantly ($p < 0.05$) affected by the applied method of investigation and the bulk density of the fillers. Instead, the differences of temperatures (ΔT) between hot plate and cold plate of the heat flow meter, simulating conditions of cold and warm season, did not show a significant effect ($p = 0.278$) on the thermal resistance (R) in this investigation where the measurements were conducted in the laboratory conditions, using LAMBDA 436/6 heat flow meter equipment.

3.3. Vapour Transfer

The determination of the heat flux through a flat wall and of the vapour flux allows the determination of temperature variations, respectively of the vapour partial pressures on the thickness of each wall layer, variations that are in both cases linear. The heat flux through the wall was first calculated with Equation (2) for an indoor temperature (t_i) of 20 °C, and outdoor temperature (t_e) of -10 °C. Following the calculations, the variation of the partial pressure (p_v) and the saturation pressure (p_s) were represented in the diagrams in Figure 8 for all investigated wall structures.

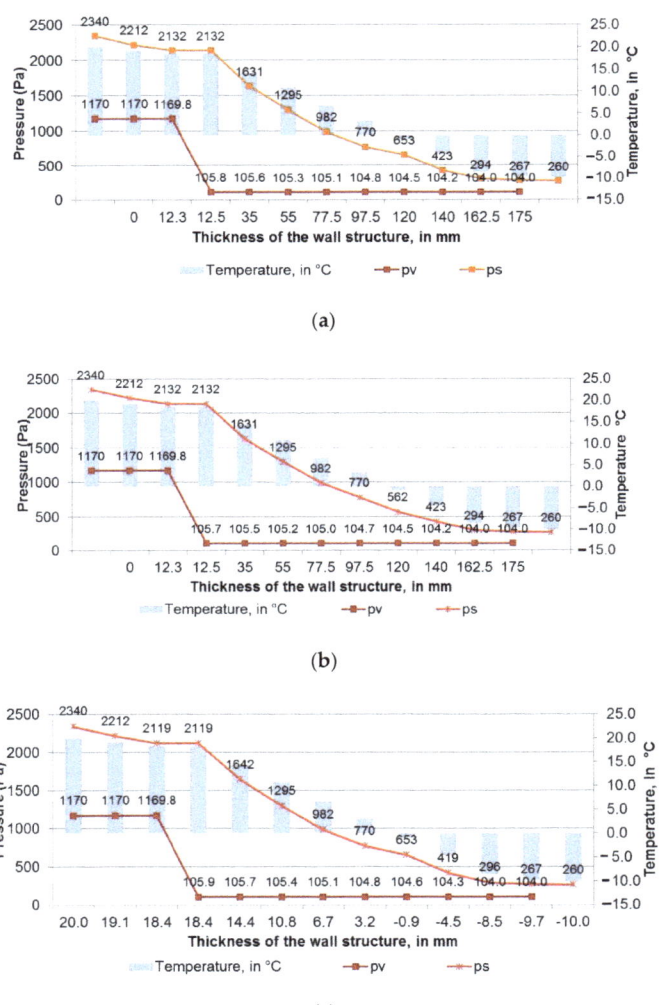

Figure 8. Cont.

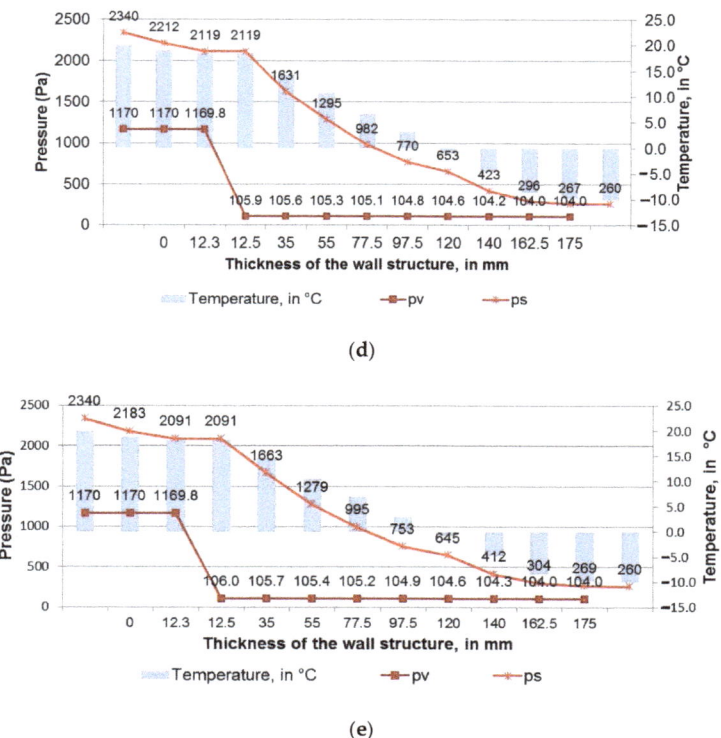

Figure 8. Graphic example of the variation of the partial pressure (p_v) and the saturation pressure (p_s) for the experimental structures with the following fillers: (**a**) wood fibers; (**b**) ground paper; (**c**) wool; (**d**) ABS waste; (**e**) reed.

If intersections of the two curves are seen in the graphs, it indicates that the condensation occurs in that element, and if there is no intersection of the two pressure variation curves, this indicates that condensation is not present in the structure. As can be seen in Figure 8, there are no intersections for any of the investigated wall structures. This proves that the presence of condensation in the structures is not observed. In this way, the thermal insulation materials are less prone to degradation and they maintain their thermal properties in time. The same trend of the graphs from Figure 8 was highlighted by the simulation process with HT-flux software [43]. The simulation exhibited a great increase in vapour pressure on the side of the wall with the highest temperature and a linear decrease towards the side with the lowest temperature, no matter the type of the insulation material.

4. Conclusions

The thermal insulation performances of the exterior timber frame walls proposed in this study are in accordance with the criteria of passive houses issued by PHI [4] for the continental temperate zone, for which the recommended thermal transmittance (U) of the exterior walls ranges between 0.30 W/m²K and 0.50 W/m²K. All investigated structures in the present research recorded U-values from 0.20 W/m²K to 0.35 W/m²K, which resulted both from numerical method and in situ one. The experimental walls were tested in five configurations using three ABS-wool thermal insulated panels placed equidistantly inside the wall and fillers, such as wood fibers, ground paper, wool, ABS waste, and reed. The laboratory tests were conducting on HFM 436 Lambda Heat Flux Meter with set indoor and outdoor temperature conditions, and the resulting thermal resistance (R) values were used to calculate the thermal transmittance (U) values. The same experimental walls were subjected to in situ measurements using thermal flow and temperature sensors placed inside and outside the experimental wall, when differences between indoor and outdoor temperature were higher than 10 °C, corresponding to cold season. The calculated U values (U_c) were higher than those measured in situ conditions (U_i) for seven days, and those differences ranged between 7.64% (structure with ABS waste filler) and 56.3% (structure with wood fibers filler). These differences between U_i and U_c could be explained by erratic climate conditions for in situ measurements, accuracy of the sensors, and the type of fillers as regard to vapour diffusion (open or closed).

The natural resources and recycled wastes used in the structure of the proposed timber frame walls achieve the sustainability objective of the European Union policy and also contribute to the low carbon emissions and energy sustainability in this sector. As natural materials, the fillers used in this study also contribute to a healthy indoor environment, allowing the walls to "breathe". The calculation of the variation of the partial pressure and of the saturation pressure proves that condensation is not present inside the five experimental structures.

Even if the structures of the external timber frame walls with wood fibers, wool, reed, and paper used as insulation materials are environmentally friendly and proven to have good thermal performances, which recommend them for passive houses, their hygroscopic properties and low anti-fungal and fire resistance limit their use without proper specific treatments, which were not the objective of the present research. Further research work can be focused on adjusting the structures of the walls investigated in this study for the cold climate, lowering the U-values below 0.15 W/m²K. This can be achieved by increasing the wall thickness or by modifying the interior structure of the wall.

Author Contributions: Conceptualization, S.-V.G. and C.C.; methodology, S.-V.G., D.Ș. and C.C.; software, S.-V.G.; validation, M.C., D.Ș. and C.C.; formal analysis, M.C.; investigation, S.-V.G.; resources, S.-V.G.; data curation, M.C.; writing—original draft preparation, S.-V.G.; writing—review and editing, C.C. and M.C.; visualization, M.C.; supervision, C.C.; project administration, S.-V.G. and C.C.; and funding acquisition, S.-V.G. All authors have read and agreed to the published version of the manuscript.

Funding: This research received no external funding.

Institutional Review Board Statement: Not applicable.

Informed Consent Statement: Not applicable.

Data Availability Statement: Not applicable.

Acknowledgments: We hereby acknowledge the structural funds project PRO-DD (POS-CCE, O.2.2.1., ID 123, SMIS 2637, No. 11/2009) for providing the infrastructure used in this work and the Contract No. 7/9.01.2014. The authors would like to thank the Transilvania University of Brasov for supporting the coverage of publication costs.

Conflicts of Interest: The authors declare no conflict of interest.

References

1. European Commission. In Focus: Energy Efficiency in Buildings. Available online: https://ec.europa.eu/info/news/focus-energy-efficiency-buildings-2020-feb-17_en (accessed on 18 February 2022).
2. European Commission. Energy Performance of Buildings Directive. Available online: https://energy.ec.europa.eu/topics/energy-efficiency/energy-efficient-buildings/energy-performance-buildings-directive_en (accessed on 18 February 2022).
3. European Commission. A European Green Deal. Striving to Be the First Climate-Neutral Continent. Available online: https://ec.europa.eu/info/strategy/priorities-2019-2024/european-green-deal_en (accessed on 18 February 2022).
4. Passive House Institute. Criteria for the Passive House, EnerPHit and PHI Low Energy Building Standard, Version 9f, Revised 15 August 2016. Available online: https://passiv.de/downloads/03_building_criteria_en.pdf (accessed on 19 February 2022).
5. Miron, I.O.; Manea, D.L.; Mustea, A. Reed and Straw-Based Thermally Insulating Panels. *ProEnvironment* **2017**, *9*, 9–15.
6. Doost-hoseini, K.; Taghiyari, H.R.; Elyasi, H.R. Correlation between sound absorption coefficients with physical and mechanical properties of insulation boards made from sugar cane bagasse. *Compos. Part B* **2014**, *58*, 10–15. [CrossRef]

7. Ardente, F.; Beccali, M.; Cellura, M.; Mistretta, M. Building energy performance: A LCA case study of kenaf-fibres insulation board. *Energy Build.* **2008**, *40*, 1–10. [CrossRef]
8. Luamkanchanaphan, T.; Chotikaprakhan, S.; Jarusombati, S. A study of physical, mechanical and thermal properties for thermal insulation from narrow-leaved cattail fibers. *APCBEE Procedia* **2012**, *1*, 46–52. [CrossRef]
9. Pinto, J.; Paiva, A.; Varum, H.; Costa, A.; Cruz, D.; Pereira, S.; Fernandes, L.; Tavares, P.; Agarwal, J. Corn's cob as a potential ecological thermal insulation material. *Energy Build.* **2011**, *43*, 1985–1990. [CrossRef]
10. Zhou, X.; Zheng, F.; Li, H.; Lu, C. An environment-friendly thermal insulation material from cotton stalk fibers. *Energy Build.* **2010**, *42*, 1070–1074. [CrossRef]
11. Agoudjil, B.; Benchabane, A.; Boudenne, A.; Ibos, L.; Fois, M. Renewable materials to reduce building heat loss: Characterization of date palm wood. *Energy Build.* **2011**, *43*, 491–497. [CrossRef]
12. Evon, P.; Vandenbossche, V.; Pontailier, P.-Y.; Rigal, L. New thermal insulation fiberboards from cake generated during biorefinery of sunflower whole plant in a twin-screw extruder. *Ind. Crop. Prod.* **2014**, *52*, 354–362. [CrossRef]
13. Kristak, L.; Ruziak, I.; Tudor, E.M.; Barbu, M.C.; Kain, G.; Reh, R. Thermophysical Properties of Larch Bark Composite Panels. *Polymers* **2021**, *13*, 2287. [CrossRef]
14. Kjeldsen, A.; Price, M.; Lilley, C.; Guzniczak, E. A Review of Standards for Biodegradable Plastics, Industrial Biotechnology Innovation Centre. Technical Report. Industrial Biotechnology Innovation Centre. 2015. Available online: https://assets.publishing.service.gov.uk/government/uploads/system/uploads/attachment_data/file/817684/review-standards-for-biodegradable-plastics-IBioIC.pdf (accessed on 19 February 2022).
15. Grigore, M.E. Methods of Recycling, Properties and Applications of Recycled Thermoplastic Polymers. *Recycling* **2017**, *2*, 24. [CrossRef]
16. Hopewell, J.; Dvorak, R.; Kosior, E. Plastics recycling: Challenges and opportunities. *Phil. Trans. R. Soc. B* **2009**, *364*, 2115–2126. [CrossRef] [PubMed]
17. Pérez, G.; Coma, J.; Martorell, I.; Cabeza, L.F. Vertical Greenery systems (VGS) for energy saving. *Renew. Sust. Energy Rev.* **2014**, *39*, 139–165. [CrossRef]
18. Sheweka, S.M.; Mohamed, N.M. Green Facades as a New Sustainable Approach Towards Climate Change. *Energy Procedia* **2012**, *18*, 507–520. [CrossRef]
19. Azkorra, Z.; Pérez, G.; Coma, J.; Cabeza, L.F.; Bures, S.; Álvaro; Erkoreka, A.; Urrestarazu, M. Evaluation of green walls as a passive acoustic insulation system for buildings. *Appl. Acoust.* **2015**, *89*, 46–56. [CrossRef]
20. Miljan, M.; Miljan, M.J.; Miljan, J.; Akermann, K.; Karja, K. Thermal Transmittance of reed-insulated walls in a purpose-built test house. *Mires Peat* **2013**, *13*, 1–12.
21. Miljan, M.; Miljan, J. Thermal Transmittance and the Embodied Energy of Timber Frame Lightweight Walls Insulated with Straw and Reed. *IOP Conf. Ser. Mater. Sci. Eng.* **2015**, *96*, 012076. [CrossRef]
22. Simpson, A. The effect of moisture on the thermal property of a reed thatch roof during the UK heating season. *Energy Build.* **2022**, *257*, 111777. [CrossRef]
23. Malheiro, R.; Ansolin, A.; Guarnier, C.; Fernandes, J.; Amorim, M.T.; Silva, S.M.; Mateus, R. The Potential of the Reed as a Regenerative Building Material—Characterisation of Its Durability, Physical, and Thermal Performances. *Energies* **2021**, *14*, 4276. [CrossRef]
24. Kain, G.; Barbu, M.C.; Hinterreiter, S.; Richter, K.; Petutschnigg, A. Using Bark as a Heat Insulation Material. *Bioresources* **2013**, *8*, 3718–3731. [CrossRef]
25. Ninikas, K.; Ntalos, G.; Hytiris, N.; Skarvelis, M. Thermal Properties of Insulation Boards Made of Tree Bark & Hemp Residues. *J. Sustain. Archit. Civ. Eng.* **2019**, *24*, 71–77.
26. Kain, G.; Tudor, E.M.; Barbu, M.C. Bark Thermal Insulation Panels: An Explorative Study on the Effects of Bark Species. *Polymers* **2020**, *12*, 2140. [CrossRef] [PubMed]
27. Euring, M.; Kirsch, A.; Kharazipour, A. Hot-Air/Hot-Steam Process for the Production of Laccase-Mediator-System Bound Wood Fiber Insulation Boards. *Bioresources* **2015**, *10*, 3541–3552. [CrossRef]
28. Gößwald, J.; Barbu, M.C.; Petutschnigg, A.; Tudor, E.M. Binderless Thermal Insulation Panels Made of Spruce Bark Fibres. *Polymers* **2021**, *13*, 1779. [CrossRef] [PubMed]
29. Akcagun, E.; Boguslawska-Baczek, M.; Hes, L. Thermal insulation and thermal contact properties of wool and wool/PES fabrics in wet state. *J. Nat. Fibers* **2019**, *16*, 199–208. [CrossRef]
30. Patnaik, A.; Mvubu, M.; Muniyasamy, S.; Botha, A. Thermal and sound insulation materials from waste wool and recycled polyester fibers and their biodegradation studies. *Energy Build.* **2015**, *92*, 161–169. [CrossRef]
31. Pennacchio, R.; Savio, L.; Bosia, D.; Thiebat, F.; Piccablotto, G.; Patrucco, A.; Fantucci, S. Fitness: Sheep-wool and hemp sustainable insulation panels. *Energy Procedia* **2017**, *111*, 287–297. [CrossRef]
32. Zach, J.; Korjenic, A.; Petranek, V.; Hroudova, J.; Bednar, T. Performance evaluation and research of alternative thermal insulation based on sheep wool. *Energy Build.* **2012**, *49*, 246–253. [CrossRef]
33. Moreno-Rangel, A.; Tsekleves, E.; Vazquez, J.M.; Schmetterer, T. *The Little Book of Bio-Based Fibre Materials in Passivhaus Construction in Latin America*; Lancaster University: Lancaster, UK, 2021; pp. 15–16.
34. Panyakaev, S.; Fotios, S. 321: Agricultural Waste Materials as Thermal insulation Dwellings in Thailand: Preliminary Results. In Proceedings of the 25th Conference on Passive and Low Energy Architecture, Dublin, Ireland, 22–24 October 2008.

35. Korjenic, A.; Petranek, V.; Zach, J.; Hroudova, J. Development and performance evaluation of natural thermal-insulation materials composed of renewable resources. *Energy Build.* **2011**, *43*, 2518–2523. [CrossRef]
36. Ninikas, K.; Mitani, A.; Koutsianitis, D.; Ntalos, G.; Taghiyari, H.R.; Papadopoulos, A.N. Thermal and Mechanical Properties of Green Insulation Composites Made from Cannabis and Bark Residues. *J. Compos. Sci.* **2021**, *5*, 132. [CrossRef]
37. Asdrubali, F.; D'Alessandro, F.; Schiavoni, S. A review of unconventional sustainable building insulation materials. *SM&T* **2015**, *4*, 1–17.
38. Zhang, S.Y.; Li, Y.Y.; Wang, C.G.; Wang, X. Thermal Insulation Boards from Bamboo Paper Sludge. *Bioresources* **2017**, *12*, 56–67. [CrossRef]
39. Zhang, L.; Luo, X.; Meng, X.; Wang, Y.; Hou, C.; Long, E. Effect of the thermal insulation layer location on wall dynamic thermal response rate under the air-conditioning intermittent operation. *Case Stud. Therm. Eng.* **2017**, *10*, 79–85. [CrossRef]
40. Garay, R.; Arregi, B.; Elguezabal, P. Experimental Thermal Performance Assessment of a Prefabricated External Insulation System for Building Retrofitting. *Procedia Environ. Sci.* **2017**, *38*, 155–161. [CrossRef]
41. Escudero, C.; Martin, K.; Erkoreka, A.; Flores, I.; Sala, J.M. Experimental thermal characterization of radiant barriers for building insulation. *Energy Build.* **2013**, *59*, 62–72. [CrossRef]
42. Susurova, I.; Angulo, M.; Bahrami, P.; Stephens, B. A model of vegetated exterior facades for evaluation of wall thermal performance. *Build. Environ.* **2013**, *67*, 1–13. [CrossRef]
43. Ninikas, K.; Tallaros, P.; Mitani, A.; Koutsianitis, D.; Ntalos, G.; Taghiyari, H.R.; Papadopoulos, A.N. Thermal Behavior of a Light Timber-Frame Wall vs. a Theoretical Simulation with Various Insulation Materials. *J. Compos. Sci.* **2022**, *6*, 22. [CrossRef]
44. Iejavs, J.; Rozins, R. Water Vapour Permeability Properties of Cellular Wood Material and Condensation Risk of Composite Panel Walls. *Pro Ligno* **2016**, *12*, 3–11.
45. Gullbrekken, L.; Geving, S.; Time, B.; Andresen, I. Moisture conditions in passive house wall constructions. *Energy Procedia* **2015**, *78*, 219–224. [CrossRef]
46. Morelli, M.; Rasmussen, T.V.; Therkelsen, M. Exterior Wood-Frame Walls—Wind–Vapour Barrier Ratio in Denmark. *Buildings* **2021**, *11*, 428. [CrossRef]
47. Janssen, H.; Blocken, B.; Carmeliet, J. Conservative modelling of the moisture and heat transfer in building components under atmospheric excitation. *Int. J. Heat Mass Transf.* **2007**, *50*, 1128–1140. [CrossRef]
48. Labat, M.; Woloszyn, M.; Garnier, G.; Roux, J.J. Dynamic coupling between vapour and heat transfer in wall assemblies: Analysis of measurements achieved under real climate. *Build. Environ.* **2015**, *87*, 129–141. [CrossRef]
49. Wang, H.; Chiang, P.C.; Cai, Y.; Li, C.; Wang, X.; Chen, T.S.; Wei, S.; Huang, Q. Application of Wall and Insulation Materials on Green Building: A Review. *Sustainability* **2018**, *10*, 3331. [CrossRef]
50. *ISO 8301:1991*; Thermal Insulation—Determination of Steady-State Thermal Resistance and Related Properties—Heat Flow Meter Apparatus. International Organization for Standardization: Geneva, Switzerland, 1991.
51. *DIN EN 12667:2001*; Thermal Performance of Building Materials and Products—Determination of Thermal Resistance by Means of Guarded Hot Plate and Heat Flow Meter Methods—Products of High and Medium Thermal Resistance, European Standard, English Version. Deutsches Institut fur Normung: Berlin, Germany, 2001.
52. *ISO 6946:2017*; Building Components and Building Elements—Thermal Resistance and Thermal Transmittance—Calculation Methods. International Organization for Standardization: Geneva, Switzerland, 2017.
53. Geving, S.; Lunde, E.; Holme, J. Laboratory investigations of moisture conditions in wood frame walls with wood fiber insulation. In Proceedings of the 6th International Building Physics Conference (IBPC), Torino, Italy, 14–17 June 2015; pp. 1455–1460.
54. Asdrubali, F.; D'Alessandro, F.; Baldinelli, G.; Bianchi, F. Evaluating in situ thermal transmittance of green buildings masonries—A case study. *Case Stud. Constr. Mater.* **2014**, *1*, 53–59. [CrossRef]
55. Durica, P.; Juras, P.; Gaspierik, V.; Rybarik, J. Long-term Monitoring of Thermo-technical Properties of Lightweight Constructions of External Walls Being Exposed to the Real Conditions. *Procedia Eng.* **2015**, *111*, 176–182. [CrossRef]
56. Rafidiarison, H.; Mougel, E.; Nicolas, A. Laboratory Experiments on Hygrothermal Behaviour of Real-Scale Timber Walls. *Maderas-Cienc. Tecnol.* **2012**, *14*, 389–401. [CrossRef]

Article

Calorific Characteristics of Larch (*Larix decidua*) and Oak (*Quercus robur*) Pellets Realized from Native and Torrefied Sawdust †

Aurel Lunguleasa *, Cosmin Spirchez and Alin M. Olarescu

Wood Processing and Design of Wooden Product Department, Transilvania University of Brasov, 29 Street Eroilor, 500038 Brasov, Romania; cosmin.spirchez@unitbv.ro (C.S.); a.olarescu@unitbv.ro (A.M.O.)
* Correspondence: lunga@unitbv.ro
† This paper is partially based on the conference paper: Lunguleasa, A.; Spirchez, C. Energetic aspects of oak and larch pellets obtained from sawdust waste improved by torrefaction. In Proceedings of the 8th International Conference on Sustainable Solid Waste Management, Thessaloniki, Greece, 23–26 June 2021; pp. 1–8. http://uest.ntua.gr/thessaloniki2021/pdfs/THESSALONIKI_2021_Lunguleasa_Sprirchez.pdf.

Abstract: This research aimed to evaluate the calorific characteristics of two biomasses from larch and oak sawdust in the form of native or torrefied pellets. Some calorific features of these two kinds of biomasses, such as ash content, higher and lower calorific values, calorific density and many others were highlighted, allowing for a comparison between oak and larch torrefied/not torrefied pellets. Installations and methods used for the process of torrefaction and for highlighting some of the calorific features were also evaluated. As a result of experiments, it was demonstrated that the larch and oak pellets were different in terms of density, but that after thermal treatment, the calorific values of both increased considerably. The investigations evidenced some increases in calorific value, up to 15.8%, for both the larch and oak sawdust/pellets. One of the main conclusions of this research was that, even though the role of biomass has diminished considerably in the last few decades, its role as a sustainable fuel remains relevant. Its use will become more widespread when the world's population understands that fossil fuels are depletable and that they must be replaced by renewable fuels such as biomass.

Keywords: biomass; calorific density; calorific value; larch; oak; torrefied pellet

Citation: Lunguleasa, A.; Spirchez, C.; Olarescu, A.M. Calorific Characteristics of Larch (*Larix decidua*) and Oak (*Quercus robur*) Pellets Realized from Native and Torrefied Sawdust. *Forests* **2022**, *13*, 361. https://doi.org/10.3390/f13020361

Academic Editor: Antonios Papadopoulos

Received: 1 February 2022
Accepted: 19 February 2022
Published: 21 February 2022

Publisher's Note: MDPI stays neutral with regard to jurisdictional claims in published maps and institutional affiliations.

Copyright: © 2022 by the authors. Licensee MDPI, Basel, Switzerland. This article is an open access article distributed under the terms and conditions of the Creative Commons Attribution (CC BY) license (https://creativecommons.org/licenses/by/4.0/).

1. Introduction

Generally speaking, the calorific value of fuels differs significantly from one fuel type to another. These differences are usually measurable between gaseous and solid fuels. However, solid and liquid fuels have the same rules when it comes to measuring calorific value and there are no quantifiable differences between them [1–3]. Heat in the form of hot air, obtained by heating the cold air in a given room or heating a plateaux of some kind, can be used in the food preparation or manufacturing industries for warming up the press plateau, etc. CO_2 from the air is captured inside the wooden biomass of trees during the growth process (about 90–100 years), forming a closed circuit due to the amount of CO_2 that was absorbed by the trees during the growth process being equal with that which is obtained during complete combustion (some minutes or hours). In other words, it would be accurate to say that the same amount of CO_2 is obtained when wooden waste/biomass is decomposed in an open space over a few years. Therefore, the use of woody waste/biomass as energy for human needs is recommended [4,5]. Natural gases have obtained a privileged position as fuels because they burn more cleanly than oil or coal, namely because that the process produces a low amount of pollution in the air. More accurately, for the same amount of fuel that is burned, natural gases spread into the atmosphere half the toxicity of coal. Also, the burning of natural gases does not release as much sulphur dioxide or nitrates [6–8] than coals. However, one of the main problems in the field of fuels is that the natural gases are not renewable and their reserves will be depleted in the next years [9]. That is why, nowadays, thenatural gases merely represent a temporary solution in the transition to more regenerative fuels.

The term biomass pertains to large renewable energy resources such as solar, wind, geo-thermal and water energy [9]. The use of this kind of fuel has been increasing every year and is expected to become the most used type of fuel in the next few years (Figure 1).

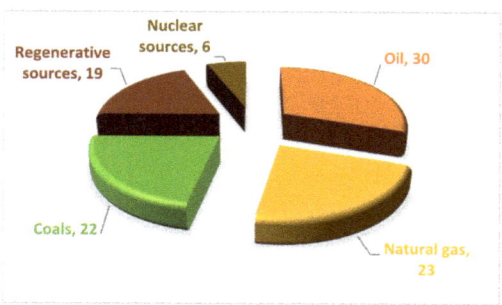

Figure 1. The world's energy resources [9].

Kambo and Dutta [1] highlighted some methods to improve raw sources of energy via thermal treatments. Natural waste taken from *Miscanthus* spp., pretreated by torrefaction and compacted in the form of pellets, was researched in relation to its mechanical and energy characteristics, for the purpose of improving some features. The effective density and calorific density of torrefied pellets increased from 834 kg/m^3 and 15.7 GJ/m^3 to 1036 kg/m^3 and 26.9 GJ/m^3, respectively. Akinrinola [7] showed that the thermal treatment of torrefied biomass from Nigeria (resulting from two wooden species and two other waste crops) was able to improve some fuel features, regardless of whether they were used for industrial or domestic energy sources. Other authors [10] took only waste of only two wood species, *Liriodendron tulipifera* L. and *Larix kaemferi* C., in order to make pellets for the purpose of improving resistance and calorific value by using lignin powder as an additive. Peng et al. [11] stated that good pellets can be realized from particles with small dimensions. A complete analysis of native and torrefied pellets was made by Kumar et al. [6], who identified the benefits of using wooden waste in combination with crushed coal.

Some parameters of the torrefaction/pelletizing process were made by Rudolfsson et al. [12]. This study was focused on the moisture content, particle dimensions and the extrusion channel length. Oh et al. [13] researched the pelletizing/torrefying process of sawdust waste from larch (*Larix Kaempferi* C.) and poplar (*Liriodendron Tulipifera* L.) species. Nhuchhen and Basu [14] analyzed a new methodology for replacing nitrogen, which is very expensive, with pressurized air. Meanwhile, Kudo et al. [15] replaced inert nitrogen gas, which results in the poor adhesion of coal shale in pellets, with wet saturated steam. The resistance of the pellets obtained because of this change increased by five times. Peng et al. [16] observed that the biomass should first be torrefied and then compacted by pelletizing. Thus, pellets with superior features, namely a high resistance to water, can be obtained.

Eseyin et al. [17] stated that torrefied biomass in the form of pellets/briquettes or native biomass has a high calorific energy, is resistant to water and fungus and has a longer storage time. Additionally, the advantages of using biomass include the fact that biomass is an environmentally friendly fuel that produces substantially lower CO_2 emissions compared to fossil fuel. Granados et al. [18] studied the influence of the torrefaction installation on the process, performed in two stages, for small wood particles of 0.5–1 mm from poplar species. During the torrefaction process, instead of an inert nitrogen atmosphere, the resulting gases were used for the heat treatment process. The mass loss was 34%

and the calorific value increased by 40% when the torrefaction temperature was 300 °C. Okoro et al. [19] studied the role of wooden biomass obtained from two species of pine versus fossil resources, especially when they were thermally treated to obtain syngas via pyro-gasification. The optimum thermal treatment was at 300 °C for 45 min. The calorific value increased by 57%. Olugbade and Ojo [20] determined a number of the energy characteristics of biomass and in particular for heat-treated agricultural biomass by torrefaction in an inert nitrogen atmosphere. Torrefaction is considered to be a process of medium pyrolysis. Alokika and Singh [21] optimized the parameters of the torrefaction process for an acacia species. TGA, SEM and FTIR analyzes showed an 18% higher calorific value, a 75% increase in fixed carbon, and an energy efficiency at 252 °C and a treatment time of 60 min, compared to raw materials. Pérez et al. [22] treated four fast-growing species from Colombia, namely Gmelina, two species of pine and one type of eucalyptus, respectively. No major differences were found between the energetic properties of these species (a volatile content of 70%, calorific value 18.7 MJ/kg and ash content below 1%) but they behaved differently during the thermal treatment process. It has also been shown via chemical analysis that only hemicellulose was degraded during torrefying, whereas lignin and cellulose remaining undegraded. Lee et al. [23] used sewage sludge for torrefaction to enrich energetic properties. The obtained product was very similar to coal. The spontaneous ignition of this fuel has been shown to occur at a temperature of 211 °C. Further research has focused on both the use of other residues obtained from other industries, such as from the coffee [24] and paper [25] industries, or the internationally correlation of standardized properties [26].

An initial deduction that can be made from the above studies in the field of sawdust thermal torrefaction is that, although torrefaction is a very efficient treatment, the nitrogen flow used during biomass torrefying is not cheap and requires complex installation. This is the reason why some simpler torrefaction methods have been tested to achieve the same effects [14,15,18]. A second conclusion that can be made is that more and more studies refer to the use of lignocellulosic biomass as a safe and sustainable method of transitioning from fossil fuel energy to the energy obtained from renewable sources.

Objectives: The aim of this paper was to improve the calorific features of the sawdust of two species (one of the softwoods and another of the hardwoods, in order to observe their different behaviors), torrefied over 200 °C without the admission of air during the thermal process. Some calorific characteristics were tasted for the native/torrefied pellets. In particular, the influence of moisture content on the calorific value of the two analyzed species was studied.

2. Materials and Methods

For the experimental, two lignocellulosic biomasses in the form of native sawdust were used. Sawdust was obtained from larch and oak timber. Waste sawdust was collected from a circular saw machine at a university wood processing workshop (Transilvania University of Brasov, Brasov, Romania) when the two wood species, in the form of timber, were processed. As the sawdust purchased had large particles and spalls, it was sorted using a 6 mm × 6 mm mesh sieve. After sorting, only the fraction that fell through the sieve was taken, with the remainder that remained above the sieve (about 12%) being eliminated. Next, the moisture of the sawdust was determined, obtaining values within the limits of 10 ± 0.6% (because the dry timber was used due to the needs of the workshop). The experimental process involved three steps: a first step, in which the characteristics of the raw material were determined; a second step, in which the small sawdust was torrefied and the native/torrefied pellets were made; and a third step for testing the obtained pellets.

Granulometry. In order to determine the characteristics of the sawdust, its granulometry was determined using sieves with 4 mm × 4 mm, 3.15 mm × 3.15 mm, 2 mm × 2 mm, 1.25 mm × 1.25 mm and 0.8 mm × 0.8 mm meshes arranged on an electric device with vibration. Six sawdust samples were used for each species analyzed, and the results obtained in the form of masses were transformed into percentage values by reporting the value of the mass remaining on the sieve to the total mass of the sample. As an example, Equation (1) shows how the percentage of chips was determined for the fraction 3.15 mm × 3.15 mm.

$$P_{3.5} = (m_{3.15} : m_s) \times 100 \ [\%] \tag{1}$$

where: $P_{3.15}$—the participation percentage of the fraction of 3.15 mm × 3.15 mm from the total chips in %; $m_{3.15}$ is the mass of the fraction of 3.15 mm × 3.15 mm which remains above the respective sieve in g; and m_s is mass of the whole sample in g.

Similar to Equation (1), the other five percentage values were determined for sieves with meshes of 4 mm × 4 mm, 2.5 mm × 2.5 mm, 1.25 mm × 1.25 mm, 0.8 mm × 0.8 mm, 0.4 mm × 0.4 mm and the rest (which remained below the sieve of 0.4 mm × 0.4 mm and was collected in the existing collector cylinder). The values for all six groups of tests were then averaged, and, based on these, the graphs of variation of the obtained values were realized.

Bulk sawdust features [26]. In the case of sawdust, its bulk density was determined by using a graduated cylinder to determine the volume of sawdust, the calculation relation being as follows Equation (2):

$$\varrho = (m_t - m_c) : (\pi \times d^2 \times h) \times 10^6 \ [kg/m^3] \tag{2}$$

where: m_t is the total mass of the sawdust with the cylinder in g; m_c is mass of the empty cylinder in g; d is inner diameter of the cylinder in mm; and h is height of the sawdust layer inside the cylinder in mm.

This bulk sawdust density was used to determine the expanding or compression coefficient of the sawdust used, using the following two equations, Equation (3):

$$Ke = \varrho w : \varrho s; \ Kc = \varrho ws : \varrho s \tag{3}$$

where: Ke is the expanding coefficient; Kc is the compression coefficient; ϱ_w is density of wood in kg/m^3; and ϱ_s is the bulk density of sawdust in kg/m^3.

The density of the solid wood was determined by taking 10 specimens with dimensions of 20 mm × 20 mm × 30 mm from the remains of the timber pieces, with the mass in lreation to their volume being reported. This Ke coefficient was used to determine the compressibility of the sawdust, comparing the density of the solid wood from which it came, with this of the sawdust with respect to its dimensional peculiarities.

Obtaining pellets from native sawdust. Pellets with a diameter of 10 mm were obtained from the sorted sawdust, using a laboratory pelletizing device (as part of calorimeter apparatus). The main characteristics that were determined for these native pellets were size, moisture content, unit density, bulk density, higher and lower calorific value (HCV, LCV), calorific efficiency, combustion time and calorific density, etc.

Ash content determination. A well-known and standardized method was used to determine the ash content of the sawdust (ASTM E1755-01) [25]. The wood sawdust was dried in a laboratory oven at 105 °C for 2 h in order to eliminate the influence of moisture content on the ash content. For this purpose, metal crucibles made of nickel–chromium alloys that are resistant to high temperatures were used. These were cleaned thoroughly and then burned on a butane gas flame, before being cooled in a desiccator and weighed up to three decimal places on a Kern, Germany electronic scale. A small amount of less than 1 g of sawdust was then placed on the surface of the crucible in 2–3 layers and then the crucibles were weighed again. In order to protect the inner cavity of the calciner (Protherm, Ploiesti, Romania), the sawdust crucible was burned on a flame of butane gas until there was no more flame and smoke, and the ash obtained had a black color. At this point, the crucible with the ashes was weighed again with the help of the high-precision balance up to 3 decimal. Next, the crucibles were introduced into the calcination furnace for a period of about 30 min at a temperature of 650 °C. After that, in order to determine if the calcination process was finished, the crucibles were checked to see if they had a light greyish ash color and that there was no longer any spark and carbon. When the ash content was determined, the sawdust samples were completely dried out in an oven and the crucible mass was also kept into consideration, as seen in Equation (8) [19–21]. The black ash and the calcined ash masses were determined as proportions with the following relationships using Equations (4) and (5):

$$BAc = (m_{ba+c} - m_c) : (m_{s+c} - m_c) \times 100 \, [\%] \qquad (4)$$

$$CAc = (m_{ca+c} - m_c) : (m_{s+c} - m_c) \times 100 \, [\%] \qquad (5)$$

where: m_{ba+c} is mass of the black ash and the crucible in g; m_{ca+c} is the mass of the calcinated ash and the crucible in g; m_{s+c} is the mass of specimen and the crucible in g; and m_c is the empty crucible mass in g.

Torrefaction of sawdust. Lignocellulosic biomass in the form of larch and oak sawdust, taken from a circular saw and sorted, was subjected to a torrefaction process inside of the calcination furnace STC 18.26 (Ploiesti, Romania) without air admission. Different temperatures, of 200, 220, 240, 260, 280 and 300 °C, and times, of 3, 5 and 10 min, were used to highlight its calorific properties. Each type of torrefying regime was performed individually (not progressively, from a lower treatment to a higher one), obtaining a total of 18 types of regimes for larch and another 18 for oak sawdust. For each regimen, 10 individual and distinct samples were used. The heat treatment was performed without oxygen intake. The torrefied samples were cooled and subjected to specific tests, mainly to determine mass loss. This material was then transformed into torrefied pellets so that they had the same dimensional characteristics as the native ones (untreated). The torrefaction process was performed on a small amount of wooden sawdust for each temperature and time [22–26]. During this period, some changes in the sawdust color occurred [27–29]. The mass loss percentage (ML) of the torrefied sawdust was determined with using Equation (6):

$$ML = (m_i - m_f) : m_i \times 100 \, [\%] \qquad (6)$$

where: m_i is the initial sawdust mass, before torrefaction, in g and m_f is the final sawdust mass, after torrefaction, in g.

The ML values represented the mean of 10 tests, each of them for all periods, temperature, and wooden species.

Density of pellets. The densities of the native and treated pellets were obtained as the ratio between their mass and volume when they had the same Mc, i.e., 10%. Because the form of the pellets had been approximated as a right cylinder (their heads were polished in order to for them to have a perpendicular to length cross section), the relationship of density was determined by Equation (7):

$$\varrho = 4 \times m : (\pi \times d^2 \times l) \times 10^6 \, [kg/m^3] \qquad (7)$$

where: m is the pellet's mass in g; d is the pellet's diameter in mm; and l is the pellet's length in mm.

High and low calorific values. The apparatus that was used to determine the calorific energy of the solid biomass in the form of sawdust pellets was the calorimeter bomb, type XRY-1C, offered by Shanghai Changji Trading Company Limited., China. The calibration of the calorimetric apparatus was performed before testing by using benzoic acid that had a calorific value of 26,463 kJ/kg. By using this value, the coefficient k was obtained with the same method of determination and the same CV relationship using Equation (8).

$$CV = (k \times (T_f - T_i) - qi) : m \, [kJ/kg] \qquad (8)$$

where: CV is the best calorific value for 0% Mc in KJ/kg; k is the calorimeter coefficient in kJ/Celsius degrees; T_f is the final temperature in °C; T_i is the initial temperature in °C; m is the mass of the specimen in kg; and q_i is the supplementary heat given by cotton and nickel wire burning in KJ.

The calorimetric installation software for determining the calorific value provided both the high and low calorific values and the burning time of the determination. In order to find the average value of the calorific value, 8–10 valid replicates were used.

Calorific density and burning rate. The purpose of the caloric density determining was to determine the amount of energy that existed in relation with biomass volume and the necessary amount to determine the transport capacities of the truck or the storage silo of the thermal power plant. The calorific density (CD) was obtained by taking into account the calorific value of the pellets and their effective density by using Equation (9):

$$CD = CV \times \varrho \ [kJ/cm^3] \tag{9}$$

where: CV is the calorific value in kJ/kg and ϱ is the density of the oven-dried pellets in kg/m^3.

The combustion speed of the biomass in the calorimeter was necessary in order to determine how fast the pellets burned, and in most cases this characteristic is used to determine the efficiency of the thermal power plant. Taking into account the burning time, the calorific value and the oven-dry mass, the burning rate was determined using Equation (10).

$$BR = CV: t \times m_0 \ [kJ/min] \tag{10}$$

where: BR is the burning rate in KJ/min; CV is the calorific value expressed on a dry basis in KJ/kg; m_0 is the mass of the oven-dried pellet in g; and t is the combustion time in min.

The mean of BR value was obtained as an arithmetic mean of eight experiments, taking in consideration different values of time and calorific value, for each sample.

Influence of moisture content on calorific value. Because the moisture content of pellets was one of the principal factors that influenced the calorific value, a relationship of dependence for 0% and other moisture content values [17] could be found by using Equation (11):

$$LCV_{Mc} = (CV \times (100 - Mc) - 2.44 \times Mc):100 \ [MJ/kg] \tag{11}$$

where: LCV_{Mc} is the low calorific value at a certain moisture content in MJ/kg; CV is the calorific value for 0% moisture content in MJ/kg; and Mc is the moisture content in %.

In order to determine this influence, the larch and oak pelleted specimens were kept in a climatic chamber until three moisture content values (10%, 20% and 50%) were obtained. The high and low calorific values (HCV, LCV) were determined from these tests. Thus, two straight lines were obtained in the x0y plane (CV0Mc). The linear equations intersected the CV axis at the same point, and the horizontal Mc axis at two other different points. The arithmetic mean of the two horizontal intersections was defined as "limitative Mc". For example, if the larch pellets had a moisture content of 20%, a calorific value of 15,499 kJ/kg was obtained, and for a moisture content of 50%, a calorific power value of 10,057 kJ/kg was obtained. In this way, two points, A (50; 10057), and B (20; 15499) could be highlighted. A line can be made through the two points, whose gross equation was found using Equation (12):

$$(y - 10{,}037):(15{,}499 - 10{,}037) = (x - 20):(50 - 20) \tag{12}$$

By performing the calculations in the previous Equation (12), the final linear equation of the form $y = n - m \cdot x$ was obtained, with the calorific value being an unknown variable and the moisture content being a known variable, as seen in Equation (13):

$$HCV = 19{,}140 - 150.8 \cdot Mc \tag{13}$$

This equation is interpreted in the sense that it has a coefficient of linear equation of 19,140 and a tangent of the slope of the linear equation of -150.8, which means that it has an angle of over 80 degrees, and the calorific value for completely dry pellets is 19,140 kJ/kg.

In the second part of the experiments, the torrefied sawdust was compressed into cylindrical pellets (with a dimension of 10 mm, a mass of 0.5–0.8 g, and a length of 9–11 mm) using the same press (as part of a calorimeter bomb) (Figure 2). At least 12 pellets were obtained from each native/torrefied sawdust lot. Also, two different categories of pellets, one made from larch sawdust (*Larix decidua*) and the other made from oak (*Quercus robur*) sawdust, were made and analyzed.

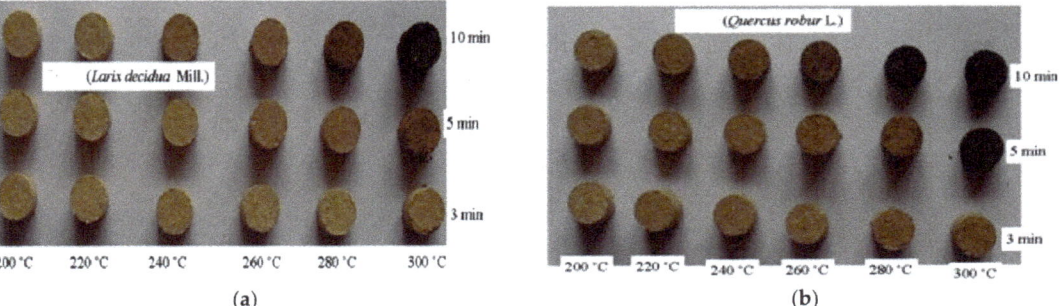

Figure 2. The torrefied pellets for larch (**a**) and oak (**b**).

The improvement in calorific value. The improving in calorific value after the thermal treatment of torrefaction was determined based on the calorific value obtained before and after treatment, depending on two parameters (temperature and treatment time), with aid of the next relation Equation (14):

$$I_{CV} = (Cvat - CVbt) : CVbt \times 100\ [\%] \tag{14}$$

where: I_{CV} is the increase in calorific value (CV) in %; $CVat$ is the calorific value after torrefaction in kJ/kg; and $CVbt$ is the calorific value before torrefaction in kJ/kg.

Calorific efficiency. The calorific efficiency was determined in relation to moisture content. This determination was based on the fact that an increase in moisture content implicitly leads to a decrease in calorific value, and thus it was important to determine the percentage from the maximum caloric power (for MC = 0%) that was actually used during combustion and how much was lost through the drying of the pellets. This determination was based on tests performed to determine the influence of moisture content on the calorific value by referring to the maximum obtained calorific value (CV). The calculation was performed as follows (Equation (15))

$$Cef = CV_{Mc} : CV \times 100\ [\%] \tag{15}$$

where: CM_{Mc} is the calorific value for a certain moisture content in kJ/kg and CV is the maximum calorific value for a moisture content of 0%.

Proximate analysis was performed by determining volatiles, fixed carbon and ash content (Ac). The same calciner was used by means of a heat-resistant vessel with a lid, so that the sawdust did not oxidize during the elimination of volatile matter (VM). The amount of fixed carbon (FC) that was obtainedwhen the sawdust had a 0% moisture content, was calculated using the following Equation (16):

$$FC + VM + Ac = 100\ [\%] \tag{16}$$

Statistical analysis. In the first stage, the obtained values were subjected to the determination of the survey median and the standard deviation in order to observe the trend and the scattering of the values. Using Microsoft Excel, the standard error was applied to the graphs obtained for a 95% confidence interval. The obtained trend equations were chosen in such a way that the determination coefficient R^2 had a value higher or was appropriated to 0.9. Analysis of variance (AVOVA one-way) was used to compare two groups of values in order to study the dependency between a dependent variable (oak density) and an independent variable (larch density). The Minitab 18 program was also used for statistical analysis in order to obtain statistical graphs and determine other statistical parameters.

3. Results

3.1. Granulometry of Native Sawdust

The granulometry of the native sawdust (Figure 3) was almost identical for larch and oak, this being determined by the fact that the raw material was taken from the same circular saw and was sorted with the same 5 mm × 5 mm sieve.

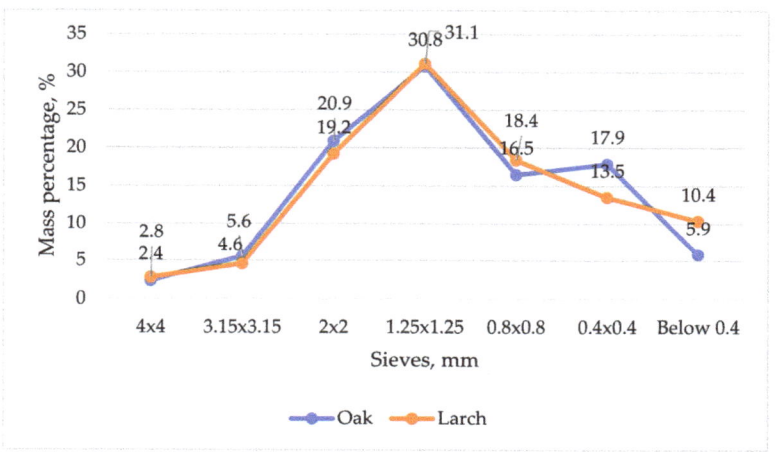

Figure 3. Granulometry of native sawdust.

The small variations observable in Figure 3 were determined by the different density of the two considered wood species

3.2. Bulk Characteristics of Sawdust

Following the procedure expressed in the previous chapter and Equation (2), a bulk density values of 162 kg/m^3 and 204 kg/m^3 were obtained for the larch and oak sawdust, respectively. The difference between the values was due to the large differences in the effective density of the two species, the calculated densities of oak and larch wood being 784 kg/m^3 and 575 kg/m^3, respectively. Moreover, the degree of compression of the sawdust of the two species was determined, taking into account the density of the wood substance (as the maximum value at which it can be compressed) for both species, which is 1450 kg/m^3. Using Equation (3), expansion coefficient values of 3.54 and 3.84 were obtained for larch and oak, respectively, as were other compression coefficient values, namely 8.9 for larch and 7.1 for oak. The compression ratio of larch was higher because the compaction started from a lower sawdust density value (162 kg/m^3) than in the case of oak (204 kg/m^3). Thus, the difference related the maximum value that can be reached in terms of compression was much larger. From this analysis, it can be concluded that heavier species (such as oak) have a lower compression than light species (such as larch).

3.3. Density of Native/Torrefied Pellets

The moisture content of the sawdust and pellets was about 10%, with this being determined by the classical gravimeter method, standardized by EN 14774-1:2009. The density of the native (un-torrefied) pellets was determined as a ratio between mass and volume and was about 1010 kg/m^3 for oak and 1012 kg/m^3 for larch. The density values were very appropriate because the dimensional characteristics of the sawdust were the same (a fraction smaller than 5 mm × 5 mm was removed from the circular sawdust) and the pelletizing press and its parameters were the same in both cases.

3.4. Calorific Features

For both the two analyzed species (*Quercus robur* and *Larix decidua*) the burning time, higher and lower values of calorific power (HCV and LCV), lineal equation for each type of calorific value and limitative MC were calculated, specifically when native (no-treated) pellets were used (Table 1).

Table 1. Calorific characteristics of native [un-treated] pellets.

Species	Time of Burning	Calorific Value			Lineal Equation	Limitative Mc, %	
		HCV	LCV	CV			
Oak	25	18,564	18,563	18,569	HCV = 18,569 − 176.4·Mc LCV = 18,569 − 195.1·Mc	105.2 95.1	100.1
Larch	29	19,135	19,132	19,140	HCV = 19,140 − 150.8·Mc LCV = 19,140 − 279.4·Mc	126.9 68.5	97.7

As a general rule, the calorific value was proportional to depended indirectly on the moisture content, with both the both higher and lower calorific values decreasing with an increase in Mc. Ideally, if the pellets under analysis were absolutely dry, i.e., they would ideally have a moisture content of 0%, then the two values, HCV and LCV, would be equal and would have the unique value CV [28]. In reality, although the pellets were dried to a constant mass (i.e., they had 0% Mc), it was seen that the higher and lower calorific values were slightly different from each other. The explanation for this is the fact that a cleaner water with a volume of 3 mL was added into the bomb in order to absorb nitrogen compounds during burning. This explains the deviation from a moisture content value of 0% (to 2–4%) and, consequently, from normal values of CV. The addition of liquid water was strongly recommended by the calorimeter provider.

3.5. Moisture Content Dependence on Caloric Power

Based on the methodology presented in the previous chapter, the graphs in Figure 4 were obtained, both for the native oak sawdust and for the larch sawdust. The graphs show once again that the calorific values of the two species were different, with the calorific value of larch (19,140 kJ/kg) being 3% higher than that of oak. The slight increase in calorific value for larch was due to the existence of a small amount of resin (which has a very high calorific value of about 34,000 kJ/kg) [24,26].

Figure 4 shows that with an increase of Mc, the calorific power will drop due to a certain amount of energy tat was used to dry the wooden pellets. The caloric value for 0% moisture content, noted with CV, was obtained by mathematically determining the regression equation. These lineal equations are observed in Table 1. Cuttings of the horizontal axis of each lineal equation determined the limitative moisture content. When there was no rise in caloric energy, the mean values of limited moisture content were 100.1% and 97.7%. The explanation for this is that the pellets' energy was equal to the energy used to consume water from wood (i.e., to dry out wood). When the moisture content is zero, both the HCV and LCV values were intersected, and a point with the same calorific value, CV, was obtained. This is why the calorific efficiency issue occurred during the burning of the pellets with a certain Mc [29].

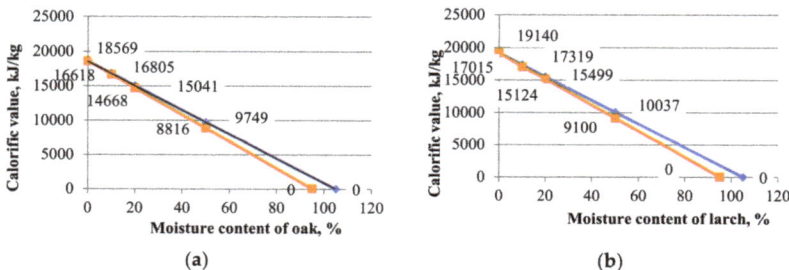

Figure 4. Calorific values for oak (**a**) and larch (**b**) in relation to moisture content.

3.6. Calorific Efficiency

With regard to calorific efficiency, Figure 5 shows that this property decreased with an increase in moisture content for oak and larch sawdust. Biomass with a 10% moisture content offered a better efficiency (95%) than that with a moisture content of 50%, the efficiency of which was 52%.

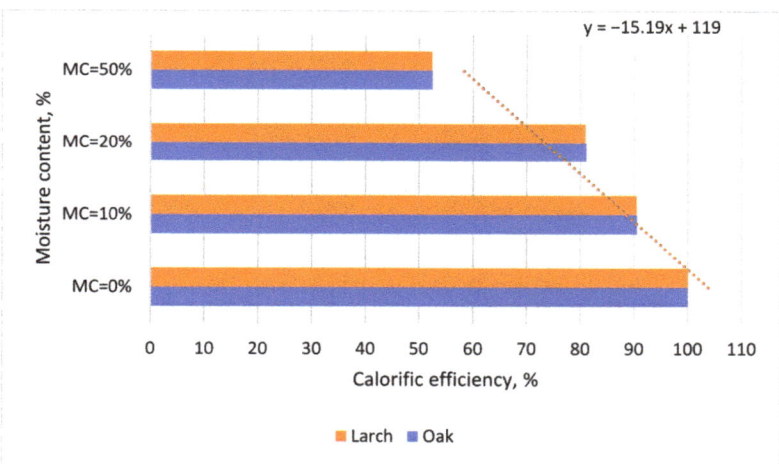

Figure 5. Calorific efficiency for larch and oak pellets.

3.7. Calorific Density

Regarding the energy properties of sawdust pellets, it was observable that the calorific density had extreme values, of 1.83–13.4 kJ/cm^3 for native larch pellets and 8.07–12.18 kJ/cm^3 for native oak pellets (Table 2). As a general rule, the calorific density decreased with an increased in pellet moisture content, with this decrease being 7.3 times for larch pellets and only of 1.5 times for oak pellets. These values show that the oak pellets were more homogeneous and had a more constant combustion. Also, the volume of the silo or means of transport would have to be slightly higher in the case of larch pellets, especially for a moisture content higher than 10%.

Table 2. Burning rate and calorific density.

Features		\multicolumn{4}{c}{Moisture Content}
		0%	10%	20%	50%
Burning rate, kJ/min	Larch	463	313	86	47
	Oak	372	346	277	164
Calorific density, kJ/cm^3	Larch	13.4	7.12	2.63	1.83
	Oak	12.18	11.06	10.89	8.07

3.8. The Burning Rate

The burning rate decreased with an increase in moisture content (Table 2). The extreme values obtained for a moisture content of 0% and 50% showed a decrease of 9.8 times for larch pellets and 2.2 times for oak pellets. It was observed with respect to burning speed, that the oak pellets were more homogeneous, that their burning would be more constant and that there would be smaller differences when the moisture content varied within the same group of pellets.

3.9. Mass Losses in Time of Torrefaction

Generally, it was determined that when the degree of torrefaction is increased (given by the values of time and temperature), the mass loss will also proportionally increase (Figure 6). Temperatures over 260 °C increased the mass loss [23–26] of oak pellets to a greater degree. A temperature of 300 °C represented the ideal temperature for torrefaction but was the highest one possible before self-burning occurs. Knowing that the mass loss for non-torrefied sawdust is zero, the mass loss over the total temperature range for 3 min was 8.84%, for 5 min was 16.95% and for 10 min was 40.24%. In the case of larch sawdust, the increase was of 7.62% for 3 min, 16.95% for 5 min and 17.59% for 10 min. It was observed that the oak sawdust had higher losses than the larch sawdust for all the treatment regimens and had a maximum increase of 128.7% for 10 min and 300 °C. Another conclusion was the observation that there was a substantial increase in losses when the temperature increased from 280 to 300 °C for both sawdust species, with a difference of 31.4% for oak and 9.9% for larch.

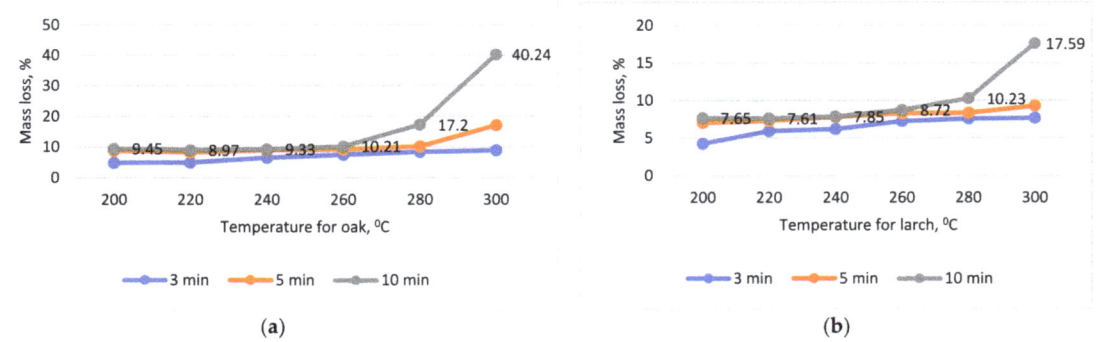

Figure 6. Mass losses for oak (*Quercus robur* L.) (**a**) and larch (*Larix decidua* Mill.) (**b**) sawdust during torrefaction.

3.10. Increasing the Calorific Value When the Torrefaction Process Occurs

With respect to the influence of temperature on the CV of *Quercus robur* pellets, Figure 7 reveals the increase in CV for treatments of 3 min, 5 min and 10 min in the case of larch pellets. Figure 8 reveals the increase in CV in the case of oak pellets.

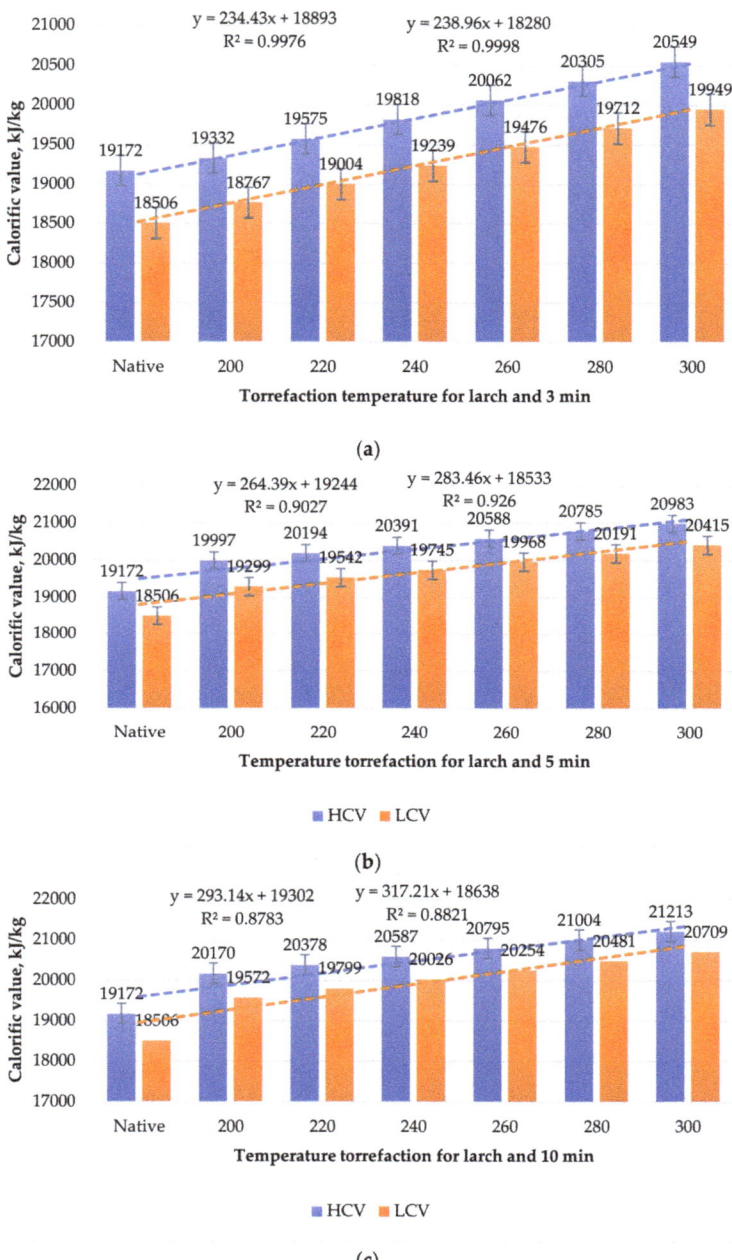

Figure 7. Calorific values for larch torrefaction for 3 min (a), 5 min (b) and 10 min (c).

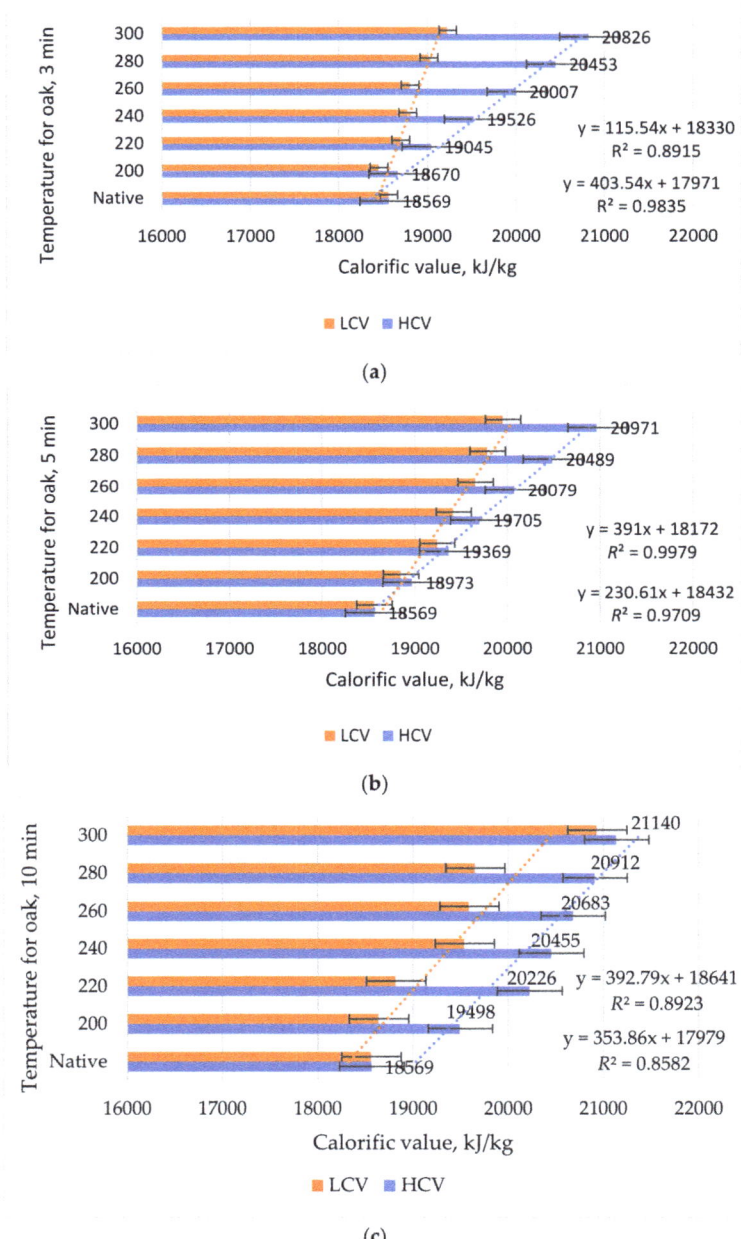

Figure 8. Calorific value of oak sawdust during torrefaction for 3 min (**a**), 5 min (**b**) and 10 min (**c**).

Regarding the influence of torrefaction temperature and time on the CV of the pellets obtained from the larch sawdust (Figure 7), it can state that the highest increase in CV, 15.8%, was obtained from the maximum torrefaction regime.

Throughout the heat treatment (0–300 °C and 10 min) of the larch sawdust, the HCV increased, a with maximum of 15.5%. It was observed that this increase does not correspond to the maximum mass loss, 17.9%, because of other volatile substances that did not increase the calorific value, with the main factor influencing the increase in calorific value being the degradation of hemicelluloses [5].

As in the case of oak sawdust (Figure 8) was seen, the growth of HCV in the case of the larch does not correspond to the loss of mass, the former being much smaller. It should be noted that the 3 min period of torrefaction resulted in an increase of less than 5% in case of larch and oak sawdust. Also, the increase in calorific power to over 21,100 kJ/kg made it possible to classify the torrefied pellets as inferior coke coals.

3.11. Ash Content, Volatile Matter and Fixed Carbon

Different values were obtained for the ash content of native sawdust, with a value of 0.42% for the *Larix decidua* and one of 0.51% for the *Quercus robur* sawdust (Figure 9). The ash content increased slightly as a result of torrefaction. On the other hand, the sawdust that was torrefied had a high ash content, but this was not totally in concordance with the mass losses found during torrefaction [26], which is usually lower. For instance, the higher values of ash content for torrefied sawdust highlighted during laboratory tests at 300 °C was 0.58% (an increase of 13.7%, whereas the mass loss was 40.2%) for oak and 0.47% (an increase of 11.9%, whereas the maximum mass loss was 17.5%) for larch. As other authors have stated before [5,13], the main constituents of ash are silicates, oxides and hydroxides, sulphates, phosphates, carbonates, chlorides and nitrates, which means that it can be used as a fertilizer in agriculture.

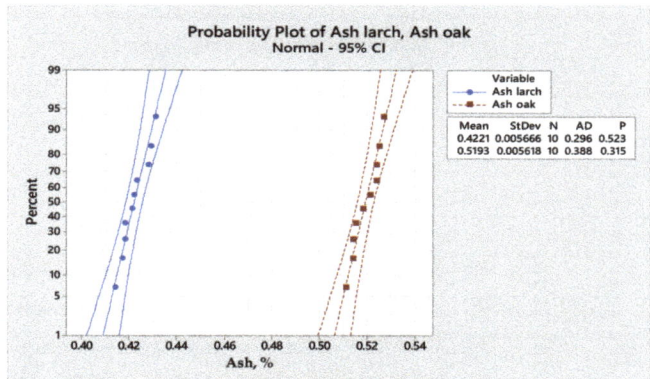

Figure 9. Probability plot for larch and oak ash content.

Proximate analysis of native and torrefied sawdust is available in Table 3.

Table 3. Proximate analysis of native/torrefied sawdust.

Specie	Type	Volatile Matter	Ash Content	Fixed Carbon
Larch sawdust	Native	78.58	0.42	21
	Torrefied	61.53	0.47	38
Oak sawdust	Native	76.49	0.51	23
	Torrefied	53.42	0.58	46

This proximate analysis highlights that by torrefying the sawdust was enriched with fixed carbon, losing some of its volatile materials. Some differences were found regarding the fixed carbon content of torrefied sawdust, with the value of 21% higher for oak.

4. Discussion

The advantages of using the pelletizing process on larch sawdust have been highlighted by other researchers, namely the high durability [10,13] achieved by increasing the calorific value [1] by about 18% and by torrefying the sawdust followed by palletization [11,16]. Replacing nitrogen with rarefied air during torrefying is just one of the alternative methods that have been proposed, along with replacing it with saturated steam [15], with pressurized air [14] or with roasting gas [18]. All these methods aim to reduce the cost of sawdust torrefying, but the price of the torrefied pellets obtained using these methods does not yet justify their benefits.

The dimensions of wood particles are essential when it comes to obtaining pellets with superior characteristics, which is why the sawdust used in this study had an average size of 1.3 mm (Figure 1). In the same way, researchers [11,16] have established that the high properties of pellets are obtained from the small particles, and [18] has studied the influence of particle size on pellet characteristics. From the point of view of granulometry, maximum values of about 31% were obtained when 1.25 mm × 1.25 mm sieve was used, which means that the maximum dimensions of the chips were found in the area of the same sieve. Minimum values of about 2% were obtained when the 4 mm × 4 mm sieve was used, because the 5 mm × 5 mm sieve was used to remove the tendrils and also removed some of the large chips.

The three characteristics of the sawdust, namely bulk density, the expansion coefficient and the compression coefficient, were dependent on the dimensions of the chips with respect to their granulometry. Therefore, the values obtained were slightly different from those obtained by other authors [11].

Calorific values for wood pellets below 19,000 kJ/kg were also obtained by [3,12], a loss of mass below 34% by [18], and an ash content below 1% by [22]. The densities of solid wood were statistically analyzed (Table 4) by means of analysis of variance (ANOVA). The values of the F-value and p-value parameters highlight the normality of the value distribution for a 95% confidence interval, as other researchers have stated before [7,17].

Table 4. Analysis of variance for wood density.

Source	DF	Adj SS	Adj MS	F-Value	p-Value
Larch density	9	346.05	38.45	0.95	0.048
Error	1	40.50	40.50		
Total	10	386.55			

In the same analysis, the pellets densities differed by less than 0.2% and the CV of the native pellets differed by less than 3.1%. Differences in the increase in calorific power after sawdust torrefaction were below 3.1% in favor of larch pellets [10]. Similar small differences were found for other features, such as caloric efficiency, energy release rate and calorific density [6].

The correlation between calorific value and degradation of main wooden compounds was achieved according to the main chemical compounds of wood, which are cellulose, hemi-cellulose and lignin. The influence of secondary chemicals (extractables and oxides from ash) was neglected. Of these major chemicals, lignin has the strongest influence on calorific value. Knowing that lignin has the highest calorific value, 25,121 kJ/kg (25.1 MJ/kg), while cellulose and hemicellulose have a value of about 17 374 kJ/kg (17.3 MJ/kg), an addictive relationship could be identified (17):

$$CV = 25{,}121 \times Li{:}100 + 17{,}374 \times (Ce + He){:}100 \ [kJ/kg] \qquad (17)$$

where CV is the calorific value of sawdust in kJ/kg; Li is the content of lignin in % wt; Ce is the content of cellulose in % wt; and He is the content of hemicelluloses in % wt.

The average values for lignin, cellulose and hemicellulose content were taken from the literature [27], and are 32%, 46% and 22% for larch, respectively, and 32%, 45% and 23% for oak, respectively.

First, it should be considered that only hemicelluloses degrade up to 300 °C and cellulose is degraded after 300 °C. Due to the total degradation of hemicelluloses, only cellulose and lignin remained in the torrefied product, with proportions of 39.2% and 60.8%, respectively. Applying Equation (16) resulted in a calorific value of 20,354 kJ/kg for larch and 19,495 kJ/kg for oak.

Second, during the torrefying process, all the cellulose can be degraded with the hemicellulose, and the maximum calorific value is 25,121 kJ/kg for both species. In reality, after performing the torrefying test, the sawdust had a maximum calorific value of 21,040 kJ/kg in the case of the oak and 21,213 kJ/kg in the case of the larch, which means that only a certain amount of cellulose was degraded during the torrefaction process, about 6.5% for the larch and 13.7% for the oak. By looking closely at the torrefaction graphs (Figure 7), it can be found that the degradation of cellulose begins in the case of larch at a temperature of 220 °C and at 215 °C in the case of oak.In this time the cellulose degradation intensity wass lower than that of hemicelluloses.

5. Conclusions

Generally, wood biomass, such as larch and oak sawdust/pellets, is environmentally friendly and offers clean and quick energy without CO_2 emissions.

The native larch and oak pellets had few differences in terms of effective density and calorific value, but after the thermal treatment of torrefaction the calorific value grew significantly, up to 15.8%.

If a comparison of the calorific properties of larch and oak sawdust/pellets is made, it can be observed that, although they are two different species (softwood and hardwood) and have different densities (775 kg/m^3 in the case of oak and 521 kg/m^3 in the case of larch at 10% moisture content), their energetic properties after torrefaction are quite appropriate.

Considering that the two analyzed species (*Quercus robur* and *Larix decidua*) are valuable species which are mainly used in furniture, decorations, timber and veneers, etc., it is recommended that only their remnants from the processing of wood (sawdust, for example) should be used in the process of manufacturing native or torrefied pellets.

Author Contributions: Conceptualization, A.L.; methodology, C.S.; software, A.M.O.; validation, A.L.; formal analysis, A.M.O.; investigation, A.L.; resources, A.L.; data curation, C.S.; writing—original draft preparation, A.L.; writing—review and editing, A.L.; visualization, A.M.O.; supervision, A.M.O.; project administration, A.L. All authors have read and agreed to the published version of the manuscript.

Funding: This research received no external funding.

Data Availability Statement: Not applicable.

Acknowledgments: The authors would like to thank the Transilvania University of Brasov for their administrative and technical support.

Conflicts of Interest: The authors declare no conflict of interest.

References

1. Kambo, H.S.; Dutta, A. Strength, storage, and combustion characteristics of densified lignocellulosic biomass produced via torrefaction and hydrothermal carbonization. *Appl. Energy* **2014**, *135*, 182–191. [CrossRef]
2. Zarringhalam, M.A.; Gholipour, Z.N.; Dorosti, S.; Vaez, M. Physical Properties of Solid Fuel Briquettes from Bituminous Coal Waste and Biomass. *J. Coal. Sci. Eng. China* **2011**, *17*, 434–438. [CrossRef]
3. Kersa, J.; Kulu, P.; Aruniit, A.; Laurma, V.; Križan, P.; Šooš, L. Determination of Physical, Mechanical and Burning Characteristics of Polymeric Waste Material Briquettes. *Est. J. Eng.* **2010**, *16*, 307–316. [CrossRef]
4. Chen, N.; Rao, J.; He, M.; Mei, G.; Huang, Q.; Lin, Q.; Zeng, Q. Preparation and Properties of Heat- treated Masson Pine (*Pinus Massoniana*) Veneer. *BioResources* **2015**, *10*, 3451–3461. [CrossRef]
5. Esteves, B.; Pereira, H. Wood Modification by Heat Treatment: A Review. *BioResources* **2009**, *4*, 370–404. [CrossRef]

6. Kumar, L.; Koukoulas, A.; Mani, S.; Satyavolu, J. Integrating Torrefaction in the Wood Pellet Industry: A Critical Review. *Energy Fuel* **2017**, *31*, 37–54. [CrossRef]
7. Akinrinola, F.S. Torrefaction and Combustion Properties of Some Nigerian Biomass. Ph.D. Thesis, School of Chemical and Process Engineering, The University of Leeds, Leeds, UK, 2014. Available online: https://etheses.whiterose.ac.uk/8867/1/Femi%20Akinrinola_%20thesis_white%20rose_01052015.pdf (accessed on 19 January 2022).
8. Boutin, G.; Gervasoni, P.; Help, R.; Seyboth, K.; Lamers, P.; Ratton, M. Alternative energy sources in Transition Countries. The case of Bio-energy in Ukraine. *Environ. Engine. Manag. J.* **2007**, *6*, 3–11. [CrossRef]
9. McKendry, P. Energy Production from Biomass (part 2): Conversion Technologies. *Bioresour. Technol.* **2002**, *83*, 47–54. [CrossRef]
10. Ahn, B.J.; Chang, H.S.; Lee, S.M.; Choi, D.H.; Cho, S.T.; Han, G.S.; Yang, I. Effect of binders on the durability of wood pellets fabricated from *Larix kaemferi* C. and *Liriodendron tulipifera* L. sawdust. *Renew. Energ.* **2013**, *62*, 18–23. [CrossRef]
11. Peng, J.H.; Bi, H.T.; Sokhansanj, S.; Lim, J.C. A Study of Particle Size Effect on Biomass Torrefaction and Densification. *Energ. Fuel* **2012**, *26*, 3826–3839. [CrossRef]
12. Rudolfsson, M.; Borén, I.; Pommer, L.; Nordin, A. Combined effects of torrefaction and pelletisation parameters on the quality of pellets produced from torrefied biomass. *Appl. Energ.* **2017**, *191*, 414–424. [CrossRef]
13. Oh, S.-W.; Park, D.H.; Lee, S.M.; Ahn, B.J.; Ahn, S.H.; Yang, I. Torrefaction of *Larix Kaempferi*, C. and *Liriodendron Tulipifera* L. Cubes: Impact of Reaction Temperature on Microscopic Structure, Moisture Absorptivity, and the Durability of Pellets Fabricated with the Cubes. *Energ. Fuel* **2018**, *32*, 431–440. [CrossRef]
14. Nhuchhen, D.R.; Basu, P. Experimental Investigation of Mildly Pressurized Torrefaction in Air and Nitrogen. *Energy Fuels* **2014**, *28*, 3110–3121. [CrossRef]
15. Kudo, S.; Okada, J.; Ikeda, S.; Yoshida, T.; Asano, S.; Hayashi, J. Improvement of Pelletability of Woody Biomass by Torrefaction under Pressurized Steam. *Energy Fuels* **2019**, *33*, 11253–11262. [CrossRef]
16. Peng, J.W.; Bi, X.T.; Lim, C.J.; Sokhansanj, S.; Peng, H.; Jia, D. Effects of Thermal Treatment on Energy Density and Hardness of Torrefied Wood Pellets. *Fuel Process. Technol.* **2015**, *129*, 168–173. [CrossRef]
17. Eseyin, A.E.; Steele, P.H.; Pittman, C.U., Jr. Current trends in the production and applications of torrefied wood/biomass—A review. *BioResources* **2015**, *10*, 8812–8858. [CrossRef]
18. Granados, D.A.; Basu, P.; Chejne, J.F.; Nhuchhen, D.R. A Detailed Investigation into Torrefaction of Wood in a Two-Stage Inclined Rotary Torrefier. *Energy Fuels* **2016**, *31*, 647–658. [CrossRef]
19. Okoro, N.M.; Ozonoh, M.; Harding, K.G.; Oboirien, B.O.; Daramola, M.O. Potentials of Torrefied Pine Sawdust as a Renewable Source of Fuel for Pyro-Gasification: Nigerian and South African Perspective. *ACS Omega* **2021**, *6*, 3508–3516. [CrossRef]
20. Olugbade, T.O.; Ojo, O.T. Biomass Torrefaction for the Production of High-Grade Solid Biofuels: A Review. *Bioenerg. Res.* **2020**, *13*, 999–1015. [CrossRef]
21. Alokika; Singh, B. Production, characteristics, and biotechnological applications of microbial xylanases. *Appl. Microbiol. Biotechnol.* **2019**, *103*, 8763–8784. [CrossRef]
22. Pérez, J.F.; Pelaez-Samaniego, M.R.; Garcia-Perez, M. Torrefaction of Fast-Growing Colombian Wood Species. *Waste Biomass Valori.* **2019**, *10*, 1655–1667. [CrossRef]
23. Lee, J.; Karki, S.; Poudel, J.; Won, K.; Cheon Oh, L.S. Fuel characteristics of sewage sludge using thermal treatment. *J. Mater. Cycles Waste Manag.* **2019**, *21*, 766–773. [CrossRef]
24. Aguilar, F.X.; Mirzaee, A.; McGarvey, R.G. Expansion of US wood pellet industry points to positive trends but the need for continued monitoring. *Sci. Rep.* **2020**, *10*, 18607. [CrossRef] [PubMed]
25. Ahmad, I.; Lee, M.-S.; Goo, H.-K.; Lee, C.-Y.; Ryu, J.-H.; Kim, C.-H. Fuel pellets from fine paper mill sludge supplemented by sawdust and by refined recovered lubricating oil. *BioResources* **2021**, *16*, 1144–1160. [CrossRef]
26. Krajnc, N. *Wood Fuel Handbook*; Food and Agriculture Organization of the United Nations: Pristina, Kosovo, 2017. Available online: http://large.stanford.edu/courses/2017/ph240/timcheck1/docs/fao-krajnc-2015.pdf (accessed on 11 May 2021).
27. *ASTM E1755-01*; Standard Method for the Determination of Ash in Biomass. Annual Book of ASTM Standards. American Society for Testing and Materials: Philadelphia, PA, USA, 2003.
28. TWH TermoWood Handbook, Edition 2.0. Finnish Thermo Wood Association, Helsinki, Finland. 2003. Available online: https://asiakas.kotisivukone.com/files/en.thermowood.palvelee.fi/downloads/tw_handbook_080813.pdf (accessed on 2 June 2017).
29. Sjostrom, E. *Wood Chemistry: Fundamentals and Applications*; Gulf Professional Publishing: London, UK, 1993; pp. 277–293.

MDPI
St. Alban-Anlage 66
4052 Basel
Switzerland
www.mdpi.com

Forests Editorial Office
E-mail: forests@mdpi.com
www.mdpi.com/journal/forests

Disclaimer/Publisher's Note: The statements, opinions and data contained in all publications are solely those of the individual author(s) and contributor(s) and not of MDPI and/or the editor(s). MDPI and/or the editor(s) disclaim responsibility for any injury to people or property resulting from any ideas, methods, instructions or products referred to in the content.

www.ingramcontent.com/pod-product-compliance
Lightning Source LLC
LaVergne TN
LVHW070157120526
838202LV00013BA/1315